RIVERBANK FILTRATION

Water Science and Technology Library

VOLUME 43

The titles published in this series are listed at the end of this volume.

RIVERBANK FILTRATION

Improving Source-Water Quality

edited by

CHITTARANJAN RAY

University of Hawaii at Mānoa,
Honolulu, Hawaii, U.S.A.

GINA MELIN

National Water Research Institute,
Fountain Valley, California, U.S.A.

and

RONALD B. LINSKY

National Water Research Institute,
Fountain Valley, California, U.S.A.

KLUWER ACADEMIC PUBLISHERS
DORDRECHT / BOSTON / LONDON

A C.I.P. Catalogue record for this book is available from the Library of Congress.

ISBN 1-4020-1133-4 (HB 2002, 2003)
ISBN 1-4020-1838-X (PB 2003)

Published by Kluwer Academic Publishers,
P.O. Box 17, 3300 AA Dordrecht, The Netherlands.

Sold and distributed in North, Central and South America
by Kluwer Academic Publishers,
101 Philip Drive, Norwell, MA 02061, U.S.A.

In all other countries, sold and distributed
by Kluwer Academic Publishers,
P.O. Box 322, 3300 AH Dordrecht, The Netherlands.

Front Cover:
The Saloppe Waterworks pumping riverbank filtrate at the Elbe River in Dresden, Germany.
It is one of the oldest bank filtration waterworks in Germany.
Provided by Thomas Grischek.

Printed on acid-free paper

Acknowledgements

This book is the direct result of the many excellent presentations and ideas brought forward at the International Riverbank Filtration Conference, held by the National Water Research Institute in cooperation with the United States Environmental Protection Agency, Louisville Water Company, and Cincinnati Water Works, in November 1999. The efforts of the following individuals are gratefully acknowledged.

Editors

Chittaranjan Ray, Ph.D., P.E.
University of Hawaii at Mānoa
Honolulu, Hawaii, United States

Gina Melin
National Water Research Institute
Fountain Valley, California, United States

Ronald B. Linsky
National Water Research Institute
Fountain Valley, California, United States

Contributors

Harish Arora, Ph.D., P.E.
O'Brien & Gere Engineers, Inc.
Landover, Maryland, United States

Kay Ball
Louisville Water Company
Louisville, Kentucky, United States

William P. Ball, Ph.D., P.E.
Johns Hopkins University
Baltimore, Maryland, United States

Philippe Baveye, Ph.D.
Laboratory for Environmental Geophysics
Cornell University
Ithaca, New York, United States

Philip Berger, Ph.D.
Ijamsville, Maryland, United States

Edward J. Bouwer, Ph.D.
Johns Hopkins University
Baltimore, Maryland, United States

Jörg E. Drewes, Ph.D.
Colorado School of Mines
Golden, Colorado, United States

Rolf Gimbel, Ph.D.
IWW Rheinisch-Westfälisches Institut für Wasserforschung
Institut an der Gerhard-Mercator-Universität Duisburg
Mülheim a.d. Ruhr, Germany

William D. Gollnitz
Greater Cincinnati Water Works
Cincinnati, Ohio, United States

Thomas Grischek, Ph.D.
Institute for Water Chemistry
Dresden University of Technology
Dresden, Germany

Alison M. Gusses, M.S.
University of Cincinnati
Cincinnati, Ohio, United States

David L. Haas, P.E.
Jordan, Jones, and Goulding
Atlanta, Georgia, United States

Thomas Heberer, Ph.D.
Institute of Food Chemistry
Technical University of Berlin
Berlin, Germany

Stephen Hubbs, P.E.
Louisville Water Company
Louisville, Kentucky, United States

Henry Hunt, CPG
Collector Wells International, Inc.
Columbus, Ohio, United States

Ronald B. Linsky
National Water Research Institute
Fountain Valley, California, United States

Hans-Joachim Mälzer, Ph.D.
IWW Rheinisch-Westfälisches Institut für Wasserforschung
Institut an der Gerhard-Mercator-Universität Duisburg
Mülheim a.d. Ruhr, Germany

Ilkka Miettinen, Ph.D.
National Public Health Institute
Division of Environmental Health
Kuopio, Finland

Gina Melin
National Water Research Institute
Fountain Valley, California, United States

Till Merkel, M.Sc.
DVGW Water Technology Center
Karlsruhe, Germany

Charles O'Melia, Ph.D., P.E.
Johns Hopkins University
Baltimore, Maryland, United States

Chittaranjan Ray, Ph.D., P.E.
University of Hawaii at Mānoa
Honolulu, Hawaii, United States

Michael J. Robison, P.E.
Jordan, Jones, and Goulding
Atlanta, Georgia, United States

Traugott Scheytt, Ph.D.
Technical University of Berlin
Berlin, Germany

Jack Schijven, Ph.D.
National Institute of Public Health and the Environment
Microbiological Laboratory for Health Protection
Bilthoven, The Netherlands

Dagmar Schoenheinz, M.Sc.
Institute of Water Chemistry
Dresden University of Technology
Dresden, Germany

Jürgen Schubert, M.Sc.
Stadtwerke Düsseldorf AG
Düsseldorf, Germany

Thomas F. Speth, Ph.D., P.E.
United States Environmental Protection Agency
Cincinnati, Ohio, United States

R. Scott Summers, Ph.D.
University of Colorado
Boulder, Colorado, United States

Ingrid M. Verstraeten, Ph.D.
United States Geological Survey
Baltimore, Maryland, United States

Jack Wang, Ph.D.
Louisville Water Company
Louisville, Kentucky, United States

W. Joshua Weiss
Johns Hopkins University
Baltimore, Maryland, United States

Graphic Design

Tim Hogan
Tim Hogan Graphics
Westminster, California, United States

Table of Contents

RIVERBANK FILTRATION: IMPROVING SOURCE-WATER QUALITY

List of Figures

List of Tables

Acronyms

AMPA	Aminomethylenephosphonic acid
ATMP	Amino-(trimethylenephosphonic) acid
BCAN	Bromochloroacetonitrile
Br-	Brominated
$CaCO_3$	Calcium carbonate
CH	Chloral hydrate
CHFP	Chloral hydrate formation potential
Cl-	Chlorinated
ClO_2	Chlorine dioxide
CP	Chloropicrin
DAA	Dischlorophenyl acetic acid
DBAN	Dibromoacetonitrile
DCAN	1,1-dichloroacetonitrile
DDA	2,2-bis(chlorophenyl)acetic acid
DDT	Dichlorodiphenyl trichloroethane
DOC	Dissolved organic carbon
DTPA	Diethylenetrinitrolopentaacetic acid
EDTA	Ethylenediaminetetraacetic acid
ESA	Ethane sulfonic acid
FP	Formation potential
GC	Gas chromatography
HAA	Haloacetic acid
HAA6	Total haloacetic acid
HAN	Haloacetonitriles
HK	Haloketones
HPC	Heterotrophic plate count
HPLC	High-performance liquid chromatography
ICR	Information Collection Rule
LT2ESWTR	Long Term 2 Enhanced Surface Water Treatment Rule
MCL	Maximum contaminant level
MCPA	(4-chloro-2-methylphenoxy) acetic acid
MS	Mass spectrometry
MTBE	Methyl tertiary butyl ether
NaCl	Sodium chloride
NASBA	Nucleic acid sequence based amplification
NOM	Natural organic matter
NTA	Nitrilotriacetic acid
PCB	Polychlorinated biphenyl
RBF	Riverbank filtration
TCAN	Trichloroacetonitrile
THCOL	Thermotolerant coliform

THM	Trihalomethane
TOC	Total organic carbon
TOX	Total organic halogen
TTHM	Total trihalomethane
UFC	Uniform formation condition
2,4-D	2,4-dichlorophenoxyacetic acid
2,4,5-T	2,4,5-trichlorophenoxyacetic acid
2,4,6-TCP	2,4,6-trichlorophenole

Units of Measure

C	Celsius
cfu	Colony forming unit
cm	Centimeter
d	Day
fnu	Formazine nephelometric units
g	Grams
ha	Hectare
h	Hour
kg	Kilogram
km	Kilometer
km^2	Kilometers squared
L	Liter
m	Meter
mg	Milligrams
mg-N/L	Milligrams as nitrogen per liter
mL	Milliliter
mm	Millimeter
MPN	Most Probable Number
m^2	Meters squared
m^3	Cubic meters
N	Newton
ng	Nanogram
nm	Nanometer
ntu	Nephelometric turbidity unit
pfu	Plaque forming unit
s	Second
t	Ton
yr	Year
μ	Micron

Conversion Table

Number	Multiply Number By	To Obtain Number In
Celsius (C)	1.8 (then add 32)	Fahrenheit (F)
Centimeter (cm)	2.54	Inch (in)
Gram (g)	454	Pound (lb)
Hectare (ha)	0.4	Acre
Kilogram (kg)	0.454	Pound (lb)
Kilometer (km)	1.6	Mile (mi)
Kilometer Squared (km^2)	2.56	Mile Squared (mi^2)
Liter (L)	3.78	Gallon (gal)
Meter (m)	0.328	Foot (ft)
Milligram (mg)	4.54×105	Pound (lb)
Milliliter (mL)	3,780	Gallon (gal)
Millimeter (mm)	304.8	Foot (ft)
Meter Squared (m^2)	10.758	Feet Squared (ft^2)
Cubic Meter (m^3)	35.288	Cubic Feet (ft^3)

Introduction

Chittaranjan Ray, Ph.D., P.E.
University of Hawaii at Mānoa
Honolulu, Hawaii, United States

Jürgen Schubert, M.Sc.
Stadtwerke Düsseldorf AG
Düsseldorf, Germany

Ronald B. Linsky
National Water Research Institute
Fountain Valley, California, United States

Gina Melin
National Water Research Institute
Fountain Valley, California, United States

1. What is Riverbank Filtration?

The purpose of this book is to show that riverbank filtration (RBF) is a low-cost and efficient alternative water treatment for drinking-water applications. There are two immediate benefits to the increased use of RBF:

- Minimized need for adding chemicals like disinfectants and coagulants to surface water to control pathogens.
- Decreased costs to the community without increased risk to human health.

But what, exactly, is RBF?

In humid regions, river water naturally percolates through the ground into aquifers (which are layers of sand and gravel that contain water underground) during high-flow conditions. In arid regions, most rivers lose flow, and the percolating water passes through soil and aquifer material until it reaches the water table. During these percolation processes, potential contaminants present in river water are filtered and attenuated. If there are no other contaminants present in the aquifer or if the respective contaminants are present at lower concentrations, the quality of water in the aquifer can be of higher quality than that found in the river. In RBF, production wells — which are placed near the banks of rivers — pump large quantities of water. The pumping action creates a pressure "head" difference between the river and aquifer, which induces the water from the river to flow downward through the porous media into pumping wells (Figure I-1). The pumped water is a mixture of both groundwater originally present in the aquifer and infiltrated surface water from the river. Depending upon the ultimate use and the degree of filtering and contaminant attenuation, additional treatments may be provided to the pumped water prior to distribution. At a minimum, RBF acts as a pretreatment step in drinking-water production and, in some instances, can serve as the final treatment just before disinfection.

1

C. Ray et al. (eds.), Riverbank Filtration, 1–15.
© 2002 *Kluwer Academic Publishers. Printed in the Netherlands.*

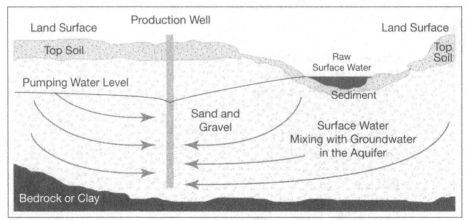

Figure 1-1. Generalized schematic of an RBF system. As production wells pump water from the aquifer, surface water flows underground to refill it. As surface water travels through the sediments, contaminants are "filtered," or removed. The filtered water (known as the "filtrate") is generally of a much higher quality than raw surface water.

For more than 100 years, RBF has been used in Europe, most notably along the Rhine, Elbe, and Danube Rivers, to produce drinking water. Although RBF is not commonly utilized in the United States, interest is increasing in using RBF as a low-cost complement or alternative to filtration systems to remove pathogens from water.

RBF has proven to be invaluable in treating drinking-water sources in Europe. Studies have shown that RBF generally removes a substantial percentage of organic compounds found in raw river water — including harmful pathogens such as *Giardia*, *Cryptosporidium*, and viruses.

2. Historical Significance

The first known utility to use RBF for water-supply purposes was the Glasgow Waterworks Company in the United Kingdom, which built a perforated collector pipe parallel to the Clyde River in 1810 to extract riverbank-filtered water. Other waterworks in the United Kingdom followed suit (e.g., Nottingham, Perth, Derby, Newark) and, in the mid-nineteenth century, RBF was officially adopted by European utilities to produce drinking water.

In Western Europe, one of the first RBF plants was established in the Lower Rhine Valley region in Düsseldorf, Germany, which is located on both sides of the Rhine River. Because of limited groundwater resources, waterworks in the Lower Rhine Valley Region preferred to use RBF to supply drinking water to a population of 600,000.

To do this, an English engineer named William Lindley created the conceptual design of the Flehe Waterworks. The construction, including wells, pump station, main pipe, and reservoir, was completed in less than 2 years, and the Flehe Waterworks started operation in 1870. Since then, the Flehe Waterworks has operated without interruption.

In 1892, there was an outbreak of epidemic cholera in Hamburg, Germany, that was caused by drinking water from a waterworks with direct intake from the Elbe River. This led to the use of artificial or natural subsoil passage of raw river water as a replacement or supplement to direct intake for public-water supply.

The actual statistics of sources used for public-water supply in Germany appear to mirror the consequences of the 1892 outbreak. They include:

- Groundwater . . . . 63.6 percent
- Springs . . . . 7.8 percent
- RBF and groundwater recharge . . . . 15.3 percent
- Surface water (lakes and dams) . . . . 13.3 percent

There are 33 waterworks along the Rhine River between 420 kilometers (km) (mouth of the Neckar River) and 840 km (German/Dutch border) that constitute the Association of Rhine Waterworks. According to statistics, the sources used for public-water supply by members of the Association of Rhine Waterworks (as of 1998) include:

- Groundwater . . . . 33 percent
- Springs . . . . 1 percent
- RBF and groundwater recharge . . . . 49 percent
- Surface water (lakes and dams) . . . . 17 percent

The amount of riverbank-filtered water used for public-water supply along the Rhine River totals approximately 250 million cubic meters per year (m^3/yr) (1998 statistics). A major part of this amount is extracted along a 120-km section of the Lower Rhine Valley between the Sieg and Ruhr Rivers.

The distance between the wells (vertical filter well galleries and horizontal collector wells) and the riverbank along the Rhine varies from 50 meters (m) to approximately 250 m, with the exception of only one gallery, which has a distance of 20 m. Aerobic (with oxygen) conditions are most common along the flow path of many RBF plants along the Rhine River. Low concentrations of biodegradable substances in river water help maintain aerobic conditions along the flow path of riverbank-filtered water. The advantage of this is a filtrate free of iron, manganese, ammonia, and nitrite. Along the Elbe River and at many sites along the Ruhr River, however, there are anoxic (without oxygen) conditions that influence the final treatment of RBF-filtered water.

Other large RBF plants in Germany are located along the Upper Elbe River around the Cities of Dresden, Meissen, and Torgau. In addition, the City of Berlin pumps its water from the bank of the Spree River and its associated chain of lakes.

There are no RBF plants along the Danube River as it flows through Germany. RBF plants in the Danube start around Vienna, Austria. The City of Bratislava in the Slovak Republic also employs RBF along the Danube. Wells have operated there for more than a century. The recent construction of a hydropower dam along the Danube River at Gabcikovo, Slovakia, and the resulting sedimentation of suspended particles in the reservoir have led to an intense investigation of the sedimentation process upon the performance of a nearby RBF plant (Mucha et al., in press).

Downstream of Bratislava, the next largest RBF plants can be found around Budapest, Hungary. These wells are located on two islands, one upstream of the City and the other downstream. The City of Belgrade in Yugoslavia is also a center for RBF along the Danube River.

In the United States, horizontal collector wells are frequently used for RBF to retrieve large amounts of water from a relatively small geographic area (e.g., Lincoln, Nebraska; Louisville, Kentucky; and Kansas City, Kansas). These horizontal collector wells are comprised of a central caisson and a series of perforated pipes connected to the caisson. Details of this can be found in Chapters 1 and 2. Horizontal collector wells are often referred to as Ranney collector wells, named after Leo Ranney, a petroleum engineer who introduced the design in the 1930s.

3. Unrecognized RBF Plants

A great number of drinking-water wells in alluvial or glacial aquifers are located in valleys near rivers, but are not considered RBF wells. Under steady-state conditions, most of the extracted water is groundwater; however, there are different events or circumstances that can change the normal capture zone of those wells and "switch on" RBF production processes, such as:

- Temporary flooding.
- Exploration of sand and gravel near the river.
- River regulation.

Temporary flooding may enhance the gradient between the river and aquifer and may allow river water to migrate towards the aquifer. This water is often referred to as "bank-stored water." Once the river stage is lowered, the bank-stored water drains back to the river. The time required for this process is a function of the hydraulic properties of the geologic medium.

Because exploration for sand and gravel near the river often requires that pits be dewatered, surface water is induced to flow toward the pits and mix with groundwater. If pumping wells are located near these pits, the pumping wells will indirectly capture a portion of the surface water.

Riparian hydrogeology can change when locks and dams are built on rivers for navigation and flood control. Higher river water levels allow surface water to move farther into the aquifers. The river might also affect wells that, originally, were some distance away from the river. One such example is the adverse impact of a hydropower dam near the City of Gabcikovo on the Danube River in Slovakia, where sedimentation of the riverbed negatively affects filtrate production.

4. Similarities Between RBF and Slow Sand Filtration

As stated earlier, the outbreak of a cholera epidemic in Hamburg, Germany, in 1892 resulted in the general rule (for Central Europe) that direct intake from rivers for public-water supply must be replaced or supplemented by the subsoil passage of raw water. The former idea was that even sophisticated treatment – regarding viruses, bacteria, and protozoa – cannot meet the efficiency of natural subsoil passage. Slow sand filtration and RBF (which exhibit many similarities, such as filter velocity and the removal of particles, microorganisms, and biodegradable substances) are alternatives to assist utilities in complying with regulatory requirements. The choice between both technologies depends on the local hydrogeological situation and may be influenced by the properties of the technologies (e.g., cost, maintenance, stability of operation); therefore, some information about slow sand filtration may be helpful.

Slow sand filtration is a purification process in which raw water (untreated source water) is passed downward through a filtering medium that consists of a layer of sand. The rate of filtration is slow, typically 0.1 to 0.2 meters per hour (m/h). Slow sand filtration is considered an inexpensive and efficient technology for removing many suspended and biodegradable contaminants as well as pathogens.

Slow sand filtration was developed in the early nineteenth century in England to produce drinking water for the City of London by filtering Thames River water through sand. Slow sand filtration continues to be used in the United Kingdom, The Netherlands, Germany, and Switzerland. In the United States, slow sand filtration is used in the New England areas. Large cities in other countries, such as the City of Chennai in India, also employ slow sand filtration for drinking-water treatment.

According to Cleasby (1990), slow sand filters do not successfully treat waters from rivers that traverse through clay-bearing formations. Most rivers carry clay and colloidal particles from clay-bearing formations found along riverbanks and riverbeds. Clay can penetrate into the filter media and clog pore spaces. Normal scraping cannot remove these clays from deeper depths. Cleasby also points out that slow sand filters are inadequate in removing color from water (<25-percent removal); however, slow sand filtration has received renewed interest in recent years because of its ability to protect against protozoa breakthrough and for use in small systems, since it does not include complex coagulation chemistry.

Cleasby (1990) conducted a literature review of slow sand filters and stated that, for a majority of slow sand filtration systems, the effective sand size is around 0.3 millimeter (mm) with a uniformity coefficient close to 2. The filter bed depth is commonly found to be 0.9 m. The gravel that supports the sand has a depth from 0.15 to 0.9 m. Filtration rates vary from 0.07 to 0.12 m/h for raw water; a higher filtration rate (0.3 m/h) may be found for waters that receive some pretreatment. Head loss through the filter varies from 0.9 to 1.5 m.

During slow sand filtration, a thin organic mat called schmutzdecke grows on the filter surface, which enhances filtration because it acts as a barrier to remove contaminants like suspended solids (since the schmutzdecke is biologically active, it also helps degrade biological particles); however, with time, the schmutzdecke layer becomes clogged with contaminants. To improve the filtration rate, it is necessary to periodically clean the filter. The schmutzdecke is removed (along with a small amount of sand) during each cleaning cycle, though it will grow back over a course of days. Ultimately, filter bed performance is improved after each cleaning. A compilation of literature by Cleasby (1990) shows that the filtration improvement rate (often referred to as the "ripening period") could vary from 6 hours to 2 weeks after cleaning, with most systems achieving peak performance within 2 days (however, if there is a significant concentration of protozoa in the surface water, then the filtrate may be unusable during the first 2 days after cleaning because the lack of the schmutzdecke layer results in poor removal). Cleasby (1990) also points out that slow sand filters were effective in removing *Salmonella typhi* in a 1980 study by the Massachusetts Board of Health. Bellamy et al. (1985) studied the performance of slow sand filters for pathogen removal. They observed that a new sand bed removed 85 percent of coliforms and 98 percent of *Giardia* cysts from source water; however, with ripening, removal rates exceeded 99 percent for coliforms and nearly 100 percent for *Giardia*. Conversely, Fogel et al. (1993) reported much lower removal effectiveness at a plant in British Columbia, Canada. They observed a 93-percent removal rate for *Giardia* cysts and an average of 48 percent for *Cryptosporidium* oocysts. Some of the low removal rates can be attributed to the high uniformity coefficient of the sand. If the filter material is poorly graded (which means that it is mostly the same size), the uniformity coefficient is low (2 to 3 range for sand). In reality, well-graded filter material is preferred, for which the uniformity coefficient should be high (5 or 6 for sand). In addition, the water temperature was around 1°Celsius (C), which enhances the viscosity of water and reduces biological activity.

Cullen and Letterman (1985) stated that filtrate turbidity typically could be <0.5 nephelometric turbidity units (ntu). In one exception, lake water that received clay and colloid-sized particles in runoff from a mountain watershed had a low removal rate.

In a 1984 survey of 27 full-scale slow sand filtration plants, Cleasby (1990) found that 74 percent used lake or reservoir water as the source water, 22 percent used river or stream water, and the remaining 4 percent used groundwater. The mean turbidity of source water was 2 ntu, with a peak around 10 ntu. The mean durations for filter runs were 42 days in the spring to 60 days in the winter. In a survey of seven plants in New York, filter run lengths varied between 1 to 6 months.

The RBF process has some similarity with the performance of slow sand filters. Organic mats, similar to the schmutzdecke, can develop at the river/aquifer interface, especially when the flow velocity is slow. River flooding may wash off this layer; however, with subsequent low-flow periods, it could be reestablished. RBF systems are vulnerable to breakthrough when the protective organic/sediment mat is washed away and river flow is high. The infiltration velocity at the river/aquifer interface varies, depending upon the location of the laterals of the collector wells. Mikels (1992) estimated that the average infiltration velocity for small-capacity collector wells at Kalama, Washington, was 1.56×10^{-5} m/h, which is substantially lower than the rates mentioned above; however, the approach velocity measured at a lateral for the collector well at Louisville, Kentucky, during initial pumping was between 0.12 and 0.16 m/h (see Chapter 7). According to Wang (2002), this is in the same range as slow sand filters (between 2 and 5 meters per day [m/d]), as given by Ellis (1985).

A slow sand filter, when used as a secondary treatment process following a primary treatment train, would receive a 2.5-log removal credit for *Cryptosporidium* removal under U.S. Environmental Protection Agency rules. In this case, the slow sand filter would be installed in front of the disinfection unit. If slow sand filtration is used as the primary mode of water treatment, then it would receive the same amount of credit (3-log removal) as conventional treatment. By comparison, properly designed and operated RBF plants reach 4-log removal efficiency.

5. Surface-Water Contaminants of Concern

For an RBF system to effectively operate, it must remove contaminants in raw surface water from lakes, rivers, or reservoirs; therefore, utilities must ensure that RBF systems are properly designed and operated to maximize contaminant removal. Utilities must also take into consideration the fact that these contaminants may fluctuate seasonally.

Physical Contaminants

Temperature and turbidity are the physical contaminants of greatest concern. In temperate climates, depending upon the season, surface-water temperature could range from freezing to ±35°C; however, groundwater temperature remains relatively unchanged (±15°C). At Louisville, Kentucky (see Chapter 7), the temperature of Ohio River water varied from a low of 2°C to as high as 32°C between the winter and summer months during a 2-year monitoring effort (Wang, 2002). According to the unpublished data of a monitoring effort in the State of Illinois, the temperature of the Illinois River between the Cities of Henry and Hardin varied by a margin similar to the temperature of the Ohio River during a monitoring study conducted by Ray et al. (1998). As shown in Wang (2002), the temperature from the collector well varied between 15 and 25°C (see Chapter 7). This variation can be a function of pumpage, monitoring point location, distance of the river to the well, well construction, or other hydrogeologic factors. Further, variations in temperature alter the performance of water-treatment plant unit operations. Groundwater provides the best moderation of temperature fluctuation. Riverbank filtrate also provides significant moderations.

Turbidity is a concern for rivers that traverse through clay-rich formations. Monitoring data for the Ohio River near Louisville, Kentucky, shows that the turbidity of river water varied anywhere between 2 ntu (July 1999) and 1,500 ntu (March 1997) during a 5-year period (1997 to 2002) (Wang, 2002); however, the filtrate from the collector well at Louisville had a turbidity of around 0.1 ntu, which is significantly below the current United States standard of 0.5 ntu. Price et al. (1999) monitored the turbidity of the Russian River near Santa Rosa, California, for coliform and other indicator organisms. River turbidity ranged between 1 and 320 ntu, correlating well with river flow. Wang (2002) observed a similar trend in the Ohio River study.

Chemical Contaminants

Chemical contaminants can be divided into four major groups:
- Inorganics.
- Synthetic organics (pesticides and volatile/semi-volatile organics).
- Natural organic matter (NOM).
- Pharmaceuticals and other emerging chemicals.

Regarding inorganics, the hardness of river water is of concern to water utilities where hardness removal is a major treatment unit operation. Hardness can be reduced during peak flow periods when the contribution from groundwater is low. For the Ohio River, Wang (2002) observed that hardness varied between 90 and 200 milligrams per liter (mg/L) as calcium carbonate ($CaCO_3$). Weiss et al. (2002) also observed that the concentrations of anion and cations in river water and filtrate from vertical and horizontal collector wells vary depending upon site conditions, well type, location, and the contaminant itself. For instance, the presence of excessive concentrations of bromide in filtrate can lead to the formation of bromate (a carcinogenic disinfection byproduct) during ozonation; therefore, monitoring is needed at RBF sites using ozone as the disinfectant to examine the relative concentrations of bromide in surface water and filtrate from RBF systems. Nitrogen and other forms of fertilizers are also of concern. Rivers traversing agricultural watersheds, especially in the Midwestern United States, can receive large amounts of nitrate in surface runoff or through tile drain discharges. Ray et al. (1998) observed peak concentrations of nitrate in the Illinois River to reach 10 mg/L as nitrogen during flood periods whereas concentrations during the winter months (mostly from sewage input) rarely exceeded 5 mg/L.

Synthetic organic chemicals and pesticides are of great concern in surface-water treatment. Rivers that traverse through agricultural watersheds receive large loads of pesticides in spring runoff, similar to that described for nitrate. The concentration peaks of many of these chemicals often coincide with flow peaks. The peak concentrations of pesticides for small rivers or watersheds that are primarily agricultural could be much higher than those for large rivers in watersheds that have diverse land use. For instance, medium- to large-size rivers, such as the Illinois, Platte, and Cedar Rivers, traverse through predominately agricultural watershed. In the case of atrazine (a herbicide and plant-growth regulator used primarily on corn and soybeans), Ray et al. (1998) observed peak concentrations close to 12 micrograms per liter (µg/L) in the Illinois River near an RBF site in the City of Jacksonville, Illinois, between 1995 and 1996. For the Platte River near Lincoln, Nebraska, Verstraeten et al. (1999) observed peak atrazine concentrations of 13 and 26 µg/L during spring runoff periods in 1995 and 1996, respectfully. These concentrations are significantly higher than the maximum contaminant levels of atrazine (3 µg/L). Verstraeten et al. (1999) also found half a dozen pesticides in the river water, including atrazine and two of its metabolites (deethylatrazine and deisopropylatrazine), alachlor, alachlor ethane sulfonic acid (ESA), metolachlor, cyanazine, and acetochlor. Wang and Squillace (1994) observed high loads of herbicides (atrazine, simazine, cyanazine, metolachlor, alachlor, propachlor, etc.) in the Cedar River near Cedar Rapids, Iowa, during spring runoff and observed a natural exchange of herbicides between the river and aquifer, along with groundwater. Stamer and Wieczorek (1996) and Stamer (1996) observed the presence of large concentrations of herbicides in water from the Platte River during monitoring efforts between 1992 and 1994, and presented potential negative health impacts when this water is directly used for drinking purposes. Navigable rivers are also subject to accidental releases of petroleum products and other industrial chemicals, such as chlorinated compounds. These all contribute to shock loads (river water with a temporary and unusual amount of pollutants). In addition to shock loads, rivers can carry residual chemicals for a significant amount of time.

NOM in surface water is a major concern for water utilities that use chlorine as the disinfectant. Chlorine combines with NOM to form disinfection byproducts, such as trihalomethanes (THMs) and haloacetic acids (HAAs), which are potentially carcinogenic. NOM concentrations and speciation vary depending upon the season, watershed characteristics, and river flow. The following water-quality parameters that are typically used as indicators of NOM in source water include, but are not limited to:

- Total organic carbon (TOC).
- Dissolved organic carbon (DOC).
- Biodegradable organic carbon.
- Ultraviolet absorbance of water at 254 nanometers (nm).
- Assimilative organic carbon.

Please refer to Standard Methods for analytical procedures and the subtle differences between TOC, DOC, and biodegradable organic carbon (American Public Health Organization et al., 1998). In Chapter 7, Wang (2002) shows that the TOC concentration of organic matter in the Ohio River varies between 2 and 4 mg/L. Higher concentrations of TOC are found during the fall season, and lower concentrations are found in the late spring and early summer months. In Chapter 8, Weiss et al. (2002) show that mean TOC concentrations were 3 mg/L for the Ohio River, 4.7 mg/L for the Wabash River, and 4.5 mg/L for the Missouri River. Unpublished data from the monitoring study of Ray et al. (1998) also indicate that the concentration of non-purgeable organic carbon in the Illinois River varied between 0.5 and 5.2 mg/L, with higher concentrations observed during late spring (mid-May). Non-purgeable organic carbon is a direct method of measuring organic carbon in which purgeable compounds, such as volatile organics and inorganic carbons like carbon dioxide, are gas-stripped. The acidification of samples converts carbonates and bicarbonates to carbon dioxide, which can be stripped by the purging gas. TOC analysis uses a differential method in which inorganic carbon is subtracted from total carbon to calculate TOC.

Brauch et al. (2001) show that the DOC level in the Rhine River varied from 3 to 6 mg/L between 1975 and 2000, with higher concentrations between 1975 and 1980. The mean DOC level in the river is now about 3 mg/L due to pollution. Other rivers in Europe also show similar concentration ranges for DOC. Grischek et al. (2001) report a mean DOC level of 5.5 mg/L for the Elbe River at the Torgau-Ost Waterworks in Germany, based upon monitoring data collected between 1995 and 1997. The discharge of sewage from wastewater treatment plants also adds to the DOC loads on rivers and other receiving waters. For example, the DOC level of Tegel Lake in Berlin (where many of Berlin's bank filtration wells are located) varies between 6 and 8 mg/L. These higher levels are partly due to wastewater discharge into the Spree River, which feeds these lakes (Zeigler et al., 2001).

Pharmaceuticals and personal care products are micropollutants (detected at microgram-per-liter to nanogram-per-liter ranges) of recent concern to drinking-water utilities. Many pharmaceuticals and personal care products are found in domestic sewage, and some pharmaceuticals and personal care products are endocrine disrupting chemicals. Only a small subset of pharmaceuticals and personal care products is suspected to be direct-acting endocrine disrupting chemicals. Pharmaceuticals and personal care products are considered to have potentially adverse effects in natural ecosystems, such as causing abnormal physiological processes and reproductive impairments of aquatic species and inducing the development of antibiotic-resistant bacteria, among others (Kolpin, et al., 2002). At present, an intense monitoring effort is underway in the United States and Europe to find the distribution and concentration ranges of these compounds in rivers, lakes, and other water sources.

Because many of these compounds are found in extremely low concentrations, an analytical determination of these compounds is difficult and requires complex instrumentation. Daughton and Ternes (1999) provide a summary of the occurrence of these compounds in the environment, their chemical structure, and other significant environmental issues. Representative classes (and members) of pharmaceuticals and personal care products reported in environmental samples can be found in a presentation by Christian G. Daughton of the U.S. Environmental Protection Agency, located at www.epa.gov/nerlesdl/chemistry/pharma/index.htm. A summary of these compounds is presented in Table I-1.

Table I-1. Selected Classes and Members of Pharmaceuticals and Personal Care Products Found in Environmental Samples

Use Class	Example Brand Name	Generic Name
Analgesics/Non-Steroid Anti-Inflammatories	Tylenol Voltaren Ibuprofen Advil	Acetaminophen Diclofenac
Antimicrobial	Many Names	Fluoroquinolones Sulfonamides
Antiepileptics	Tegretal	Carbamazepine
Antiseptics	Igrasan DP 300	Triclosan
Lipid Regulators (Antilipidemics, Cholesterol-Reducing Agents, and Their Metabolites)	Astromid-S Lopoid	Clofibrate (Clofibric Acid Metabolite) Gemfibrozil
Contraceptives	Oradiol	17α-estradiol 17α-ethinyl estradiol
Antihypertensives (Betablockers, Beta Andrenergic Receptor Inhibitors)	Concor Lopressor	Bisoprolol Metoprolol
Musks (Synthetic)	Musk Xylene Celestolide Substituted Amino Anitrobenezes	Nitromusks Polycyclic Musks Reduced Metabolites of Nitromusks
Anti-Anxiety/Hypnotic Agents	Valium	Diazepam
Sunscreen Agents	Eusolex 6300	Methylbenzylidene Camphor
Antineoplastics	Cycloblastin Holoxan	Cyclophosphamide Isofamide

Adapted from a presentation by C.G. Daughton, U.S. Environmental Protection Agency (available at www.epa.gov/nerlsd1/chemistry/pharma/index.htm).

There is very little data on the concentrations of these compounds in surface waters within the United States. The U.S. Geological Survey has initiated the monitoring of many of these compounds through its National Water Quality Assessment program. The results of this monitoring program are expected to be available in the future; however, monitoring data for specific RBF sites may be available for Germany and other European countries. Heberer et al. (1997) show that the concentrations of several of these pharmaceuticals and personal care products found in the City of Berlin's surface water near RBF well sites were 7,300 nanograms per liter (ng/L) for clofibric acid, 380 ng/L for diclofenac, 200 ng/L for ibuprofen, 1,250 ng/L for phenazone, and 690 ng/L for primidone. These compounds are probably present below detection levels (1 to 10 ng/L) throughout the year. Heberer et al. (1998) also report that, for pharmaceutical residues, the maximum reported concentrations at 30 representative sampling locations in Berlin surface waters were up to 1,900 ng/L. Maximum concentrations for nitro musks were 390 ng/L; for polar pesticides, the concentrations were always below 0.1 µg/L.

Other chemicals of interest include:

- Adsorbable organic halogen.
- Adsorbable organic sulfur.
- Nitrilotriacetic acid (NTA).
- Ethylenediaminetetraacetic acid (EDTA).
- Diethylenetrinitrolopentaacetic acid (DTPA).
- Aromatic sulfonates.

NTA, EDTA, and DTPA are widely used as chelating agents in detergents and industrial cleaners and in the textile, photo, and pulp and paper industry. Sacher et al. (2001) report peak concentrations of 4 µg/L for NTA, 16 µg/L for EDTA, and 2.75 µg/L for DTPA between January 1994 and August 2000 in the Rhine River.

Biological Contaminants

Biological contaminants in surface water include protozoa, bacteria, and viruses. *Cryptosporidium* and *Giardia* are the two major waterborne protozoa of concern. Fecal and total coliform bacteria and, in some cases, the spores of aerobic and anaerobic bacteria are also monitored. In addition, human enteric viruses and bacteriophage are monitored at some European and American RBF sites. There is great deal of data available for surface water viruses, bacteria, and individual organisms such as *Giardia* and *Cryptosporidium*. The purpose of this section is not a comprehensive presentation; rather, representative data from selected sites in Europe and the United States are presented for reference only.

In 1996, the U.S. Environmental Protection Agency promulgated the Information Collection Rule (ICR), which is an effort to collect water-quality data from public water systems serving 100,000 people or more. In compliance with the ICR, 207 water-treatment plants collected monthly samples for viruses in raw water over a period of 18 consecutive months (U.S. Environmental Protection Agency, 2000b). The 100-liter (L) samples were then analyzed for enteroviruses using the ICR-approved method. Results were reported as a most probable number (MPN) per 100 L. Viruses were detected in 24 percent of the 3,365 samples collected by participating plants. The virus concentrations ranged from 0 to 1,974 MPN/100 L. Over 80 percent of the plants detected viruses in their source water. The median and the ninetieth-percentile of the mean virus concentrations at the plant were 0.4 and 5 MPN/100 L, respectively; the maximum mean virus concentration was 112 MPN/100 L (Shaw et al., in press).

In a 5-month monitoring effort at Louisville, Kentucky, Wang et al. (2001) reported aerobic spore counts approximately between 3,000 to 15,000 colony forming units (cfu) per 100 milliliters (mL) of sample collected from the Ohio River near the Louisville RBF plant. The average aerobic spore count was about 8,700 cfu/100 mL. Total coliform concentrations in the Ohio River ranged between 9 and 33,040 MPN/100 mL for samples collected between January and July of 2000. The heterotrophic plate counts (HPC) for river water samples ranged between 10 and 8,820 cfu/100 mL. For the Russian River in Northern California, Price et al. (1999) monitored the presence of total coliform bacteria in conjunction with turbidity and particle monitoring. The turbidity and total coliform count correlated well. The peak concentration of total coliform in the river reached 16,000 MPN/100 mL in January 1993, and similar high readings were observed in December 1992 and February 1993. At the same time, the river turbidity reached a peak value of 260 ntu. The river also experienced peak flow during the winter months (December 1992 through March 1993).

Medema et al. (2001) studied the concentrations of several protozoa, bacteria, and viruses in water from the Meuse River near the City of Roosteren in The Netherlands between January 1998 and May 1999. The mean concentrations of *Giardia* and *Cryptosporidium* (four samples) were 95 and 140/L per sample with maximum concentrations of 170 and 460/L, respectively. Peak concentrations of enteroviruses and reoviruses were 0.9 and 13.5/L of water, respectively, and their mean values (five samples) over the sampling period were 0.52 and 7.1/L, respectively. For the same site, six samples were analyzed for somatic coliphages and nine samples for F-specific RNA bacteriophages. The mean concentrations were 43,900 and 10,600/L, respectively, and the peak concentrations were 74,600 and 26,400/L, respectively.

In Chapter 8, Weiss et al. (2002) found neither *Cryptosporidium* nor *Giardia* in Ohio River water. Sample volumes were 6.2 L; however, *Clostridium* had concentrations of 1,220 cfu/L while two other bacteriophages (*E. coli* C and Famp) had concentrations of 490 cfu/L and 120 cfu/L, respectively. These concentrations are somewhat lower than that reported by Medema et al. (2000) for the Meuse River in the Netherlands. As part of the ICR, the U.S. Environmental Protection Agency (2001) conducted monthly samplings at 347 sites for microbial pathogens over a period of 18 months. For this study, Messner and Wolpert (2000) reported that concentrations of pathogens in flowing streams were, in general, higher than in lakes.

6. Case Studies of Log Removal Credit in the United States

"Log removal" is a shorthand term for $\log_{10}$ removal, which refers to the physical-chemical treatment of water to remove, inactivate, or kill pathogenic organisms such as *Giardia lamblia* and viruses. For example, 1-log removal equals a 90-percent reduction of the target organism; a 2-log removal equals a 99-percent reduction; and a 3-log removal equals a 99.9-percent reduction.

Log removal credit is a regulatory term used in the United States that expresses the amount of pathogens that a water utility has removed from water using technologies like slow sand filtration and RBF. For example, some water utilities that employ RBF may receive 1-log removal credit. This means that the RBF process has removed 90 percent of the initial concentration of pathogens; however, if the target removal is 99.9 percent (3 logs), the utility must remove an additional 2 logs using conventional filtration or other alternative techniques.

According to United States law, the granting of log removal credit is, in general, negotiated between the water utility and primacy agency responsible for enforcing regulations like the Surface Water Treatment Rule. Promulgated in 1989, the Surface Water Treatment Rule established maximum contaminant level goals of zero for *Giardia lamblia*, viruses, and *Legionella*, as well as set filtration and disinfection requirements for all public water systems using surface-water sources or groundwater sources under the direct influence of surface water (U.S. Environmental Protection Agency, 1989).

The following distinctions apply to drinking-water sources in the United States:

- *Groundwater*: Subsurface water contained in porous rock strata and/or soil. It is not affected by recently infiltrated surface water.
- *Groundwater Under the Direct Influence of Surface Water*: Water defined by the U.S. Environmental Protection Agency in the Surface Water Treatment Rule as any water

beneath the surface of the ground that has a significant occurrence of insects or other microorganisms, algae, organic debris, or large-diameter pathogens like *Giardia lamblia*, or significant and relatively rapid shifts in water characteristics — such as turbidity, temperature, conductivity, or pH — that closely correlate with climatological or surface-water conditions. It is a legal definition that implies that groundwater pumped from a well has been affected by recently infiltrated surface water.

- *Surface Water*: Water from sources open to the atmosphere, such as lakes, reservoirs, rivers, and streams.

Included are three short descriptions of log removal credit negotiations between primacy agencies and water utilities with RBF wells that were subject to the Surface Water Treatment Rule. These utilities include the Sonoma County Water Agency in Santa Rosa, California, the City of Kearny in Nebraska, and the Central Wyoming Regional Water System in Casper, Wyoming.

Sonoma County Water Agency

The Sonoma County Water Agency in Santa Rosa, California, operates five horizontal collector wells and several vertical wells on the banks of the Russian River. The Agency was not required to filter the pumped water, and only chlorination was provided to the drinking water; however, in September 1991, the California Department of Health Services (the primacy agency) concluded that the five collector wells were subject to the Surface Water Treatment Rule. As a result, the Agency was required to demonstrate that these wells were removing pathogens equivalent to that for direct intake or slow sand filtration. A study was conducted between March 1992 and May 1993 to show that filtration would not be needed if the well waters were not groundwater under the direct influence of surface water. It was determined that, with the exception of Collector Well 5, all other well waters were not groundwater under the direct influence of surface water, but were actually groundwater. As a result, the California Department of Health Services assigned a 2.5-log removal credit for *Giardia* and 1.0-log removal credit for viruses. This required the Agency to provide the necessary level of disinfection to achieve the needed log removals under the Surface Water Treatment Rule. Although the Agency's Russian River plant was practicing chlorination, it was difficult to achieve the dictated level of chlorination (C) and contact time (T) (C × T) criteria for Collector Well 5 since the water from all the collector wells was quickly combined. This was particularly troublesome during the winter months when the C × T levels are higher; therefore, the Agency petitioned the California Department of Health Services, requesting that the system be considered groundwater unless river conditions result in poor water quality in Well 5. The California Department of Health Services issued a conditional permit for the use of Well 5 in June 1995 as long as river conditions do not adversely affect well-water quality. In essence, this decision required further studies to define the condition that would enable the use of the well as groundwater not considered under the influence of surface water.

The Agency initiated a study between February 1997 and April 1998 to determine the river conditions in which Collector Well 5 would be classified as groundwater under the direct influence of surface water (Price et al., 1999). It was determined that for 83 percent of the time, Collector Well 5 was considered groundwater while, for 17 percent of the time, it was determined to be groundwater under the direct influence of surface water. Most of the poor-quality water was produced during the winter months when the river level is high and water demand is low. Based upon the study, the Sonoma County Water Agency developed a well management plan in which Collector Well 5 would not operate during low-water demand winter months when river-water quality is poor. The Agency proposed to operate Collector Well 5 when the flow in the Russian

River is below 142 cubic meters per second (m³/s). This approval is still conditional and requires the management of the well with respect to river flow.

The City of Kearney, Nebraska

The City of Kearny, Nebraska, derives its water from 12 wells located on nearby Killgore Island in the Platte River. The wells are 17- to 18-m deep, and the pumping capacity of each well varies between 0.063 to 0.095 m³/s. The City's water supply was determined to be groundwater under the direct influence of surface water and, therefore, was subject to the Surface Water Treatment Rule. This meant that the City was required to build a filtration plant that would cost between $15 and $30 million. To avoid the cost of construction, the City undertook a 6-month monitoring program for *Giardia* and viruses in 1995. Based upon the results, the Nebraska Department of Health Services (the primacy agency) gave conditional approval of 2-log filtration credit for *Giardia* and 1-log credit for viruses as well as required further studies over a longer period of time. The City undertook another monitoring program between October 1997 and October 1998 and demonstrated that the system easily removed between 2 and 2.5 logs of *Giardia*. As a result, the Nebraska Department of Health Services gave the final approval for 2-log removal of *Giardia* and 1-log removal for viruses. The Nebraska Department of Health Services also asked the City to continue its monitoring efforts to observe if there any conditions in the river that would adversely affect filtrate quality. The City has contracted a consulting firm to continue this work.

Central Wyoming Regional Water System

The Central Wyoming Regional Water System in Casper, Wyoming, operates a number of vertical wells, an infiltration gallery, and three horizontal collector wells on the banks of the North Platte River. In evaluating the results of a series of monitoring activities (including microscopic particulate analysis), the Region 8 office of the U.S. Environmental Protection Agency (the primacy agency in the State of Wyoming) classified the system as groundwater under the direct influence of surface water, thus subjecting the system to the Surface Water Treatment Rule. When the Central Wyoming Regional Water System petitioned the U.S. Environmental Protection Agency to consider the well field's RBF process as an alternative treatment technology that could remove 3-logs of *Giardia* and 4-logs of viruses, the U.S. Environmental Protection Agency responded that, because of inadequate data, it would not provide the log removal credit. The U.S. Environmental Protection Agency also imposed other criteria, such as implementing a regional wellhead protection program, providing secondary disinfection, monitoring chlorine and turbidity as per the Surface Water Treatment Rule, and having redundant disinfection units; however, on December 10, 2001, the U.S. Environmental Protection Agency reversed its position and granted conditional approval to the Central Wyoming Regional Water System to consider its system as an alternative filtration technology that achieves 99-percent removal of *Cryptosporidium* oocysts. This was contingent, in part, upon the design and completion of a study to demonstrate that 99-percent of *Cryptosporidium* is actually removed by filtration. The Central Wyoming Regional Water System will submit the findings of the study to the U.S. Environmental Protection Agency by November 1, 2003.

7. The Value of Applying RBF as a Pretreatment Technology

Based upon the above discussions, it is clear that RBF systems can help utilities in various ways. Most importantly, RBF is an asset to these utilities. The purpose of investing in such an asset (e.g., the wells, delivery system, and treatment trains) is to benefit from the services that make an RBF system valued by both the utility and consumers receiving the product of that investment: drinking water. The various services that high-quality drinking water provide often have unrecognized values. One of the unrecognized values of high-quality drinking water is avoided medical cost. Additional unrecognized values include longer life span, cancer risk reduction, and enhanced environments such as the wetlands, lakes, or rivers where recreational activities are centered. Table I-2 shows several values that can result from the services provided by RBF.

Table I-2. The Value of RBF

Services and Benefits	Value
Contaminant (Pathogen/Chemical) Removal	Reduced Medical Costs Longer Life Span Improved Productivity Capital Cost Reduction Cancer Risk Reduction Enhanced Environment
Reduced Maintenance	Capital Cost Reduction
Improved Reliability (as Source Water)	Drought Protection
Nutrient (Organics) Removal	Reduced Treatment Costs Lower Regulatory Scrutiny Lower Monitoring Costs
Enhanced Community Supply (Due to Total Dissolved Solids Reduction)	Greater Customer Satisfaction Lower Corrosion of Household Plumbing

As can be seen, there are numerous advantages in using RBF as a pretreatment technology. The value of RBF is not just found in reduced treatment and delivery costs, but also in the many invaluable services it provides to the consumer, environment, and future generations.

References

American Public Health Association, American Water Works Association, and Water Environment Federation (1998). *Standard Methods for the Examination of Water and Wastewater,* 20th Edition, American Public Health Association, Washington, D.C.

Bellamy, W.D., G.P. Silverman, D.W. Hendricks, and G.S. Logsdon (1985). "Removing *Giardia* cysts with slow sand filtration." *Journal American Water Works Association,* 77(12): 52.

Brauch, H.J., U. Müller, and W. Kühn (2001). "Experiences with riverbank filtration in Germany." *Proceedings, International Riverbank Filtration Conference,* Internationale Arbeitsgemeinschaft der Wasserwerke im Rheineinzugsgebiet (IAWR), Amsterdam, The Netherlands, 33-40.

Cleasby, J.L. (1990). "Filtration." *Water quality and treatment,* F.W. Pontius, ed., McGraw Hill, Inc., New York.

Cullen, T.R., and R.D. Letterman (1985). "The effect of slow sand filter maintenance on water quality." *Journal American Water Works Association,* 77(12): 48.

Daughton, C.G., and T.A. Ternes (1999). "Pharmaceuticals and personal care products in the environment: Agents of subtle change?" *Environmental Health Perspective,* 107(6): 907-938.

Ellis, K.V. (1985). "Slow sand filtration." *CRC Critical Reviews in Environmental Control,* 15(4): 315-354.

Fogel, D., J. Issac-Renton, R. Guaspirini, W. Moorehead, and J. Ongerth (1993). "Removing *Giardia* and *Cryptosporidium* by slow sand filtration." *Journal American Water Works Association,* 85(11): 77-84.

Grischek, T., E. Worch, and W. Nestler (2001). "Is bank filtration under anoxic conditions feasible?" *Proceedings, International Riverbank Filtration Conference,* Internationale Arbeitsgemeinschaft der Wasserwerke im Rheineinzugsgebiet (IAWR), Amsterdam, The Netherlands, 57-65.

Heberer, T., U. Dünnbier, C. Reilich, and H.J. Stan (1997). "Detection of drugs and drug metabolites in groundwater samples of a drinking water treatment plant." *Fresenius' Environmental Bulletin,* 6: 438-443.

Heberer, T., K. Schmidt-Bäumler, and H.J. Stan (1998). "Occurrence and distribution of organic contaminants in the aquatic system in Berlin-Part I: Drug residues and other polar contaminants in Berlin surface and groundwater." *Acta Hydrochimica et Hydrobiologica*, 26: 272-278.

Kolpin, D.W., E.T. Furlong, M.T. Meyer, E.M. Thurman, S.D. Zaugg, L.B. Barber, and H.T. Buxton (2002). "Pharmaceuticals, hormones, and other organic wastewater contaminants in U.S. Streams, 1999-2000: A national reconnaissance." *Environmental Science and Technology*, 36(6): 1202-1211.

Medema, G.J., M.H.A. Juhàsz-Holterman, and J.A. Luijten (2001). "Removal of micro-organisms by bank filtration in a gravel-sand soil." *Proceedings, International Riverbank Filtration Conference*, Internationale Arbeitsgemeinschaft der Wasserwerke im Rheineinzugsgebiet (IAWR), Amsterdam, The Netherlands, 161-168.

Messner, M.J., and R.L. Wolpert (2000). "Occurrence of *Cryptosporidium* in the nation's drinking water sources – ICR data analysis." *Proceedings, Water Quality Technology Conference*, American Water Works Association, Denver, Colorado.

Mikels, M.S. (1992). "Characterizing the influence of surface water on water produced by collector wells." *Journal American Water Works Association*, 84(9): 77-84.

Mucha, I., D. Rodak, Z. Hlavaty, and L. Bansky (2002). "Ground water quality processes after the bank infiltration from the Danube at Cunovo." *Riverbank filtration: Understanding biogeochemistry and pathogen removal*, C. Ray, F. Laszlo, and A. Bourg, eds., Kluwer Academic Publishers, Dordrecht, The Netherlands (in press).

Price, M.L., J. Flugum, P. Jeane, and L. Tribbet-Peelen (1999). "Sonoma County finds groundwater under direct influence of surface water depends on river conditions." *Proceedings, International Riverbank Filtration Conference*, National Water Research Institute, Fountain Valley, California.

Ray, C., D.K. Borah, D. Soong, and G.S. Roadcap (1998). "Agricultural chemicals: Impacts on riparian municipal wells during floods." *Journal American Water Works Association*, 90(7): 90-100.

Sacher, F., H.J. Brauch, and W. Kühn (2001). "Fate of hydrophilic organic micropollutants in riverbank filtration." *Proceedings, International Riverbank Filtration Conference*, Internationale Arbeitsgemeinschaft der Wasserwerke im Rheineinzugsgebiet (IAWR), Amsterdam, The Netherlands, 139-148.

Shaw, S., S. Regli, and J. Chen (2002). "Virus occurrence and health risks in drinking water." *Information collection rule data analysis*, M.J. McGuire, J.L McLain, and A. Obolensky, eds., American Water Works Association Research Foundation, Denver, Colorado, in press.

Stamer, J.K. (1996). "Water supply implications of herbicide sampling." *Journal American Water Works Association*, 88(2): 76-85.

Stamer, J.K., and M.E. Wieczorek (1996). "Pesticide distribution in surface water." *Journal American Water Works Association*, 88(11): 79-87.

U.S. Environmental Protection Agency (1989). "Drinking water; National primary drinking water regulations: Disinfection; Turbidity, *Giardia lamblia*, viruses, legionella, and heterotrophic bacteria; Final rule." *Fed. Reg.* 54:27486, U.S. Environmental Protection Agency, Washington, D.C.

U.S. Environmental Protection Agency (1992). "Consensus method for determining groundwaters under the direct influence of surface water using Microscopic Particulate Analysis (MPA)." *EPA 9009-92-029*, U.S. Environmental Protection Agency, Washington, D.C.

U.S. Environmental Protection Agency (1998). "National primary drinking water regulations. Interim Enhanced Surface Water Treatment Rule." *Fed. Reg.* 63:69478, U.S. Environmental Protection Agency, Washington, D.C.

U.S. Environmental Protection Agency (2000). "ICR Auxiliary 1 Database, Version 5.0." *EPA 815-C-00-002*, U.S. Environmental Protection Agency, Washington, D.C.

U.S. Environmental Protection Agency (2002). "National primary drinking water regulations. Long Term 1 Enhanced Surface Water Treatment." *Fed. Reg.* 67:1811, U.S. Environmental Protection Agency, Washington, D.C.

Verstraeten, I.M., J.D. Carr, G.V. Steele, E.M. Thurman, K.C. Bastian, and D.F. Dormedy (1999). "Surface water/ground water interactions: Herbicide transport into municipal collector wells." *Journal of Environmental Quality*, 28(5): 396-405.

Wang, J. (2002). "Riverbank filtration case study at Louisville, Kentucky." *Riverbank filtration: Improving source-water quality*, C. Ray, G. Melin, and R.B. Linsky, eds., Kluwer Academic Publishers, Dordrecht, The Netherlands.

Wang, J., R. Song, S. Hubbs (2001). "Particle removal through riverbank filtration process." *Proceedings, International Riverbank Filtration Conference*, W. Julich and J. Schubert, eds., Internationale Arbeitsgemeinschaft der Wasserwerke im Rheineinzugsgebiet (IAWR), Amsterdam, The Netherlands, 127-138.

Wang, W., and P. Squillace (1994). "Herbicide interaction between a stream and an adjacent alluvial aquifer." *Environmental Science and Technology*, 28: 2336-2344.

Weiss, W.J., E.J. Bouwer, W.P. Ball, C.R. O'Melia, H. Aurora, and T.F. Speth (2002). "Reduction in disinfection, byproduct precursors and pathogens during riverbank filtration at three Midwestern United States drinking-water utilities." *Riverbank filtration: Improving source-water quality*, C. Ray, G. Melin, and R.B. Linsky, eds., Kluwer Academic Publishers, Dordrecht, The Netherlands.

Ziegler, D., C. Hartig, S. Wischnak, and M. Jekel (2001). "Behaviour of dissolved organic compounds and pharmaceuticals during lake bank filtration in Berlin." *Proceedings, International Riverbank Filtration Conference*, W. Julich and J. Schubert, eds., Internationale Arbeitsgemeinschaft der Wasserwerke im Rheineinzugsgebiet (IAWR), Amsterdam, The Netherlands, 151-160.

Part I:
Systems

Chapter 1. Conceptual Design of Riverbank Filtration Systems

Henry Hunt, CPG
Collector Wells International, Inc.
Columbus, Ohio, United States

Jürgen Schubert, M.Sc.
Stadtwerke Düsseldorf AG
Düsseldorf, Germany

Chittaranjan Ray, Ph.D., P.E.
University of Hawaii at Mānoa
Honolulu, Hawaii, United States

1. Introduction

When considering the application of RBF, it is incumbent upon the engineer to recognize that the following parameters may affect the performance of an RBF system:

- Available river water that can be induced to flow into the aquifer.
- Quality of river water.
- Commercial river traffic (a source of pollution; dredging may also be necessary).
- Flow velocity and bed load characteristics.
- Seasonality of river flow.
- Stability of the river channel.

Most RBF systems are constructed in alluvial aquifers located along riverbanks. These aquifers can consist of a variety of deposits ranging from sand, to sand and gravel, to large cobbles and boulders. Ideal conditions typically include coarse-grained, permeable water-bearing deposits that are hydraulically connected with riverbed materials. These deposits are found in deep and wide valleys or in narrow and shallow valleys. RBF systems in deep and wide valleys may have a wider range of options since wells (vertical and horizontal collector wells) can be placed at greater depths (which can provide higher capacities) and can be placed farther away from the river to increase the degree of filtration. In a narrow, shallow valley, horizontal collector wells may be more advantageous than vertical wells since well screens can be placed at the lowest elevation (maximizing the available drawdown) and extended out beneath the riverbed, and longer lengths of screen can be installed to minimize entrance velocities.

RBF systems can even be constructed in low permeability zones (typically, clay and silt layers) within an alluvial aquifer. If the confining layers are extensive and continuous, well screens can be placed above the confining layer to infiltrate water from the surface source; well screens can be placed below the confining layer to obtain maximum filtration, whereby the well may not be classified as groundwater under the direct influence of surface water; and well screens can also be placed both above and below the confining layer to maximize the capacity available. If low permeability zones are discontinuous on a

19

C. Ray et al. (eds.), Riverbank Filtration, 19–27.

local or regional scale, it should be possible for water to infiltrate from the surface-water source and migrate downward to the screen around these semi-confining layers. Some utilities have reported that redox zones form down-gradient of these low permeability zones. Redox zones are areas where the oxygen concentration is very low and reduced species of chemicals are present.

The conditions of each project site must be suitable for vertical wells, directionally-drilled horizontal wells, or horizontal collector wells, including cost comparisons. Typically, a horizontal collector well can develop a capacity equivalent to multiple vertical wells and multiple (directionally-drilled) horizontal wells or galleries; therefore, total system costs must be developed to facilitate this comparison to include both capital and long-term operation and maintenance costs.

2. Well Types and the Suitability of Vertical Versus Horizontal Collector Wells

Historically, three types of wells have been used for RBF since the technology was first established in the 1800s. They include:

- *Horizontal Collector Well:* A circular central collection caisson sunk into the ground with horizontal lateral well screens pushed out into unconsolidated aquifer deposits, in many cases into alluvial deposits beneath a river or lake. It is typically used by United States water utilities to produce drinking-water supplies from groundwater sources or from riverbanks through filtration (Figure 1-1). Horizontal collector wells are also called "collector wells" in the United States.
- *Vertical Well:* A tubular well that is drilled vertically downward into a water-bearing stratum or under the bed of a lake or stream (Figure 1-2).
- *Pit Well:* A shallow, large-diameter well that, in most instances, is manually dug into the ground (a pit well is either constructed by excavating with power machinery or by hand tools rather than drilling or driving). Typically, a pit well (also known as a "dug well" in the United States) is constructed for an individual residential water supply.

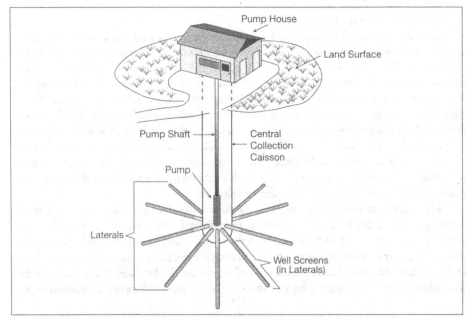

Figure 1-1. Horizontal collector well with a pump house.

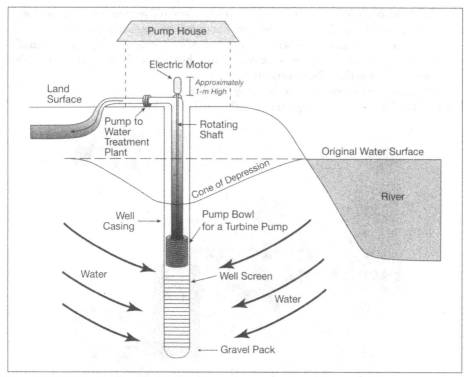

Figure 1-2. Vertical well with a pump house.

While vertical and horizontal collector wells have been used primarily for RBF throughout the years, some open pit wells were used in the nineteenth century. These wells reflect the technical means available at the time. Pit wells (also referred to as "dug wells") were cylindrical in shape and had diameters between 7 to 10 m (dug well diameters are smaller in the United States), with perforated walls in the lower part of the brick masonry. Groundwater could penetrate through the open ground and the perforated wall. The water yield of those simple wells was about 500 cubic meters per hour (m³/h). Another historical well used for RBF was a perforated collector pipe located in a shallow aquifer, which functioned as an infiltration gallery.

Drilling and construction technologies were developed at the end of the nineteenth century that allowed pit wells to be replaced with vertical filter wells; however, the equipment used to pump water from the subsurface was limited. Only piston pumps, driven by steam engines, were available. Because water yields of the former vertical filter wells (particularly in shallow aquifers) were low compared to pit wells, the siphon tube concept was introduced in Germany to extract water from a great number of vertical filter wells at one time using only one pump. Siphon well systems are generally connected via a discharge manifold to one or more suction pumps. They are used in shallow aquifers where the water level is lower than the suction lift of the pumps. In the United States, siphon wells are used in construction dewatering where fewer pumps are needed and, through connection to a discharge manifold, many siphon wells (which are usually smaller diameter than conventional vertical wells) can be connected in series. In general, siphon tube systems can support the construction of well galleries parallel to a riverbank, with up to 100 vertical filter wells connected together.

The siphon tube concept has been used for approximately 100 years to connect vertical well galleries. As a technique, it has survived because of numerous advantages, including:

- Low operation and maintenance costs (only one collector caisson with pumps is needed).
- Easy adaptation to varying raw water demand (e.g., by a few variable-frequency or variable-speed pumps in the collector caisson).
- Uniform stress of production wells during operation.

Figures 1-3 to 1-5 show a typical siphon tube well setup used in Germany.

Figure 1-3. Well gallery in Düsseldorf, Germany.

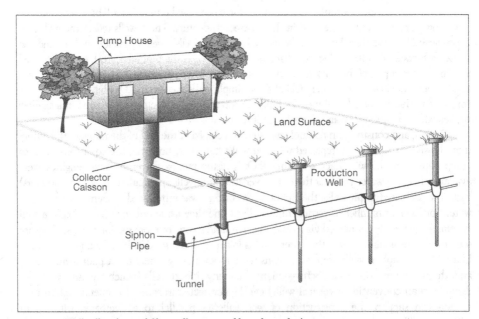

Figure 1-4. Well gallery (vertical filter wells connected by siphon tubes).

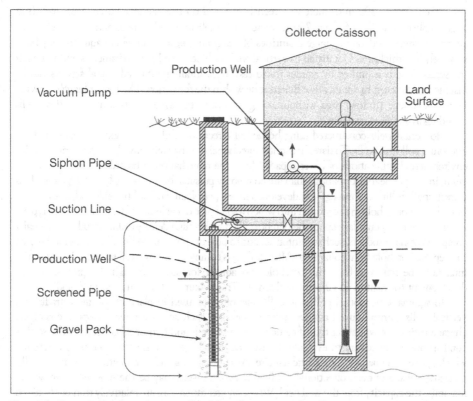

Figure 1-5. Cross-section of a production well and the collector caisson from Figure 1-4.

Traditionally, vertical wells are used for developing groundwater supplies in alluvial aquifer systems. The screen lengths of vertical wells are controlled by the saturated thickness of the aquifer and desired pump setting. In general, the pumps are set above the screen zone to prevent the introduction of air directly into the formation through the well screen and to improve flow conditions. In some cases, a drop pipe is added to the pump suction to further improve the flow hydraulics within the screen. Screen lengths and diameters are selected to control entrance velocities and to avoid pumping excessive sand and fine particles from the aquifer. Since the pumps must be placed above the screens, the available drawdown would be equal to the height of the original water surface minus the location of the pump, plus a safety factor to provide adequate submergence for the pump and to avoid breaking suction.

During the last 70 years, horizontal collector wells have been developed for the production of groundwater in unconsolidated, water-filled sediments (Radke and Hüper, 2001). Some of the distinct advantages that horizontal collector wells offer are:

- More of the available drawdown can be used since well screens can be installed at a lower elevation in the aquifer.
- More well screen can be exposed to the aquifer at a given site since screen length is not limited by the saturated thickness of the formation. This increased screen length lowers the entrance velocity of the water through the well screen, reducing the rate of clogging and minimizing head loss between the aquifer and the well; however, the screen length itself may be limited by the hydraulic losses inside the screen pipe.

Often, 150 to 300 m or more of screen is installed on the laterals of a collector well (the largest collector well so far has 750 m of screen). Collector wells are a particularly effective alternate where moderate to large quantities of water are needed and where aquifers may have relatively low thickness. Compared to vertical wells (which have limited diameters and in which the screen length is limited by aquifer thickness), horizontally positioned lateral screens can be installed in the most hydraulically efficient zone within the formation. The longer lengths provide increased surface (inflow) area without being subject to the same limitations as vertical wells regarding available drawdown.

Horizontal wells constructed using horizontal directional drilling methods are different from horizontal collector wells. To date, horizontal directional drilling wells have been primarily used for environmental applications of lower capacity (e.g., the remediation of hazardous waste sites) and, often, in less permeable formations than are commonly pursued for water-supply development. This (directional) drilling technology was developed in the petroleum and utility fields and, as such, was intended to bore a hole in the ground into which a solid pipe or cable would be installed. The typical drilling technology entails keeping the borehole filled with a high density drilling mud slurry, usually composed of clay products, such as bentonite, until a pipe or cable can be installed. This mud is used to keep the borehole from collapsing. After drilling is complete and the well screen is installed, the mud must be removed from the borehole and aquifer formation through the process of well development to result in an efficient well screen. This maximizes the well capacity (yield).

In summary, horizontal collector wells may often be used in place of the more traditional vertical wells within alluvial aquifer systems; however, each site will have site-specific geologic characteristics that will affect the efficiency of vertical wells and horizontal collector wells, so that comparisons are necessary to evaluate the respective yields and construction costs to determine which system may be best for a particular site. At many sites, some combination of vertical wells, collector wells, and even directionally-drilled horizontal wells may be the most effective way to maximize the capacity from the well field. Where aquifer formations are relatively thin or of limited extent, horizontal collector wells may be advantageous in maximizing the yield possible from available sites and in minimizing the number of pumping facilities needed. Obviously, where a suitable hydraulic interconnection exists between alluvial aquifer systems and an adjacent surface-water source, yields can be maximized. A schematic of horizontal collector wells and vertical wells designed in the United States is shown in Figure 1-6, while Figure 1-7 shows an alternate schematic in which horizontal filter wells (also referred to as "collector wells" in Europe) are constructed some distance away from the river with their laterals fully extended within the aquifer.

The importance of horizontal collector wells for RBF may be demonstrated by the number of these wells along European rivers:

- Along the Rhine River, more than 50 collector wells are under operation. The Düsseldorf Waterworks in Germany operates 12 horizontal collector (filter type) wells with a capacity between 900 to 3,000 m^3/h each.
- More than 200 collector wells are operational in the Danube region.
- There are also collector wells along the Save, Main, Maas, Ruhr, Enns, Elbe, and Oder Rivers.

There is continuous debate whether vertical filter wells or horizontal collector wells should be selected for RBF plants. The decision in each particular case must regard site conditions, most notably the hydrogeological situation of the aquifer and the hydraulic conditions in the river, especially concerning riverbed clogging.

The saturated thickness of the aquifer should not be less than 6 m and the transmissivity in the range of 1,500 meters squared per day (m^2/d) or higher. An assessment of the tendency for riverbed clogging must consider:

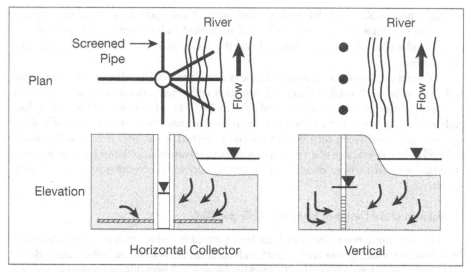

Figure 1-6. Representation of both a horizontal collector well and a vertical well (United States design). Modified and reproduced with permission from Ray et al. (2002).

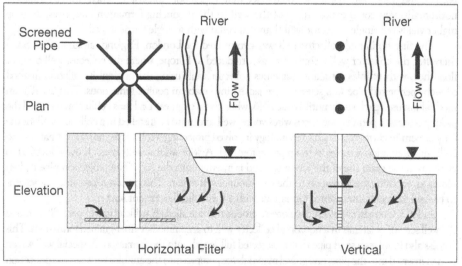

Figure 1-7. Representation of both a horizontal filter well and a vertical RBF well (alternate design).

- Regional situation of the river.
- Flow regime.
- Bed load transport situation.
- Site of the production well at the inner or outer bend of the river.
- Induced infiltration velocities near the riverbed.

When wells are over-pumped (i.e., pumping water from the ground faster than it can be recharged), several problems arise. Over-pumping results in higher infiltration velocities at the river/aquifer interface as well as amplified clogging of the interstitial space beneath the riverbed, making it inaccessible for rehabilitation and restoration. These results adversely affect long-term infiltration capacities and lower the well yield. A detailed hydrogeologic investigation should determine the optimal pumping capacity during the design phase of the RBF projects to avoid such problems.

If the site conditions do not restrict the use of a collector well (based upon first design of the wells or a feasibility evaluation), the capital, operation, and maintenance costs of both alternatives (a series of vertical wells versus a horizontal collector well) should be compared for resulting life-cycle costs. It is common for collector wells to have an advantage over vertical wells when operation and maintenance costs are compared for a specified period, and collector wells can also be found to be cost-effective when total system costs are compared (a collector well versus a series of vertical wells that will produce the equivalent capacity with connecting pipelines, electrical service, etc.).

3. Evolution of the Design of Horizontal Collector Wells

The use of horizontal collector wells for developing infiltrated water supplies originated in the 1930s after petroleum engineer Leo Ranney found that falling prices for oil made directionally-drilled horizontal oil wells less cost-effective. He modified his approach from drilling horizontal boreholes into oil-bearing rock formations to a hydraulic jacking process in which perforated pipes are installed into unconsolidated sand and gravel water-bearing aquifer formations. His theory, for both oil and water, was that if you could place wells (open boreholes or screened) into a formation horizontally, you could expose more of the well to the producing formation and, thus, develop higher yields per single collector well than you could with a single vertical well.

The first horizontal collector well was constructed for London, England, around 1933. Soon thereafter, the collector well technology was introduced to Europe, where the collector well concept flourished, with utilities installing numerous collector wells using the original installation method whereby perforated pipe well screens were jacked into place in aquifer formations. This installation method was used exclusively until about 1946, when Swiss engineer Dr. Hans Fehlmann modified the jacking process to permit continuous wire-wound well screens to be installed in a collector well for the City of Bern in Switzerland. This technology involved projecting a solid pipe into the formation and collecting formation samples as the pipe is projected. A wire-wound well screen is then designed to conform to the grain size of the formation and is inserted into the pipe. The projection pipe is then retracted, exposing the formation to the wire-wound well screen. This process allows fine slot screens to be used to match fine-grained formations with a hydraulically efficient screen.

In 1953, German engineers developed a process that installed an artificial gravel-pack filter around the well screens of laterals in a horizontal collector well to accommodate finer-grained formations. This process also involves a solid pipe that is projected full-length into the formation. A special well screen is then inserted into the pipe, and gravel materials are pumped into the annulus between the projection pipe and the screen while the projection pipe is retracted. The use of an artificial gravel-pack filter provides a transition between fine-grained formation deposits and more efficient screen openings.

These two advances in collector well technology improved the hydraulic efficiency of collector wells and permitted collector well laterals to be installed in a wider range of geologic formations. Both the Fehlmann and gravel-packing technologies were brought into the United States in the mid-1980s and have been used extensively since. In 1985, two collector wells were built in the Midwest using the process developed by Dr. Fehlmann. Concurrently, the first gravel-packed collector well screens used in the United States were installed in New Jersey.

As the design and construction process for the horizontal collector well evolved, it became evident that wells installed adjacent to and, sometimes, underneath surface-water sources were able to develop large quantities of water. As water levels were lowered by pumping, the hydraulic gradients in the aquifer permitted water to be infiltrated from an adjacent river or lake, providing recharge into the aquifer to replenish water removed by pumping. This infiltration process pre-filters river water as it percolates through the riverbed sediments toward the aquifer (recharging it) and, ultimately, into the well screens, typically removing objectionable characteristics of the river water, such as turbidity and microorganisms. Because the "recharge water" from the river is infiltrated over such a large area, infiltration rates are extremely low, providing a high degree of filtration in most cases.

During the first 50 years that collector wells were installed, they were often built immediately adjacent to surface-water sources to:

- Be in close proximity to the apparent source of recharge.
- Take advantage of the filtering capacity of the riverbed sediments and aquifer to provide high-capacity infiltrated water supplies.

During that time, these well systems promoted the fact that they could induce the infiltration of moderate to very high quantities of filtered surface water using RBF principles.

As regulatory agencies began evaluating groundwater under the direct influence of surface water issues in the 1990s, siting and design philosophies for collector wells were revised to:

- Improve the filtration of surface water.
- Locate wells to minimize the potential for contamination from surface-water sources.
- Improve caisson installation methods to minimize disturbance to the aquifer.
- Improve surface-sealing techniques around the caisson.

This involved the proper selection of the horizon (elevation) for projecting the lateral screens and sometimes locating the wells a sufficient distance back from the river to increase the degree of filtration and travel time for recharge water. The ability (or efficiency) of the streambed and aquifer to filter out objectionable microorganisms and to reduce the turbidity from surface-water sources will vary from region to region and, certainly, from site to site. In most alluvial settings, it should be possible to achieve some degree of filtration to improve water quality.

References

Ray, C., T. Grischek, J. Schubert, J. Wang, and T. Speth (2002). "A perspective of riverbank filtration." *Journal American Water Works Association*, 94(4): 149-160.

Radke, B., and G. Hüper (2001). "Riverbank filtration with horizontal collector wells." *Proceedings of the International Riverbank Filtration Conference.* W. Jülich and J. Schubert, eds., Internationale Arbeitsgemeinschaft der Wasserwerke im Rheineinzugsgebiet (IAWR), Amsterdam, The Netherlands.

Chapter 2. American Experience in Installing Horizontal Collector Wells

Henry Hunt, CPG
Collector Wells International, Inc.
Columbus, Ohio, United States

1. Introduction

RBF has been used to develop moderate to very high capacities of infiltrated water along United States waterways using horizontal collector wells since the mid-1930s. For many years, these wells were considered to be an "alternative" approach to standard or conventional water wells; however, continued advances in the technology (coupled with an increased focus on RBF as a treatment strategy) increased the awareness, acceptance, and popularity of this well design. As new water sources are being sought (for new well fields or to replace direct surface-water intake withdrawals), the use of collector wells can be considered as a viable alternative where hydrogeologic conditions are favorable.

2. Timeline

The following timeline presents notable dates regarding the use of collector wells:
- 1927 First horizontal oil well (Texas).
- 1930s First collector well for oil (Ohio).
- 1933 First collector well for water (London, England).
- 1936 First collector well in the United States for water (Ohio).
- 1940 First collector wells in the United States for public drinking water using RBF principles (New York and New Jersey).
- 1944 First use of a collector well for artificial recharge (Kentucky).
- 1946 First collector well using Fehlmann technology (Switzerland).
- 1953 First collector well using gravel-packing for lateral well screens (Germany).
- 1954 First collector well for filtered seawater (California).
- 1985 First collector wells in the United States using Fehlmann technology (Ohio and Michigan).
- 1985 First collector well in the United States using gravel-packing for lateral screens (New Jersey).
- 1997 Highest capacity (>1.75 m³/s) collector well in the world (Kansas City, Kansas).

3. Historical Progression

The first collector well installed in the United States was for industrial use, as were 46 of the first 50 collector wells. Only about one-third of the first 100 collector wells built over the first 20 years were used for public-drinking water; these wells were sited along rivers to take advantage of induced infiltration to support yields. The early construction procedures for collector wells were better suited to large capacity users; therefore, these wells were more attractive to large industrial

29

C. Ray et al. (eds.), Riverbank Filtration, 29–34.

users than to public drinking-water systems. As the construction procedures were refined and improved, collector wells became more appropriate for municipal use. The sustained development of public drinking-water supplies in the United States did not occur until the late 1940s.

This trend continued through the years, using induced infiltration to recharge the aquifer to support well yields from both horizontal collector wells and conventional vertical wells sited along rivers and streams. Induced infiltration has long been recognized as a viable process for recharging alluvial aquifers to support moderate to very high well yields; however, the use of RBF as a viable treatment process had not been widely recognized or acknowledged in the United States until recently. In 1990, two horizontal collector wells were installed for the City of Lincoln in Nebraska to provide an infiltrated water supply of between 1.53 and 1.75 m^3/s from the Platte River. This was followed by the installation of a 0.66- to 0.88-m^3/s horizontal collector well for the Louisville Water Company in Kentucky in 1999. Both cases serve as examples of modern-day RBF installations and as forerunners of applying this technology for the pretreatment of surface water in the United States. The Des Moines Water Works in Iowa also installed a horizontal directionally drilled well and several horizontal collector wells to produce an infiltrated water supply of 1.1 m^3/s from the Raccoon River alluvial aquifer.

The high visibility of these projects has helped promote the use of horizontal collector wells in alluvial aquifers to take advantage of the filtration capabilities of natural riverbed and riverbank sediments. Utilities located along alluvial valleys that now treat surface water are considering the use of RBF as a cost-effective alternative to retrofit treatment systems to meet future water-treatment requirements for parameters like turbidity and temperature and for removing pathogens, such as *Giardia* and *Cryptosporidium*. It is expected that other utilities will investigate the use of RBF as a pretreatment option for meeting future water-supply demands and treatment goals.

A number of communities across the United States develop groundwater from alluvial aquifer systems along rivers and streams using collector wells. Current well siting practices include locating the caisson and lateral well screens away from the surface-water source. This increases the travel time for water that may infiltrate as recharge into the aquifer. At many sites, riverbed sediments and aquifer deposits provide filtration to remove turbidity and other microscopic particulates from water before the water reaches the collector well caisson.

4. Collector Well Construction

The construction of collector wells involves two main components: the concrete wet well caisson and lateral well screens. The central caisson is made of reinforced concrete that is constructed using the open-end caisson method, whereby each section (called "lifts") is formed and poured at ground surface and is sunk into place by excavating soils from within the caisson. The lifts are tied together with reinforcing steel, and water stops as each lift settles to ground level. The lower section is fitted with openings that will be used for projecting the lateral well screens. As these openings reach the design depth for the lateral screens, excavation is stopped and a bottom-sealing plug is poured in to the caisson (Figure 2-1). The concrete caissons are typically constructed with an inside diameter ranging from 3 to 6 m or larger, if necessary. The caissons can be installed to depths of 46 m using normal construction methods, and possibly deeper using special hydraulic equipment. The average depth of the caissons in the United States is 21 m and the average diameter is 4 m.

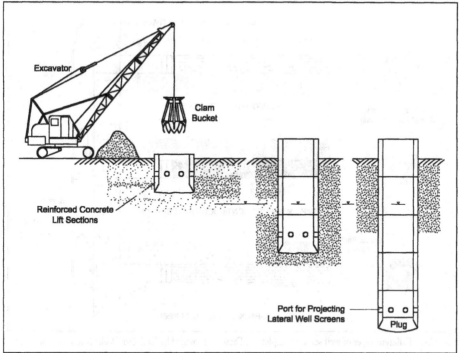

Figure 2-1. Schematic of the sinking of caisson during the construction of a collector well. Drawing provided by Collector Wells International, Inc.©

Water is pumped from inside the caisson, and the laterals are projected through the openings in caisson walls. There are three standard methods for installing lateral well screens:

- Original method using perforated pipe sections.
- Projection pipe method (developed by Dr. Hans Fehlmann).
- German method for installing an artificial gravel-pack filter around well screens.

This multi-design capability allows the lateral well screens to be matched most efficiently with the formation materials to be screened. Figure 2-2 shows the three methods (Hunt, 1985).

Original Method

The original method to install lateral well screens involves projecting pipe sections that have been perforated by punching or sawing. The pipe sections are directly attached to a digging head that is used to direct the projection of the lateral pipe. In this approach, the pipe sections are projected into the aquifer and left in place. The openings on the pipe typically provide a maximum open area of 20 percent, which is limited since the pipe needs to be strong enough to accommodate the jacking forces used during projection. The perforated pipe well screen has been projected in diameters of 20, 30, and 40 centimeters (cm) out to a maximum length of about 107 m. Using this method, the perforated pipe is most commonly made of standard carbon-steel materials due to cost. Occasionally, stainless steel or special alloy materials are used. Because of the method for perforating the pipe, the minimum slot size that can be made is sometimes too large to sufficiently retain fine-grained formation materials for efficient well development.

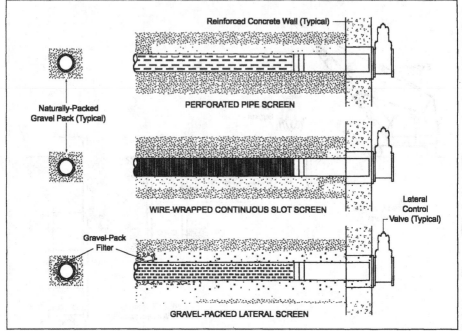

Figure 2-2. Different types of well screen completion. Drawing provided by Collector Wells International, Inc.©

Projection Pipe Method

The projection pipe method involves the use of a special heavy-duty projection pipe that is pushed out into the aquifer formation. During the projection process, formation samples are collected and analyzed for grain-size distribution. Once the pipe has been placed in the aquifer to the desired distance, a wire-wrapped continuous-slot well screen (with slot openings selected to conform to the aquifer deposits encountered) is inserted inside the projection pipe. The projection pipe is then withdrawn so that it may be used in projecting the next lateral. The lateral lengths range from about 30 to 75 m using this method, with 20- or 30-cm diameter screens installed. Because of this method, the well screen gains the following advantages:

- More open area (up to 40 percent or more).
- More durable construction (stainless steel is normally used).
- More flexibility in slot size to accommodate a wide range of formation deposits.

This method also provides the ability to use well screen materials (other than steel) that are applicable in saline and brackish environments. It is also possible to install laterals in formations containing large cobbles and boulders using a modification of this technique.

Gravel-Packing Method

The gravel-packing method uses a projection pipe for the initial projection. For this method, formation samples are also collected as the pipe is projected. Once the projection pipe has been pushed to the full design length, specially designed well screens (usually stainless steel) are inserted and an artificial gravel-pack filter is placed around the well screens as the projection pipe is withdrawn. This permits the installation of a gravel filter to act as a transition zone between a fine-grained aquifer formation and the slots in the well screen to prevent ongoing sand intrusion into the well. This method has been used for both seawater and fresh water (inland) applications.

5. Hydrogeological Investigation/Testing

Prior to selecting the appropriate well design alternative and potential well sites, a hydrogeological investigation is typically conducted at prospective sites. The investigation process is designed to:

- Evaluate the hydraulic characteristics of each formation to select the most appropriate horizons within the formation for installing well screens.
- Select the most efficient method for installing well screens to maximize the possible yield from each site.

During the project planning and engineering phase, the City of Lincoln, Nebraska, considered installing a series of 13 vertical wells (the City had operated a well field comprised of 38 vertical wells for a number of years) or two radial collector wells to meet future needs. During the evaluation, site-specific hydrogeological investigations to examine exploratory test drilling and detailed aquifer testing to determine the hydraulic characteristics of the aquifer were necessary to predict well yields and prepare well design. The testing determined that several potential sites were available on an island in the Platte River, and hydraulic interval testing within several test boreholes identified the most hydraulically efficient horizon within the aquifer formation for placing the lateral well screens. Subsequently, two horizontal collector wells were built on the island to take advantage of river infiltration to support the desired well yield.

6. Design and Construction Details

For reference, some design and construction details for the horizontal collector wells in Lincoln, Nebraska; Louisville, Kentucky; and Prince George, British Columbia, Canada, are shown in Table 2-1.

Table 2-1. Design and Construction Details for the Horizontal Collector Wells in Lincoln, Nebraska; Louisville, Kentucky; and Prince George, British Columbia, Canada

	Lincoln Well 1	Lincoln Well 2	Louisville Well	Prince George Well
River Valley	Platte River	Platte River	Ohio River	Nechao River
Capacity (m³/s)	0.88	0.88	0.88	0.66
Depth (m)	24	26	30.5	30.5
Caisson Diameter (m)	4	4	4.9	4.9
Number of Laterals	7	7	7	14
Lateral Diameter (cm)	30	30	30	20
Total Lateral Length (m)	393	427	512	545

Figure 2-3 illustrates a typical collector well, including the structure for the pump house. Typically, the well caisson is carried above known or anticipated flood elevations, where it can be completed with a pump house building, or as an open-air slab, if weather or security conditions

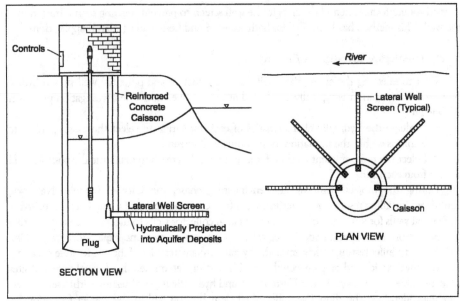

Figure 2-3. Cross-section and plan view of a horizontal collector well. Drawing provided by Collector Wells International, Inc.©

allow. The individual well yields from collector wells in the United States have ranged from about 0.0044 to 1.75 m³/s. The diameter of the concrete caisson usually ranges from 3 to 6 m, and the depth ranges from about 10 m to over 38 m. The number of lateral well screens varies according to anticipated yields, historically ranging from 2 to 14 per well (one has 23 laterals), and screen diameters range from 20 to 30 cm. The length of individual lateral lines depends on expected capacity, and will vary up to about 75 m, with the total footage installed in a collector well ranging from 100 to 750 m in an individual well. The geology and project needs at each site are different, requiring that a site-specific design be prepared (which generally falls within the ranges shown above). This flexibility permits collector wells to be considered for a wide range of applications in diverse geographic settings.

Collector wells continue to be considered for producing water supplies from RBF, groundwater, and filtered seawater (for desalination) in the United States. Their popularity has increased with the development of highly-publicized systems throughout the Midwest, vast improvements in the construction technology available over the past 15 years (which have expanded their applicability), and increased scrutiny and demand for treatment improvements of surface-water supplies that have brought RBF in as a viable treatment process for consideration at many sites. With these advances, collector wells have produced capacities ranging from 0.0044 to 1.75 m³/s of infiltrated water supplies at sites all across the United States for over 60 years.

References

Hunt, H.C. (1985). "Design and construction of radial collector wells." Presented in Water Wells Design & Construction Course, University of Wisconsin, Madison, Wisconsin.

Chapter 3. German Experience with Riverbank Filtration Systems

Jürgen Schubert, M.Sc.
Stadtwerke Düsseldorf AG
Düsseldorf, Germany

1. Introduction

Natural rivers meander in flat regions, primarily at their middle and lower parts, and form bends. A cross-section of a bend may show a stabilized and, sometimes, paved bed at the outer section of a bend, with moveable ground at the inner section (Figure 3-1). Clogging is more common among RBF wells near the outer section of a bend; however, if clogging is controlled by bed load transport, it will not restrict filtration. The yield of riverbank-filtered water in the inner part of a bend is normally higher because of the moveable ground of the riverbed, as well as the natural underground cross-flow due to the river gradient.

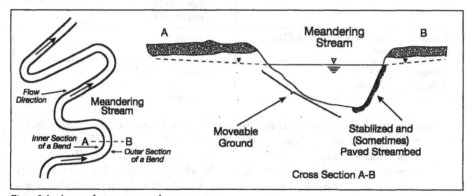

Figure 3-1. A meandering stream with cross-section.

2. River Characteristics for Siting RBF

Using a simple model, three different regions can be distinguished along a river:
- Upper part (with erosion).
- Middle part (with bed load transport).
- Lower part (with deposition).

The model's assumption does not truly reflect the natural design of a river, which normally does not have a single erosion basis (the mouth of the river) along its flow path but, due to geological conditions, has several; however, this model can help avoid mistakes in selecting future RBF sites.

The grain-size distribution of riverbed material is valuable in understanding the geomorphology of the river, which in turn determines the suitability of the site for RBF. On principle, erosion regions as well as regions where very fine particles like silt and fine sand are deposited (such as upstream of

35

C. Ray et al. (eds.), Riverbank Filtration, 35–48.
© 2002 *Kluwer Academic Publishers. Printed in the Netherlands.*

dams and near the mouth of the river) should be avoided when selecting an RBF site. Besides the geological data, the hydraulic gradient of the river can provide rough information about grain-size distribution (which helps build up the aquifer over time), mean flow velocity in the river, and the capability of bed load transport. Bed load transport is the movement (rolling, skipping, or sliding) of sediment, such as soil, rocks, particles, or other debris, along or very near the riverbed by flowing water. It is instrumental in the self-cleaning process of the riverbed.

About 80 percent of the RBF sites along the Rhine River in Europe are located in the Lower Rhine Valley, between 660 and 780 km. Table 3-1 lists data on river characteristics in this region.

Table 3-1. Characteristics of the Rhine River in the Lower Rhine Valley Region
(Between 660 and 780 km) in Europe

River Characteristics	Data
Average Hydraulic Gradient	0.21 to 0.18 m/km
Average Flow Velocity	1.0 to 1.4 m/s
Average Shear Force on the Riverbed	About 10 Newtons Per Meters Squared (N/m^2)
Average Grain-Size Diameter in the Riverbed	33 mm at 640 km 26 mm at 730 km 10 mm at 860 km
Average Grain-Size Diameter of the Bed Load (at 660 to 780 km)	13 to 8 mm
Average Hydraulic Conductivity of the Adjacent Aquifer	2.2×10^{-2} to 3.3×10^{-3} m/s

In addition, the bed load rate function at 845 km (the next downstream monitoring point) is:
- Flow rate (m^3/s)   1,000   2,000   3,000   4,000   5,000   7,000
- Bed load rate (kg/s)   0   7   15   23   31   47

The average flow rate at 845 km is about 2,300 m^3/s, while the minimal flow rate is 800 m^3/s. The flow at the initiation of bed load transportation is about 1,050 m^3/s. The bed load discharge at this monitoring point is about 220,000 m^3/yr. Additional suspended sand is transported up to 330,000 m^3/yr. From decades-long operational experience, it is clear that clogged areas in the upstream region have sufficient self-cleaning capabilities.

3. Flow Dynamics of Rivers and the River/Aquifer Interaction

A discharge hydrograph is a graphic representation of the discharge (flow rate) of a stream at a given point in time. A discharge hydrograph not only characterizes the sources of a river and its main tributaries (such as an alpine region, hill region, open country, or mixed), but it also reveals the dynamic behavior along the flow path of a river year-round. The hydrograph may be regarded as a "fingerprint" of the upstream catchment area of a river. Figure 3-2 shows an example of a discharge hydrograph for a section of the Rhine River in Düsseldorf, Germany.

The dynamic behavior of the river level not only influences flow and transport in RBF, but also influences water quality in the river as well as in raw water in the wells.

Ubell (1987) investigated river/aquifer interactions from a quantitative point of view in the Neuwieder Becken, a section of the Middle Rhine region near Koblenz, Germany. The geological data of this region are well known. The aquifer has a hydraulic conductivity between 2×10^{-2} and

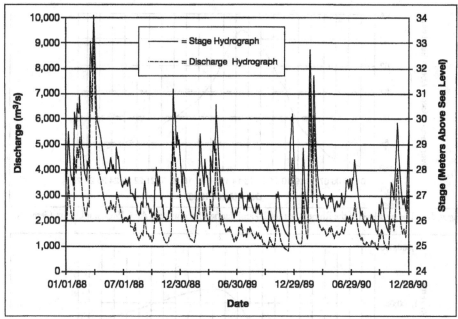

Figure 3-2. Discharge and level hydrograph of the Rhine River at 744.2 km (Düsseldorf, Germany) from January 1988 to December 1990.

4×10^{-3} m/s, a pore volume rate of 0.2, and a thickness of 10 to 15 m; this is also representative of the Lower Rhine Valley. The average groundwater recharge rate from precipitation is 5.0 liters per second per square kilometer ($L/s/km^2$).

The main task of these investigations was to understand and quantify the bank storage process, which occurs when groundwater is temporarily stored in sediments adjacent to a stream channel because of a rise in stream elevation during flooding. A second task was to develop tools to simulate this process to help determine how important bank storage is on the groundwater balance of aquifers in the vicinity of rivers and how it may influence the flow of the river itself. The field studies began in 1982 and continued for nearly a decade (Giebel et al., 1990).

A gallery of seven monitoring wells was installed perpendicular to the flow direction of the Rhine River at River Kilometer 602.37 to collect data on the groundwater level. Based on the gauge observations (including the river and relevant data of the aquifer), a time series of the specific volume of bank storage and infiltration/exfiltration rates was determined over the years. Figure 3-3 shows the river level during a flood wave in April 1983 near Urmitz, Germany, on the Rhine River at 602.4 km.

The volume of the specific bank storage during the flood event is shown in Figure 3-4. Approximately 1-million cubic meters (m^3) of riverbank-filtered water entered the aquifer in a few days over a riverbed length of 1 km.

An interesting effect of the flood event is reflected in Figure 3-5, which shows the specific values of infiltration and exfiltration, with a maximum infiltration rate of 2,400 L/s/km length of the river (this infiltration rate, caused by a temporary flood wave, is about three to five times higher than the infiltration rates of existing RBF plants). A significant amount of riverbank-filtered water is stored for weeks or months in the aquifer. The observed maximum distance of

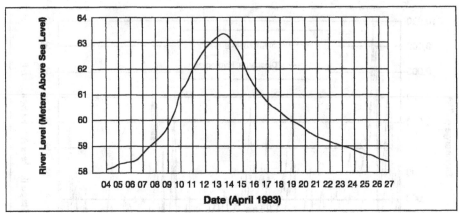

Figure 3-3. The river level of the Rhine River (602.4 km) during a flood wave in April 1983 near Urmitz, Germany.

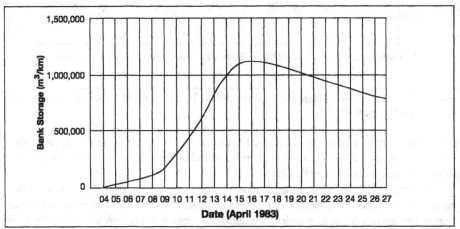

Figure 3-4. Cumulative volume of bank storage along the Rhine River (602.4 km) during a flood wave in April 1983 near Urmitz, Germany.

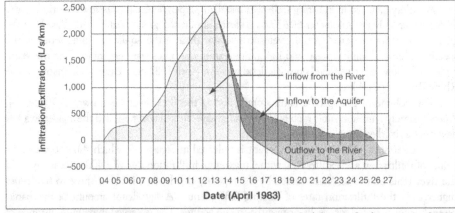

Figure 3-5. Infiltration and exfiltration rates along the Rhine River (602.4 km) during a flood wave in April 1983 near Urmitz, Germany.

riverbank-filtered water that penetrates into the aquifer during this event is 300 m. The succession of even smaller flood events increases the volume of bank storage and the depth of penetration.

Field studies on bank storage induced by fluctuating river levels give insight into the hydraulics of river/aquifer interactions. These could never be substituted by small-scale pilot experiments. Additional water-quality monitoring can answer many questions concerning the effectiveness of RBF in attenuating pollutant loads. The natural bank storage process, in connection with the river-level hydrograph (fingerprint), clearly shows that RBF must be regarded as a highly dynamic process.

4. Field Studies on RBF – Hydraulic Aspects

To understand river/aquifer interactions during RBF, monitoring concepts must consider the dynamic behavior of the system as a whole. This means that monitoring data must be collected over a long time period and monitoring wells must be suitable for depth-orientated samples, including flooding.

The Düsseldorf Waterworks in Germany has used RBF since 1870 to procure water. During the first 80 years, RBF alone, without additional treatment (disinfection only), sufficed to obtain safe drinking water. After World War II, the quality of Rhine River water began to deteriorate. A few years later, the consequences of this became evident by odor and taste problems in well water (since then, the necessity arose to treat raw water). But, at that time, there were neither standards nor experience available on how to remove the mostly unknown compounds. A prototype of advanced treatment steps with ozone, biological filtration, and granular activated carbon adsorption was developed and has operated successfully since 1961.

Following the Sandoz Accident in Switzerland in 1986, in which chemicals like insecticides were released into the Rhine River (see Chapter 10), the flow and transport phenomena that occurred between the river and the wells became the subject of an RBF research project (Sontheimer, 1991). The results of this project included a three-dimensional, dynamic flow and transport simulation model (Schmid et al., 1990/1991), which describes the effect of shock loads on raw water in the wells resulting from accidental river pollution. Another result is a tailor-made monitoring system (Schubert, 1997) that has proven invaluable in determining and reporting any pollution in the Rhine River due to accidents.

Over the last few years, additional investigations have been undertaken regarding the results of the project (Schubert, 2001), among them:

- The outbreak of waterborne diseases in some countries (due to pathogens in drinking water) questioned the appropriate elimination rates of suspended solids and pathogens during RBF.
- The concentrations of most known organic micropollutants in Rhine River water are decreasing; however, the effects of unknown mixtures of pollutants (even in very low concentrations) upon human health and the environment are uncertain. As a result, the Ames Test was employed to describe changes in mutagenic activity during the process of converting river water to drinking water via RBF.

Flow and Transport Phenomena in RBF

Figure 3-6 shows the fluctuating chloride concentration in the Rhine River due to an industrial effluent discharge at the upper Rhine region. RBF balances out this fluctuating concentration, as shown by the chloride concentration in the production well (dotted line). This "mixing" (or compensation effect) of RBF is well known. But why does this phenomenon occur?

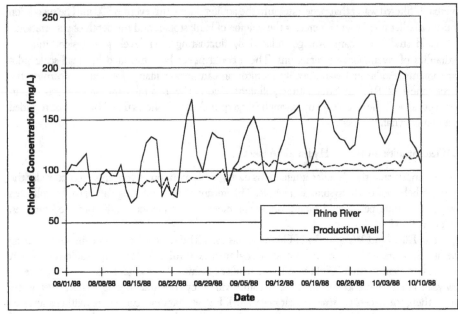

Figure 3-6. Chloride concentration in Rhine River water compared with that in adjacent well water (from August to October 1988).

If an ideal tracer is monitored during subsoil passage, the processes of chemical reactions, sorption, and biological degradation are excluded. The propagation of dissolved tracers with the groundwater is only governed by convection, hydro-mechanical dispersion, and molecular diffusion. The behavior of the ideal tracer, chloride, during RBF is shown in Figure 3-6. Compared to passive transport with groundwater flow (convection), molecular diffusion and hydro-mechanical dispersion are both low-scale effects; therefore, the mixing capability must be mainly the result of convection.

To prepare the monitoring devices for related field studies and the basic assumptions for a three-dimensional dynamic flow and transport model, a simple hypothesis was used (Figure 3-7).

A rough estimation on the behavior of such a model, regarding three different infiltration points (1,2,3) in the riverbed, leads to the following conclusions:

Flow path length s: $s_1 < s_2 < s_3$

Flow velocity v: ($v = k \times \Delta h/s$) $v_1 > v_2 > v_3$

Flow time t: ($t = s/v$) $t_1 << t_2 << t_3$

where k is the hydraulic conductivity; and Δh is the difference between the river water level and well water level.

The variations in flow time between different infiltration points in the riverbed and Production Well 45 must be much greater than the variations within flow path length. To verify these rough conclusions, a monitoring strategy was chosen with three rows (A, B, C) of observation wells: two rows (A, B) between the well gallery and the river and one row (C) on the opposite side of the well gallery. Each row consists of three monitoring wells with short filter screens at different depths to allow depth-orientated sampling (Figure 3-8). A second monitoring profile was installed some 600 m downstream; it is monitored the same way.

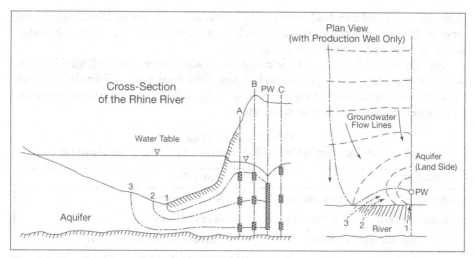

Figure 3-7. Hypothesis of groundwater flow during RBF. A, B, and C are rows of observation wells. PW represents the Production Well. 1, 2, and 3 represent infiltration points of the river water.

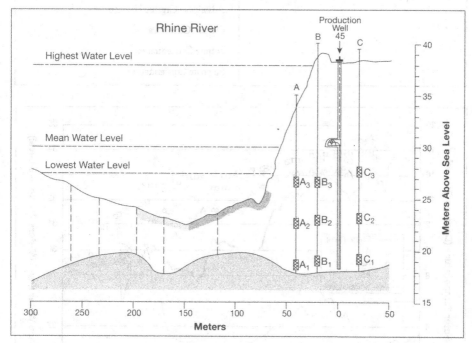

Figure 3-8. Monitoring wells at the Flehe Waterworks (731.5 km). A, B, and C indicate rows with three monitoring and sampling wells each. Indices 1, 2, 3 indicate the different depths of the screen pipe.

From former investigations, it was known that even comprehensive "snapshots" cannot guarantee insight into RBF flow and transport phenomena. To understand river/aquifer interactions during RBF, monitoring strategies must consider the dynamic behavior of the entire system, which is mainly governed by fluctuating river levels. This means that monitoring data (hydraulic, physical/chemical, and microbial) must be collected over a long time period. The monitoring time period for the RBF research project was chosen between 7 and 11 months for each study session. Specific monitoring data are presented in Table 3-2.

Already, some early results of the field experiments with temperature data (Figure 3-9) were used to confirm the earlier hypothesis and to design and develop an RBF simulation model. Moreover, the long-term data of the chloride tracer (Figure 3-10) were a good basis to test and calibrate the RBF flow and transport model afterwards.

Table 3-2. Monitoring Data
Collected from Daily Samples Used to Characterize River/Aquifer Interactions During RBF

Hydraulic Data	Quality Data
River Water Level (Three Monitoring Sites)	Water Temperature
Groundwater Level (40 Monitoring Sites)	Electrical Conductivity
Intake of Raw Water at Pump Stations	Chloride Concentration (Tracer)
	Ultraviolet Absorbance at 254 nm
	DOC
	Adsorbable Organic Halogen
	Oxygen Concentration
	Nitrate Concentration
	Sulphate Concentration

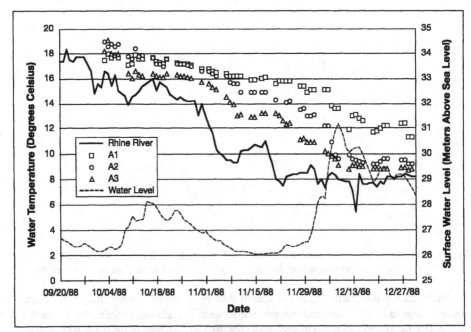

Figure 3-9. Water temperature in Rhine River water and in Sampling Wells A1 (lower layer), A2 (middle layer), and A3 (upper layer) at the Flehe Waterworks (731.5 km) from September to December 1988.

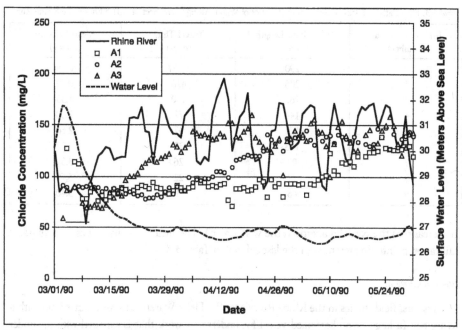

Figure 3-10. Chloride concentration in Rhine River water and in Sampling Wells A1 (lower layer), A2 (middle layer), and A3 (upper layer) at the Flehe Waterworks (731.5 km) from March to May 1990.

An important finding was the significant age stratification of riverbank filtrate between the river and wells. Age stratification represents the difference in the residence time of water in the aquifer. Based on water samples collected from the rows of monitoring wells (A, B, C), young water (with a residence time of just days) was found in the upper layers of the aquifer and old water (with a residence time varying from weeks to months) was found in the lower layers of the aquifer. Because of age-stratification, water withdrawn from an RBF well entered infiltration areas in the riverbed at widely differing times. This explains the almost total compensation of fluctuating concentrations between the river and wells.

An existing and well-calibrated two-dimensional model of the whole catchment area of the Düsseldorf Waterworks cannot simulate three-dimensional flow in the vicinity of the riverbed and the wells because of the curvature of flow lines near the river/aquifer interface and well screens; therefore, the two-dimensional model in these locations was refined to have the three-dimensional component for flow and transport simulation. This covered the wells of the Flehe Waterworks and the riverbank. The model to simulate flow and mass transport processes was designed as a dynamic, three-dimensional numerical model, and was developed and calibrated in cooperation with Ruhr University in Bochum, Germany.

To calculate the travel time/distance of river water to a production well near River Kilometer 731.5, several hypothetical infiltration points were considered in the three-dimensional model. The travel time and flow path distances varied, as shown in Table 3-3. For example, if a water particle enters at Infiltration Point 1 (farthest from the well) of the river/aquifer interface, it will travel 290 m before reaching the well (with a travel time of 1,157 days). By contrast, if a particle enters the river/aquifer interface at Infiltration Point 9 (closest to the well), the flow path length will only be 68 m and the travel time will be 20 days. The flow velocity along the flow path varies

Table 3-3. Flow Path and Travel Time Characteristics of Water Along Cross-Section 731.5 km on the Rhine River

Infiltration Point (number)	Flow Path Length (m)	Travel Time (days)	Mean Velocity (m/day)
1	290	1,157	0.25
2	206	420	0.49
3	162	120	1.35
4	144	72	2.00
5	125	47	2.66
6	108	33	3.27
7	92	26	3.54
8	79	22	3.59
9	68	20	3.40

(it increases as the particles approach the well screen). The mean velocity values for individual infiltration points are presented in the last column of Table 3-3.

Riverbed Clogging

The first field studies in the Rhine River near the Flehe Waterworks were carried out with a diving bell (a large, open-bottomed vessel for underwater work that is supplied with air under pressure) between 1953 and 1954 to investigate riverbed clogging during high loads of organic contaminants and suspended solids in river water. Two clogged layers could be determined: one on the surface of the infiltration area (mechanical clogging) and the other stretching about 1 decimeter below (chemical clogging) (Gölz et al., 1991).

After these early investigations, a unique experiment was conducted to determine the influence of the clogged area on the water yield of the wells. A "window" (large pit) was dredged in the riverbed within the clogged areas in front of the well gallery, with a length of 300 m and width of 70 m and, as expected, the water yield increased significantly, but the effect was temporary. A few weeks later, the dredged window was clogged again.

Clogging is unavoidable. The grain size of the material in the silt layer ranges from <0.002 to 0.2 mm. The larger particles plug the pore channels in short time and, together with the smaller particles, build up a nearly impermeable layer. As a result, it is only at the beginning of the clogging process that some smaller particles may penetrate into the aquifer. When suspended silt cannot penetrate the aquifer, it is removed and deposited in the upper layer of the aquifer. Clogged areas tend to expand from the well side of the riverbank to the middle of the riverbed. If the sheer force on the riverbed is high enough, the expansion of these clogged areas will be restricted or limited by bed load transport, which washes out sand and silt from the riverbed.

In 1987, a second investigation with a diving bell was conducted in the same area. The purpose behind investigating the bottom of the Rhine River (Figure 3-11) was to take water samples directly below the riverbed and to obtain information about:

- Geological information under the riverbed (from exploratory borings and samples).
- Type and extent of clogging in the infiltration areas.
- Regions of different permeability.

The results of this investigation revealed that three distinct zones exist on the riverbed (Figure 3-12).

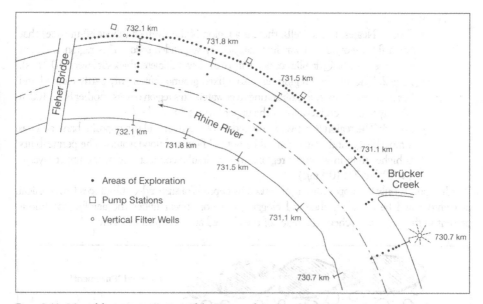

Figure 3-11. Map of the investigation area: Flehe Waterworks in Düsseldorf, Germany.

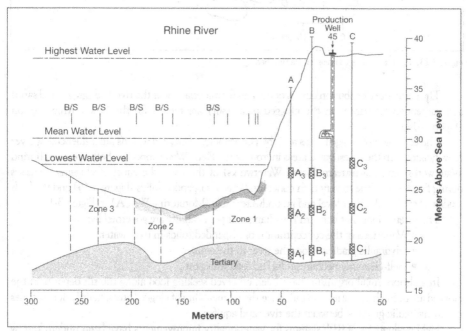

Figure 3-12. Zones of classification of the riverbed in front of the Flehe Waterworks (731.5 km). B = Borehole core, 200-mm diameter. S = Slot probe, 25-mm diameter. A, B, and C indicate rows within three monitoring and sampling wells each. Indices $_{1, 2, 3}$ indicate the different depths of the screen pipe.

The zones include:

- *Zone 1:* Nearest to the wells, there is a region of the riverbed (about 80-m wide) that has a fixed ground and was fully clogged by suspended solids. This region is almost impermeable (the hydraulic conductivity of the few-millimeter thick silt layer is 10^{-8} m/s).
- *Zone 2:* The attached region has also a fixed ground, but is only partly clogged and has good permeability for infiltrating river water. This region covers another 80 to 100 m (the hydraulic conductivity of the upper layer is 3×10^{-3} m/s).
- *Zone 3:* The region is between the middle of the river and the opposite bank and has a movable ground that is shaped by normal flow and flood waters. The permeability is higher than in all other regions (the hydraulic conductivity of the upper layer is 4×10^{-3} to 2×10^{-2} m/s).

In Zone 1, only the upper thin layer caused by deposits of suspended substances (mechanical clogging) could be detected; chemical clogging did not appear under the aerobic conditions present in the aquifer. A schematic image of the clogged area is shown in Figure 3-13.

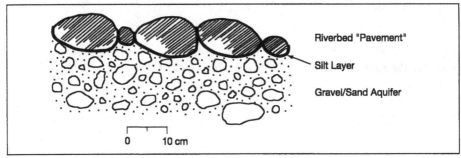

Figure 3-13. Pattern of clogged areas in the riverbed.

The grain-size distribution curves of the aquifer material under the riverbed (gravel and sand) and suspended substances in the clogged areas (silt) are typical for the Lower Rhine region (Figure 3-14).

The extension of clogged areas and the permeability of clogged regions are influenced by river flow dynamics. In the investigated areas in front of the Flehe Waterworks and both upstream (Grind Waterworks) and downstream (Staad Waterworks) of the river, the effect of clogging increases during flood events due to very high concentrations of suspended solids (Figure 3-15) and the high gradient between the river level and groundwater table (Monitoring Well A1 in Figure 3-16).

The interactions that concern the changes of permeability are governed by:

- Variations of the concentration of suspended solids in river water.
- Hydraulic gradient from the river to the aquifer.
- Self-cleaning mechanism by bed load transport.

In addition, small organisms have been observed seeking food along the silt deposits in the clogged areas. Such activities also influence the permeability of clogged areas, mainly during times of low hydraulic gradient between the river and aquifer.

Field evaluations of RBF systems for water-quality improvements have been undertaken by many European water utilities, including Düsseldorf Waterworks in Germany. RBF at Düsseldorf has contributed positively to the removal of suspended particles, turbidity, biodegradable compounds, and pathogens, as well as to a decline in mutagenic activity. Chapter 12 provides details of these processes at Düsseldorf.

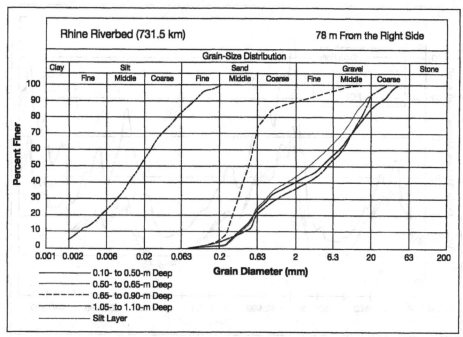

Figure 3-14. Grain-size distribution of the aquifer under the riverbed and the silt in the clogged areas.

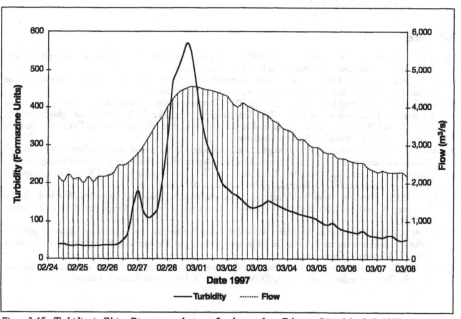

Figure 3-15. Turbidity in Rhine River water during a flood wave from February 24 to March 8, 1997.

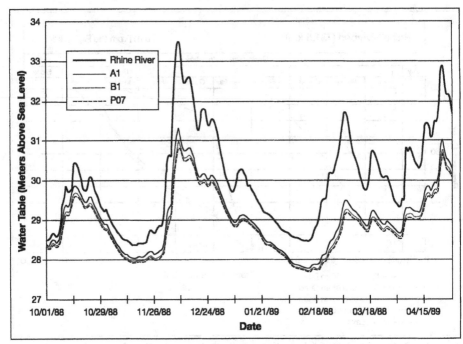

Figure 3-16. Water level of the Rhine River and the water table at Sampling Wells A1, B1, and P07 at 731.5 km from October 1988 to April 1989. The locations of Sampling Wells A1 and B1 are shown in Figure 3-12. Monitoring Well P07 is located between Sampling Well B1 and Production Well 45 (from Figure 3-12) at 731.5 km

References

Giebel, H., E. Gölz, H.J. Theiss, and K. Ubell (1990). *Hydrogeologie und Grundwasserhaushalt im Neuwieder Becken.* Bundesanstalt für Gewässerkunde, Mitteilung Nr. 54, Koblenz, Germany.

Gölz, E., J. Schubert, and D. Liebich (1991). "Sohlenkolmation und Uferfiltration im Bereich des Wasserwerks Flehe (Düsseldorf)." *Gas- und Wasserfach Wasser/Abwasser,* 132, Jahrgang, 2: S.69-76.

Schmid, G., P. Obermann, and J. Gotthardt (1990/91). BMFT-*Forschungsvorhaben Sicherheit der Trinkwassergewinnung aus Uferfiltrat bei Stoßbelastungen, Teilprojekt 3, Teil I: Grundwasserströmung (Dez. 1990), Teil II: Schadstofftransport (März 1991) (Research Project: Safety of RBF against shock loads, Project No. 3, Part 1: Groundwater flow, Part 2: Mass Transport [March 1991]).* Ruhr-Universität, Bochum, Germany.

Schubert, J. (1997). "Monitoring strategy to safeguard the water supply in the case of accident-related river pollution." Proceedings, Monitoring Tailor-made II, International workshop on information strategies in water management. J.J. Ottens, F.A.M. Claessen, P.G. Stoks, J.G. Timmerman, R.C. Ward, eds., Institute for Inland Water Management and Waste Water Treatment (RIZA), Lelystad, The Netherlands, p. 467 - 468.

Schubert, J. (2001). "How does it work? Field Studies on Riverbank Filtration." *Proceedings, International Riverbank Filtration Conference.* W. Jülich and J. Schubert, eds., International Association of the Rhine Waterworks (IAWR), Amsterdam, p. 41-55.

Sontheimer, H. (1991). *Trinkwasser aus dem Rhein ? Bericht über ein vom Bundesminister für Forschung und Technologie gefördertes Verbundforschungsvorhaben zur Sicherheit der Trinkwassergewinnung aus Rheinuferfiltrat bei Stoßbelastungen.* Academia Verlag, Sankt Augustin, Germany.

Ubell, K. (1987). "Austauschvorgänge zwischen Fluss- und Grundwasser" (*Interactions between river water and groundwater*). Deutsche Gewässerkundliche Mitteilungen (DGM) 31, Edition 4, 119-125 and Edition 5, 142-148.

Chapter 4. Riverbank Filtration Construction Options Considered at Louisville, Kentucky

Stephen Hubbs, P.E.
Louisville Water Company
Louisville, Kentucky, United States

Kay Ball
Louisville Water Company
Louisville, Kentucky, United States

David L. Haas, P.E.
Jordan, Jones, and Goulding
Atlanta, Georgia, United States

Michael J. Robison, P.E.
Jordan, Jones, and Goulding
Atlanta, Georgia, United States

1. Introduction

A productive aquifer fills the ancient riverbed of the Ohio River, including a 11-km stretch between two treatment plants that are operated by the Louisville Water Company in Louisville, Kentucky (United States). This area of the riverbed, which has been considered for water-supply development since the late 1940s, includes a stretch of scenic highway and is regarded as a significant visual landscape within the community. Preservation efforts within this area have resulted in limited commercial development; therefore, few activities have the potential to contaminate the underlying aquifer.

When the Louisville Water Company evaluated water-supply construction options in the late 1990s, an effort was made to design facilities that addressed the following concerns:

- Proximity to existing water transmission systems.
- Aquifer productivity.
- Ease of maintenance.
- Visual impacts of the facilities.
- Impact of construction activities.
- Construction costs.

Two conventional construction options were initially considered to provide the 9.86-m³/s capacity needed for converting to RBF:

- The first option involved constructing individual horizontal collector wells of 0.66- to 0.88-m³/s capacity, spaced about 610 m apart, and concentrated around existing river pump stations. It was determined that approximately 15 wells would be needed to provide a capacity of 9.86 m³/s and that the system would have to accommodate future growth.

49

C. Ray et al. (eds.), Riverbank Filtration, 49–59.

- The second option involved constructing approximately 100 conventional vertical wells, with an average capacity of 0.095 m³/s each; however, this option was rejected because of operational difficulties and right-of-way issues resulting from 100 separate facilities and construction sites in the corridor.

To better identify the construction issues and to define the capacity of the aquifer for long-term, high-yield production, a full-scale horizontal collector well was constructed on Louisville Water Company property at the 2.63-m³/s capacity B.E. Payne Water Treatment Plant. The well was designed for an output of 0.66 to 0.88 m³/s, and was architecturally detailed to be compatible with the surrounding area. This facility went on-line in the summer of 1999, with a total project cost (including engineering and supervision) of approximately $5 million.

Interaction with community groups led to the evaluation of alternative ways of extracting water from the aquifer, focusing on hydraulic efficiency, cost, and visual impact. Several creative design options were considered for the second phase of the project, which would expand RBF capacity at the B.E. Payne Water Treatment Plant to 2.63 m³/s. Three options were pursued to preliminary design:

- A soft-soil tunnel constructed in the sand and gravel aquifer, with well screens extending horizontally from a tunnel under the river.
- A hard-rock tunnel system connecting several horizontal collector wells to a common pump station.
- A hard-rock tunnel system connecting a series of 30 vertical wells without pumps to a centralized pump station.

2. Site Conditions

The demonstration well is located on the property of the Louisville Water Company at the B.E. Payne Water Treatment Plant. This property lies along the bank of the Ohio River on the Kentucky side, about 19 km north of downtown Louisville. The project site is located in a flood plain on glacial alluvial deposits of the Ohio River. At the project site, these deposits consist of 10 m of silt and clay overlying 20 m of sand and gravel. The bedrock beneath the alluvial deposits consists of horizontally stratified limestone, shale, and dolomite of Ordovician age. The sand and gravel layer acts as a leaky confined aquifer, with the groundwater table located 4.6 m below ground surface at elevation 128 m above mean sea level (controlled by the locks and dams of the Ohio River).

The subsurface investigation for the second phase of the project included the following elements:

- Large-diameter bucket auger borings to bedrock.
- Grain-size analysis of composite samples obtained from the bucket auger borings.
- Core borings of bedrock materials.
- Examination of rock outcrops near the site.

Large-diameter bucket auger borings were used to collect representative samples of the alluvium. Because alluvium contains significant amounts of coarse gravel, cobbles, and possibly boulders, the large-diameter bucket auger was considered the most appropriate sample collection method. Grain-size distribution curves were developed based on sieve analyses of the bucket auger samples.

The results of the rock-core boring program indicated the presence of limestone and shale. A particularly good layer of limestone occurred from 39.6- to 47.9-m below ground surface. Limestone beds in this unit tend to be on the order of 0.6- to 0.9-m thick, interbedded with 7.5- to 10-cm thick layers of shale. All of the limestone tested in this layer has a rock quality designation in the range of 90 to 100 percent, which according to Deere and Deere (1989) is classified as "excellent."

Based on the results of the subsurface investigation, feasible construction options existed for either:

- A tunnel system in the rock, or
- A soft-soil tunnel constructed in the sand and gravel aquifer.

The following sections discuss specific considerations for implementing RBF for the soft-soil and hard-rock tunnel options.

3. Site Hydraulic Characteristics

The site hydraulics characteristics were initially estimated by evaluating historical data of pumping activities in the area. These values were refined after the demonstration well was constructed through a series of controlled pumping tests performed in October 1999, March 2000, October 2000, April 2001, and September 2001.

Prior to the construction of the 0.88-m^3/s collector well, the transmissivity of the 21.3-m thick sand and gravel aquifer in this area was estimated at 2,485 m^2/d, based upon previous data collected during the construction of the treatment plant. This corresponds to an average aquifer hydraulic conductivity of approximately 116 m/d.

After the 0.88-m^3/s collector well was completed, the horizontal hydraulic conductivity was measured at 119 m/d and the vertical conductivity was measured at 39.6 m/d at a flow of 0.88 m^3/s. The leakance (a measure of the ease of flow between the river/aquifer interface) was measured at 2.4 1/day in the October 1999 pumping test. Subsequent pumping tests showed that the leakance decreased to:

- 0.72 1/day in March 2000.
- 0.25 1/day in October 2000.
- 0.20 1/day in April 2001.
- 0.15 1/day in September 2001.

The decrease in leakance indicates that clogging of the river/aquifer interface is a function of time.

Altogether, these data indicated that the design-limiting factor in the demonstration well at Louisville was the movement of water between the river/aquifer interface. The decrease in leakance with time indicates that particles are blocking the flow path and are extending the area of recharge out into the river bottom. Piezometric measurements taken in August 2000 and August 2001 indicated the cone of influence for the well had extended as far as 305 m out into the 610-m width of the river.

It is believed that this type of clogging is typical of streams with moderate scouring velocities (<0.9 m/s), like the Ohio River. Streams with much higher scouring velocities (1.8 m/s), like the Rhine River in Europe, typically have the top layers of the aquifer renewed with each high-velocity event and would not see a decrease in leakance with time to the same extent as a low-velocity stream. These observations led the Louisville Water Company to consider a facility design that distributes the stress on the aquifer evenly along the riverbank, as opposed to concentrating this stress at radial collector caissons spaced 305- to 610-m apart.

Considering the concern for numerous aboveground facilities and the desire to use the maximum amount of river frontage, the Louisville Water Company initiated the design of groundwater collection systems based on several tunnel designs. These designs followed the basic concepts of infiltration galleries, but allowed a much higher capacity because of the depth at which the laterals and tunnels could be constructed.

4. Soft-Soil Tunnel Option

The ideal RBF groundwater extraction system would draw water evenly from the entire riverbank along a given stretch of river long enough to provide adequate volume without over-stressing the aquifer system. With this in mind, a design concept was developed based on soft-soil tunneling capabilities. The technology for soft-soil tunnels has developed over the past 20 years and is now a feasible construction technique for large-diameter tunnels in sand and gravel aquifers. The technique provides for unlimited lengths of tunnels with diameters ranging from 4.3 to over 12 m.

The soft-soil tunneling technique provides for the construction of a large-diameter conduit within the water-bearing stratum of the aquifer, through which horizontal well (e.g., lateral) screens could be installed. To our knowledge, this combination of soft-soil tunneling with lateral screens has not yet been attempted. Construction companies that had installed lateral screens in conventional caissons were contacted to determine if such a facility could be constructed. It was determined that a tunnel diameter of 4.3 m or greater would be required for efficient construction.

The need to perform maintenance on laterals without taking a major portion of the facility out of service resulted in a design that used a 4.3-m diameter tunnel with two 1.20-m conduits laid in the bottom for water conveyance. Laterals would be pushed out from the side of the tunnel towards the river and tied into the 1.20-m mains. The upper half of the tunnel would be dry, allowing any one lateral to be removed from service and maintained, as needed. This design eliminated the need for multiple redundant tunnel facilities. Figures 4-1 through 4-3 illustrate the concept of this design.

Under the conditions found at the B.E. Payne Water Treatment Plant plant, a total of 1,830 m of tunnel would be required for 1.97-m^3/s capacity. The design included 23 laterals, each 61-m long, spaced at 79.3-m intervals, directed out towards the river. The preliminary project cost estimate for this facility was approximately \$50 million, including 8 percent for engineering and construction management, and 20 percent for contingency.

The cost of the soft-soil tunneling was influenced by the relatively short distance of tunnel required. A significant cost element for soft-soil tunneling is the earth-pressure balancing machine required to construct the tunnel, as well as the mobilization, entrance caisson, and exit caisson required for the tunnel. These costs are the same for any length of tunnel.

This design option represented the best facility design, yielding a high degree of operation flexibility and least overall impact. The cost estimate for this construction option, however, was significantly greater than other options.

5. Hard-Rock Tunnel Option with Horizontal Collector Wells

A second option considered for extracting water from the aquifer included a traditional hard-rock tunnel system below the riverbank, tying traditional collector wells to a common pump house. This option has the advantage of combining proven technology for both tunneling and well screen installation. The system design includes three collector caissons spaced 610-m apart, connected via a hard-rock tunnel to a central pump house for a capacity of 1.97 m^3/s. The most economical tunnel design was a 2.44-m diameter tunnel, designed to easily accommodate additional collectors from any direction. This provided flexibility for future collector caissons to be added. Figures 4-4 and 4-5 provide an illustration of this design concept.

The cost estimate for this system was based on using a tunnel boring machine for 1,830-m long excavation. Three collector wells capped at grade would be installed. The pump house was designed for 2.63-m^3/s capacity. The total project cost estimate for this installation, including engineering and construction management at 8 percent and a 20-percent contingency, was approximately \$30 million.

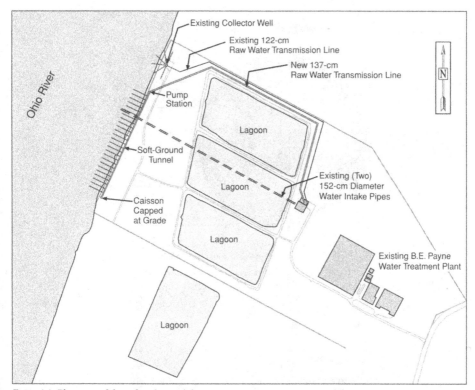

Figure 4-1. Plan view of the soft-soil tunnel design option.

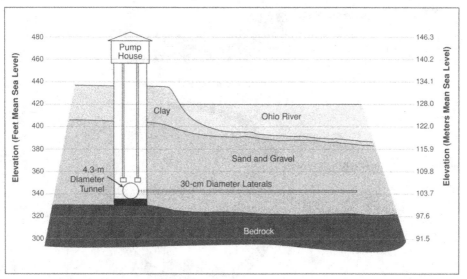

Figure 4-2. Profile of the soft-soil tunnel design option.

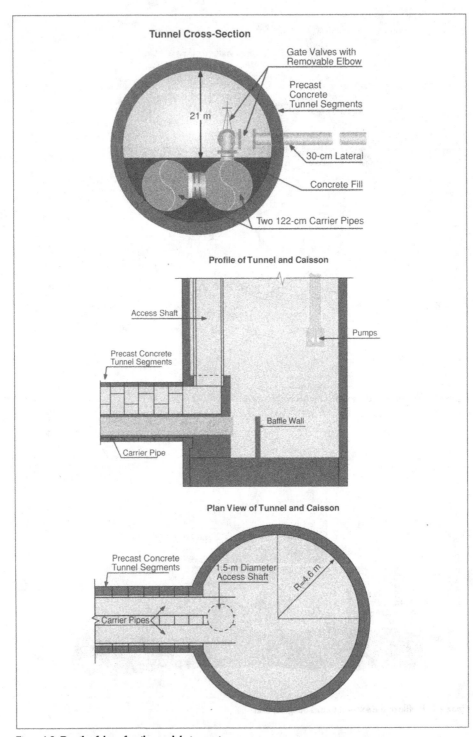

Figure 4-3. Details of the soft-soil tunnel design option.

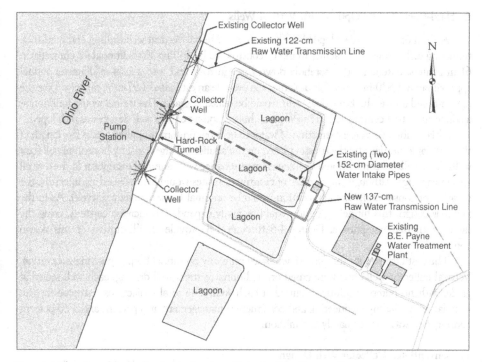

Figure 4-4. Plan view of the hard-rock tunnel option with horizontal collector wells.

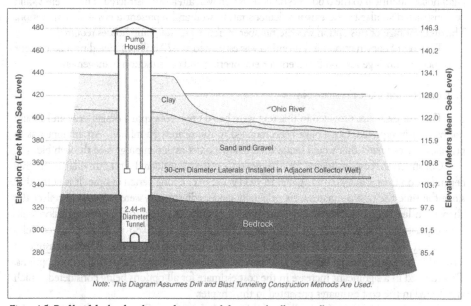

Figure 4-5. Profile of the hard-rock tunnel option with horizontal collector wells.

6. Hard-Rock Tunnel Option with Vertical Wells

A third option was developed based upon conventional vertical well drilling and hard-rock tunneling technology. This facility would be constructed by drilling 30 traditional 40-cm wells on 61-m centers, extending the borehole down through rock and into a 2.44-m diameter tunnel approximately 1,830-m long. This design would provide an estimated 1.97-m³/s capacity. Like the option noted above, the individual wells would be sealed at grade. The tunnel would function as a conveyance to a centralized 2.63-m³/s pump house, and a ventless well design was developed.

This option allows the extraction of water to be more uniformly applied across the length of the riverbank, minimizing the stresses placed on the aquifer by large-capacity horizontal collector wells. It also allows conventional well screen maintenance with minimal disruption to the overall system capacity. Although the number of construction sites would be increased significantly, the duration of construction at each location would be minimal (approximately 1 week). As in the previous design, this tunneling option could be easily expanded to increase capacity, using the same centralized pumphouse. Figures 4-6 through 4-8 provide an illustration of this design concept.

The cost estimate for this system was based primarily on tunnel boring machine excavation. Vertical collector well costs were based on traditional vertical well drilling, each well sealed at grade, with no pump or utilities required at each well. The total project cost estimate for this installation, including engineering and construction management at 8 percent and a 20-percent contingency, was approximately $25 million.

7. Conventional Collector Well Design

The tunnel design options were compared to the construction of three collector wells and pumphouses identical to the 0.66-m³/s demonstration well already constructed. These wells would be constructed similar to the existing demonstration well and represent a cost-effective option. The disadvantage of this option was the number of above ground pump houses required.

The cost of construction for this option was estimated at $20 million, including a 20-percent construction contingency and 8 percent for engineering and construction management.

8. Construction Cost Estimate Notes

The cost estimates provided in this text were based upon conceptual design only and should not be over-interpreted. Costs have been rounded to the nearest $5 million. An attempt was to provide costs estimates that would retain their relative order as design progressed through bid.

Construction cost estimates for the design options considered at Louisville are highly dependent on the capacity of the aquifer to yield water over a long period of time. It is noted that since the time the cost estimates were originally made, adjustments were required to reflect the changes in leakance observed through the first 24 months of pumping the demonstration well. The capacity of the collector well was originally calculated at 0.88 m³/s, but this capacity calculation was reduced to 0.66 m³/s after 2 years of operating experience. These adjustments resulted in a greater spacing between the individual laterals and caissons in the various options. This resulted in a significant increase in the cost estimate for all options being considered, which is reflected in the cost estimates presented in this chapter.

Experience gained at Louisville and historical data from Germany indicates that the maximum amount of water that can reasonably be extracted from the riverbank at Louisville ranges from 0.22 to 0.31 m³/s per 305 m of riverbank. This design factor influences the length of

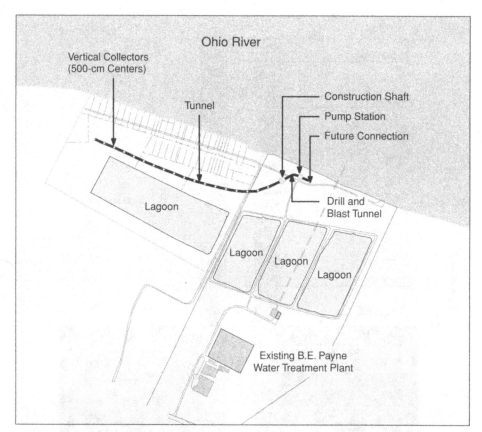

Figure 4-6. Plan view of a hard-rock tunnel option with vertical wells.

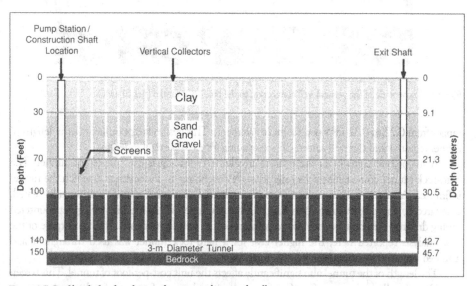

Figure 4-7. Profile of a hard-rock tunnel option with vertical wells.

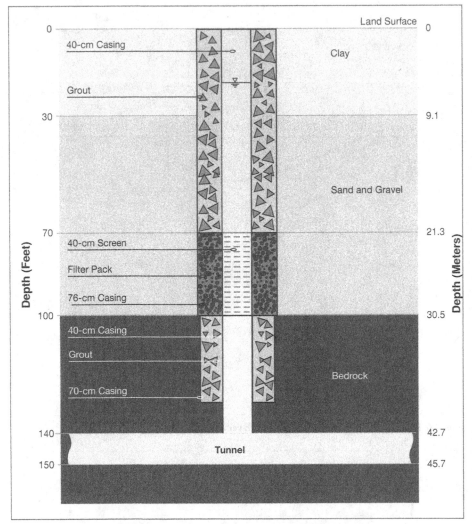

Figure 4-8. Schematic of the vertical well connection to the hard-rock tunnel (not to scale).

ranges from 0.22 to 0.31 m³/s per 305 m of riverbank. This design factor influences the length of tunnel required and the well screen/caisson spacing for each particular design.

A major cost element of the options involving tunneling is the cost of the tunnel. The hard-rock tunnel cost estimates vary significantly depending on the manner in which the tunnel is finished. A substantial cost reduction can be realized if the tunnel interior is minimally treated, compared to the cost estimate of a lined tunnel. Because of the specific geology encountered during drilling, the costs for the hard-rock tunnel could vary significantly. For the purpose of the cost analysis reported here, the higher cost for a hard-rock tunnel was used throughout and assumed the need for a liner.

The length of the tunnel also significantly affects the unit cost-per-foot of tunnel. This is seen most dramatically in the cost of a soft-soil tunnel, with a large up-front cost for the earth-pressure balance drilling machine. Costs for a 3,050-m tunnel could approach half the unit costs of a 305-m tunnel.

In each of the alternatives involving a tunnel and a centralized pump station, the 0.88-m^3/s pump station was estimated at $4 million.

9. Evaluation of Alternatives

In evaluating the alternatives for construction, the factors that drove the deliberations included the ease of well screen maintenance, capital cost, the number of above-ground facilities, and the development of technology that could be later applied to the 2.64-m^3/s installation planned for the larger treatment plant at Crescent Hill. The advantages provided by the soft-soil tunnel option included access to each individual lateral for maintenance, ability to be scaled-up by extending the length of the collector tunnel, and a minimal number of aboveground structures. The obvious disadvantage was cost. A second disadvantage was the untried nature of this type of construction and the costs associated with contractors reflecting this increased risk in their bid prices.

The advantage of the rock tunnel/horizontal collector well option was the relatively lower cost, expandability by adding additional caissons attached to the pump station by tunnel, and the limited number of aboveground facilities. The disadvantage of this option was the requirement to remove an entire caisson from service whenever maintenance was required on the laterals. This would result in a significant decrease of capacity whenever maintenance was required, forcing off-peak scheduling and an increased risk of inadequate water supply during maintenance. As a result, the increased operating risk must either be accepted or an additional caisson must be constructed for redundancy.

The hard-rock tunnel/vertical well option provided the advantages of lower cost, conventional well construction and maintenance techniques, and minimal aboveground disruption, but at a greater number of construction sites. This option would impact a larger number of landowners. It also has the benefit of providing for a more even extraction of water along the riverbank than the collector well option, reducing the stresses and resulting in a decrease in capacity experienced with the 0.66-m^3/s demonstration collector well.

Previous experience with designing, bidding, and constructing the conventional collector well provides an advantage for this option. The disadvantages included the lack of flexibility for maintenance described for the rock tunnel/horizontal collector option, and the number of aboveground structures required. This option was not considered feasible for the Crescent Hill facility; therefore, an additional disadvantage is that this option would provide no additional technical design information to be applied towards applying RBF at the larger plant site.

Taking these options into consideration, preliminary design was initiated on the hard rock tunnel option with vertical wells in January 2002. Final design is anticipated in 2003, with construction beginning in 2004 and extending through 2006.

References

Deere, D.U., and D.W. Deere (1989). "Rock quality designation (RQD) after twenty years." *Contract Report No. GL-89-1*, U.S. Army Corps of Engineers, Washington, D.C.

Chapter 5. Operation and Maintenance Considerations

Henry Hunt, CPG
Collector Wells International, Inc.
Columbus, Ohio, United States

Jürgen Schubert, M.Sc.
Stadtwerke Düsseldorf AG
Düsseldorf, Germany

Chittaranjan Ray, Ph.D., P.E.
University of Hawaii at Mānoa
Honolulu, Hawaii, United States

1. Introduction

To properly manage an RBF well, an operator must ensure routine maintenance and understand the environmental factors that impact well performance. For example, the manager of a utility would prefer that the production rate remains unchanged with time when the well is operating at full capacity. Similarly, if the well operates at less than maximum capacity, the well must be able to deliver water at peak capacity when such a need or emergency arises; however, the production capacities of RBF wells are bound to change due to environmental conditions in the well, aquifer, and river. Common problems include biofouling and well-screen clogging. Aquifers could clog with fine particles or biological matter, and the clogging or scouring of a riverbed could severely alter the hydraulic connection between the river and aquifer. Riverbed clogging could also reduce the amount of flow and increase the drawdown needs for a given pumpage. Conversely, scouring can enhance the flow to wells but, at the same time, could increase the turbidity or bacteria count of pumped water, especially for collector wells with laterals beneath the rivers. Furthermore, operation and maintenance needs vary due to:

- Size of the RBF facility.
- Types of wells employed.
- Continuous or intermittent operation of wells.
- Materials used for well construction.
- Geologic environment.
- River conditions.

Most utilities are required to monitor the quality of pumped water. The amount of water pumped or the number of people served will determine how often samples are collected; however, many utilities rarely keep track of ambient water-level and water-quality data, which may help determine the performance and future maintenance needs of RBF wells. Such data would include:

- Water-level and water-quality records of the river near the RBF facility.
- Water-level and water-quality records of RBF wells.
- Water levels in monitoring wells between the river and well.
- Water levels in monitoring wells that are on the land side of the RBF system.
- Ambient groundwater quality.

C. Ray et al. (eds.), Riverbank Filtration, 61–70.
© 2002 *Kluwer Academic Publishers. Printed in the Netherlands.*

While most small- to medium-sized utilities do not have the resources to frequently monitor (monthly or less) these parameters, it would be useful to have monthly records of these parameters during significant events, such as droughts or floods.

2. Select Operating Wells in the United States

There is no comprehensive record of RBF wells built and operated for drinking water or industrial water supply in the United States. It is almost impossible to develop a list of such wells unless a survey is conducted of all water utilities. In particular, many utilities using vertical wells, as stated in the Introduction of this book, might not know that their wells could be classified as RBF wells. Lacking such data, only an overview of public-water systems in the United States and their sources are presented in Table 5-1. In the United States, public-water systems are classified into categories based upon the number of population served. Any water system serving 25 people or having 15 service connections is considered to be a public-water system. There are approximately 54,000 public-water systems that serve most of the U.S. population. Public-water systems that serve less than or equal to 3,300 people are termed "small systems." Medium systems serve between 3,301 and 10,000 people. Large systems serve above 10,001 people. There are 350 very large systems in the United States that serve over 100,000 people each. In 1995, the total pumpage for public supply (including commercial, industrial, domestic consumption, and losses) was 152 million m³/day, of which 37 percent was groundwater and 63 percent was surface water.

Table 5-1. Water Sources for Public-Water Systems in the United States in 2001

	Groundwater	Surface Water	Total
Number of Systems	42,212	11,571	53,783
Population Served	85,743,562	178,401,567	264,145,129
% of Systems	78	22	100
% of Population	32	68	100

Most RBF collector wells constructed in the United States over the past 50 years were installed by employees trained by the original company that invented this well-drilling technology in the 1930s, who in turn played a part in the many improvements and advances in the specialized construction methods used today. Since a comprehensive list of all wells installed in the United States is not available, Table 5-2 provides the design data and other operational informations for selected wells. As shown in Table 5-2, the design capacity of the wells vary. The recent trend has been towards installing medium to large capacity wells. The frequency of maintenance appears to be low (more than 10 years). Most RBF systems have the ability to maintain service during scheduled maintenance through the activation of back-up vertical wells, use of other collector wells in the system, or through various rehabilitation schedules that permit continued service (including in-line storage pumping, manifolding, or specialized rehabilitation procedures concurrent with production). Included are short descriptions of system and maintenance issues for RBF systems in Jacksonville, Illinois, and Cedar Rapids, Iowa. Both of these systems are listed in Table 5.2.

Table 5-2. Design and Maintenance Details for Select Horizontal Collector Wells in the United States

Site	Year Installed	Design Capacity (m³/s)	Number of Wells	Caisson Diameter (m)	Depth (m)		Laterals		Total Lateral Lengths per Well (m)	Year of First Maintenance/ Type of Work	Frequency of Maintenance/ Period of Shutdown	Backup Well Type
					Well	Lateral	Number	Diameter (cm)				
Boardman, Oregon	1976, 2002	0.39 and 0.66	2	4, 4.9	15	13	14	20 and 25	244	None	NA	None
Casper, Wyoming	1958	0.09	3	4	10.7	8.6	9	25	274	None	NA	Vertical
Cedar Rapids, Iowa	1995, 2002	0.3, 0.44, 0.44, 0.37	4	4, 4.9, 4.9	21.3	20 Average	5	30 Average	251	None	NA	Vertical
Evansville, Indiana	1978	0.4	1	4	24.4	22.3	8	30	488	None	NA	Intake
Independence, Missouri	1990	0.44	1	4	38	36	8	25	488	None	NA	Vertical
Jacksonville, Illinois	1955	0.2	1	4	27.4	25.6	7	30	357	1983	10 Years/1 Month	Vertical
Kalama, Washington	1975	0.1	1	4	12	10	3	25	98	2000	25 Years/None	None
Kansas City, Kansas	1997	1.1 (1.75 Maximum)	1	6	37	34.5	14	30	747	None	NA	None
Kennewick, Washington	1957	0.13 Average Each	5	4	13.7	12 Average	8	25	91	2000	43 Years/None	None
Lake Havasu City, Arizona	2000	1.1	1	4.9	32	29	14	30	537	None	NA	Vertical
Lincoln, Nebraska	1994	1.5 to 1.75 Total	2	4	23 Average	21 Average	14	30	381	None	NA	Vertical
Louisville, Kentucky	1999	0.88	1	4.9	30	28.5	7	30	488	None	NA	Intake
Mankato, Minnesota	1970	0.22	1	4	17.5	16	9	30	300	1982	Self-Performed	Vertical
Perth Amboy, New Jersey	1985	0.26	1	4	24.4	22	5	25	250	2001	16 Years/1 Month	Well Points
Sioux Falls, South Dakota	1956 to 1999	0.09 Average Each	14	3 to 4	16.8 Average	14.6 Average	3	20 Average	99	Varies*	Self-Performed	Vertical
Sonoma County, California	1957, 1975, 1982	3.72 Total	5	4	34 Average	31 Average	41	20 and 25	331	2001	44 Years/1 Month	Vertical
St. Helens, Oregon	1955, 1969, 1999	0.13, 0.13, 0.22	3	4.4, 4.9	27.5, 29, 19	25.3, 27, 17	34	9 and 25	247	None	NA	Vertical
Terre Haute, Indiana	1991	0.44	1	4	22.9	20.7	8	30	366	None	NA	None

NA = Data not available. *Due to the large number of wells, the frequency of maintenance varies for the site in Sioux Falls, South Dakota.

Jacksonville, Illinois

The RBF well at Jacksonville, Illinois, is located on the bank of the Illinois River near the City of Naples in Scott County. A 37-km long pipeline transports pumped water from the well to a storage reservoir in the City of Jacksonville. The water from the reservoir undergoes conventional treatment (coagulation, flocculation, sedimentation, filtration, and disinfection) prior to being pumped into the distribution system. There are seven laterals, each 30-cm in diameter, for a total length of 357 m. The laterals are located at a depth of 25.6 m from ground surface, and part of these laterals are under water during normal flow. Under high-flow conditions, most of these laterals remain beneath the river. There is a minimum of 15-m porous media between the river and the laterals. The caisson has three pumps, and the normal pumpage is on the order of 0.2 m³/s. Two backup wells, located 300 to 400 m from the riverbank, are pumped when the collector well is shut down for maintenance. The last time the collector well was shut down was during the summer of 1995; the maintenance work lasted for 1 month. The well was back in service towards the end of the summer. The typical frequency of maintenance is no less than 10 years. Floods in the Illinois River and the subsequent clogging of the riverbed (after flow recession) have been reported to affect the production capacity of the collector well at this site.

Cedar Rapids, Iowa

The City of Cedar Rapids in Iowa has three municipal well fields located on the banks of the Cedar River, a tributary of the Iowa River, which ultimately joins the Mississippi River. There are 53 vertical wells and four collector wells. Two of the four collector wells were constructed in 1995 and the rest were installed in 2002. The pumping rates of the collector wells range from 0.3 to 0.44 m³/s each. Current pumpage is on the order of 1.5 m³/s. Several studies have been conducted in these well fields to assess the surface and groundwater interaction near the pumping wells and to evaluate the potential contamination of pumped water from chemicals present in surface water. As shown in Table 5-2, each collector well has five 30-cm laterals, and the total lengths of the laterals screens are about 250 m per well. The depths of the vertical wells range between 12.8 to 23 m. The collector wells are approximately 21-m deep. Since the wells are relatively new, none have undergone maintenance yet. The site also has an adequate supply of vertical and collector backup wells for when a collector well is taken out of service for maintenance.

3. Select Operating Wells in Germany

Along the Rhine River, more than 200 waterworks from seven European countries (Switzerland, Liechtenstein, Austria, France, Germany, Belgium, and The Netherlands) supply drinking water to about 28-million inhabitants. The waterworks are represented by the International Association of the Waterworks in the Rhine Region, which coordinates monitoring programs and assists research programs, among other activities, for 120 water supply companies in the Rhine River region. The total amount of withdrawn raw water from the Rhine River in 1998 was $2,430 \times 10^6$ m³ (Furrer et al., 2000). The sources of raw water, which are treated for domestic and industrial use, can be distinguished in Table 5-3.

More than 80 percent of riverbank-filtered water is withdrawn (1998) for domestic and industrial use in the Lower Rhine Valley between the mouth of the Sieg River at 660 and the Rhine River at 790 km. Table 5-4 represents a summary of RBF plants, along with their locations and pumpage, in this region.

Different well types are employed in the Lower Rhine Valley. While there are still some cylindrical pit wells (7- to 10-m diameter) from the nineteenth century, the majority of wells are

Table 5-3. Distribution of Drinking-Water Sources in the Rhine River Region

Source of Raw Water	Amount Withdrawn (per year)	Distribution (%)
Spring Water	$65 \times 10^6 \text{ m}^3$	3
Groundwater	$877 \times 10^6 \text{ m}^3$	36
Riverbank-Filtered Water	$220 \times 10^6 \text{ m}^3$	9
Recharged Groundwater	$378 \times 10^6 \text{ m}^3$	16
River Water (for Groundwater Recharge)	$389 \times 10^6 \text{ m}^3$	16
Lake Water	$395 \times 10^6 \text{ m}^3$	16
Dam Water	$104 \times 10^6 \text{ m}^3$	4
Total Amount	$2,428 \times 10^6 \text{ m}^3$	100

Table 5-4. Volume of Water Withdrawn from the Rhine River Between 660 and 789 km (1998)

Water Supply Company	Well Field Name	Location Along River	Water Volume Withdrawn
Wahnbachtalsperrenverband	Siegniederung	660 km, r	$22.6 \times 10^6 \text{ m}^3$
WBV Wesseling-Hersel		662 km, l	$2.5 \times 10^6 \text{ m}^3$
GEW-Köln AG	Weißer Bogen	679 km, l	
	Hochkirchen	683 km, l	
	Langel/Worringen	703 km, l	$30.2 \times 10^6 \text{ m}^3$
RGW AG	Stammheim	697 km, r	$18.5 \times 10^6 \text{ m}^3$
Bayer AG	Leverkusen	704 km, r	$7.4 \times 10^6 \text{ m}^3$
WW Baumberg GmbH		712 km, r	$3.9 \times 10^6 \text{ m}^3$
Stadtwerke Düsseldorf AG	Auf dem Grind	722 km, l	
	Flehe	731 km, r	
	Am Staad	750 km, r	$49.8 \times 10^6 \text{ m}^3$
Wuppertaler Stadtwerke AG	Auf dem Grind	722 km, l	$10.5 \times 10^6 \text{ m}^3$
Stadtwerke Duisburg AG	Wittlaer	757 km, l	$9.4 \times 10^6 \text{ m}^3$
RWW GmbH	Duisburg-Mündelheim	768 km, r	$2.1 \times 10^6 \text{ m}^3$
NGW GmbH	Duisburg-Baerl	785 km, l	$28.2 \times 10^6 \text{ m}^3$
WVN GmbH	Binsheimer Feld	789 km, l	$7.8 \times 10^6 \text{ m}^3$
Total Amount (1998)			$192.9 \times 10^6 \text{ m}^3$

r = Right bank   l = Left bank.

represented by vertical filter wells (400- to 600-mm diameter). Most wells are grouped into well galleries parallel to the riverbank. Individual wells in the well galleries are connected together by siphon tubes. The pumping equipment is normally present in only one well. When this well is pumped, water is forced through the siphon tubes from the other wells to refill the pumped well. Horizontal collector wells (Fehlmann-type with a gravel-pack filter) may be designed to function with either of two different capacities:

- Small-size type with a capacity between 900 and 1,200 m³/h; central caisson diameter of 3.2 m; 4 to 6 radial laterals (20- to 30-cm diameter, 30- to 40-m length).
- Medium-size type with a capacity between 2,700 to 3,000 m³/h; central caisson diameter of 5 m; 10 to 12 radial laterals (25- to 30-cm diameter, 60- to 90-m length).

Utilities may have either small-capacity horizontal filter wells or medium-capacity horizontal filter wells, or a combination of both. The choice of the capacity of collector wells depends on water demand and site conditions. About 60 to 70 percent of the operating horizontal collector/filter wells in the Lower Rhine Valley are medium-sized (2,700 to 3,000 m³/h).

Well Sites for RBF in the Lower Rhine Valley in Europe

Figure 5-1 shows several RBF sites along the Rhine River between 715 and 760 km with accompanying groundwater protection zones in the region near Düsseldorf, Germany. A cross-section through the aquifer in this region is shown in Figure 5-2. The confining impervious layer consists of very fine Tertiary sands with a permeability of less than 10^{-5} m/s. The aquifer consists of sandy, gravely Pleistocene sediments. Its average permeability varies (due to pumping tests) between about 10^{-3} and 10^{-2} m/s. The saturated thickness of the alluvial deposit varies with the site and is controlled to some extent by topography and riverbed level. The thickness of these deposits ranges between 10 and 20 m. The surface of the formation is covered by clay and fine-sand layers, with a depth of a few decimeters up to 3 m. The average groundwater recharge rate in the region is 7 L/s/km².

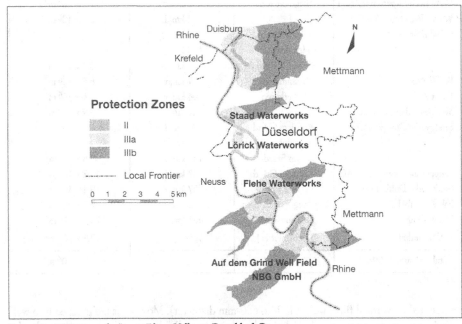

Figure 5-1. RBF sites at the Lower Rhine Valley in Düsseldorf, Germany.

The Rhine River is about 400-m wide and can be characterized by the following data:
- Median discharge . . . . 2,100 m³/s.
- Hydraulic gradient . . . . . . 0.20 m/km.
- Flow velocity . . . . 1.0 to 1.4 m/s.

Included are detailed descriptions of the Flehe Waterworks and Auf dem Grind Well Fields, which are located in this region of the Rhine.

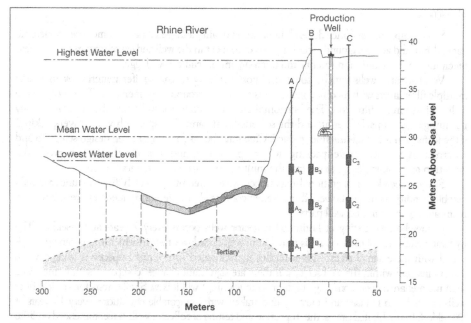

Figure 5-2. Cross-section through the aquifer in the Düsseldorf, Germany, region at 731.5 km. A, B, and C indicate rows with three monitoring and sampling wells each. Indices $_{1, 2, 3}$ indicate the different depths of the screen pipe.

a. Flehe Waterworks Well Field

The Flehe Waterworks Well Field is situated in the outer side of a river bend at the eastern bank of the Rhine River between 730.7 and 732.5 km. The thickness of the aquifer varies between 15 to 25 m. The data below describes a gallery of 70 wells located between 730.9 and 732.2 km (600-mm diameter, 22- to 28-m depth, 15-m screen length, connected by siphon pipes). The gallery is situated at a distance of 60 m from the bank. The average retention time of river water in the subsoil is 3 weeks. The design capacity of this part of the well field is 44,000 cubic meters per day (m^3/d). In long terms, 70 to 75 percent of the extracted water is riverbank-filtered water and 25 to 30 percent is groundwater rising from the 17-kilometer squared (km^2) catchment area on the land side. The average specific yield of riverbank-filtered water is 25 cubic meters per meter per day (m^3/m/d).

b. Auf dem Grind Well Field

The Auf dem Grind Well Field is situated on a "peninsula" formed by a distinct river bend at the western bank of the Rhine River between 719 and 726 km. The thickness of the aquifer varies between 25 to 30 m. The surface of the peninsula is flooded during very high river water levels. Seven horizontal collector wells with a capacity of 2,800 m^3/h each are situated at distances between 200 to 300 m from the bank of the Rhine River. The average retention time of river water in the subsoil is 10 weeks. The design capacity of the well field is 400,000 m^3/d in total. The maximum demand was 330,000 m^3/d in 1976. Due to the cross-flow of river water through the underground (induced by the gradient of the river level) and the location of the well field in the inner bend of a meander, 90 to 95 percent of the extracted water is riverbank-filtered water and only less than 10 percent is groundwater rising from the 19-km^2 catchment area on the land side. The average specific yield (intake per meter length of the bank) of riverbank-filtered water is 54 m^3/m/d.

Maintenance

Screen-pipe maintenance depends upon water quality and well type. In some cases, anaerobic bank filtrate and aerobic groundwater (or vice versa) meet in the well and cause iron and manganese precipitation. In addition, nutrients and microorganisms can cause clogging.

Vertical filter wells can be adapted almost perfectly to the aquifer material by installing multiple filter layers with modified grain-size distribution around the screen pipe. This protects the well against sand infiltration. The horizontal laterals of collector wells can also be protected by installing a gravel-pack filter around the screens (Fehlmann-type approach with gravel-packing) if the formation is sufficiently fine-grained. Care must be taken during screen design selection and construction to select gravel-packing, where appropriate, and to select the proper gavel-pack gradation and screen design to control the infiltration of sand from the formation.

Several decades of operational experience with different types of RBF wells under chiefly aerobic conditions in the Lower Rhine Valley confirm that different frequencies are suitable for maintaining underground well parts.

Comprehensive testing of horizontal collector wells occurs every 7 years at Düsseldorf. The hydraulic efficiency of each lateral is tested from step drawdown tests in which the pumping rate is varied with incremental steps. This is further supplemented with visual inspections using a video camera inserted within the well screens. If there are significant changes compared with former data, then maintenance operations (e.g., high-pressure washing) will follow. Similar tests for vertical filter wells are easier to manage and must be undertaken with comparable conditions every 15 years at Düsseldorf. In the United States, the inspection and testing of collector wells is recommended on 5- to 10-year intervals to monitor well performance and to plan for any maintenance that may be required.

4. Other Applications

Horizontal collector wells have also been installed in the United States for water production in settings other than RBF. Three such applications are:
- Seawater (beach) collector wells.
- Groundwater collector wells.
- Artificial recharge wells/aquifer storage and recovery wells.

Seawater (Beach) Collector Wells

Seawater collector wells are used to produce filtered seawater for specialized purposes, such as reverse osmosis or cooling. The central pump station caisson can be installed some distance away from the beach, with the lateral well screens projected out horizontally into the beach deposits (Figure 5-3). In this way, suspended debris and surface-water organisms are typically filtered out before water reaches the pumps, providing pre-filtration of raw water to improve the quality of water entering the treatment plant (e.g., desalination) or for point-of-use.

This well design allows well screens to be installed in the most hydraulically efficient aquifer layers to maximize the possible yield from the formation. Since fewer wells are needed (compared to vertical wells) and as the caissons can be constructed away from the beach or completed at or below grade, visual impacts are minimized (Hunt, 1996).

Groundwater Collector Wells

Groundwater collector wells consist of a central reinforced concrete caisson that serves as a wet well pump station. Well screens are projected out horizontally from within the caisson into the surrounding aquifer deposits (Figure 5-4), and are projected into the most hydraulically efficient zone within the aquifer to maximize yield. If more than one aquifer is identified, well screens can

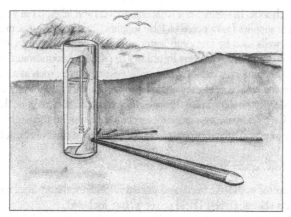

Figure 5-3. Conceptual representation of a filtered seawater (beach) collector well. Drawing provided by Collector Wells International, Inc.©

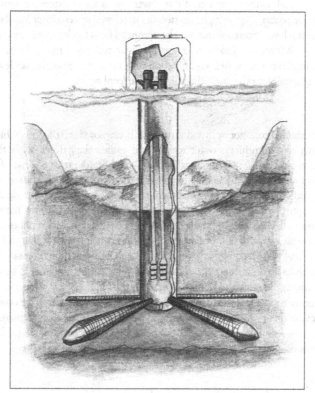

Figure 5-4. Conceptual representation of a groundwater collector well. Drawing provided by Collector Wells International, Inc.©

be projected into each zone. In settings not adjacent to rivers and lakes, these wells are designed to produce groundwater supplies (as opposed to RBF supplies), and they share some similarities with the horizontal filter wells used by European water utilities; however, in the United States, these wells have not been used in large numbers by utilities to produce "groundwater" solely.

Where alluvial deposits exist adjacent to a surface-water source, such as a river, lake, or even the ocean, well screens can be projected into deposits that are hydraulically connected to the surface water to achieve natural filtration through RBF processes. For example, the Board of Public Utilities in Kansas City, Kansas, uses a groundwater collector well on the bank of Missouri River, where the laterals extend radially at two levels in the productive parts of the aquifer.

Artificial Recharge Wells/Aquifer Storage and Recovery Wells

Horizontal collector wells have been used since the 1940s for the artificial recharge of aquifers in a range of projects (Hunt, 1984). These projects have included:

- *Louisville, Kentucky* – A horizontal collector well was used to recharge treated city water into the alluvial aquifer to replenish declining groundwater levels in the downtown area, which was caused by many years of over-pumping.
- *Canton, Ohio* – Three collector wells were used to recharge RBF water derived from a shallow surficial aquifer past a confining layer into a lower extensive aquifer for storage using passive recharging through two tiers of lateral well screens installed above and below the confining layer. Stored water was then pumped from the lower aquifer into the system.
- *Manitowoc, Wisconsin* – Lake water was passively recharged into a local aquifer system using an intake-collector well combination unit to restore groundwater levels to support higher capacities for other wells installed in the local aquifer.

5. Conclusion

Numerous horizontal collector wells and vertical wells, employed at RBF sites in the United States and Europe, extract large quantities of water from alluvial aquifers along rivers. While the tendency in Europe is to use vertical wells, most utilities in the United States prefer the use of large-capacity horizontal collector wells. Most of these wells have been operating for decades. The successful operation of RBF wells depends upon both regularly maintaining the well screen and pumps and operating the system within design parameters. For the United States, the frequency of maintenance for most horizontal collector wells appears to be on the order of 10 years or longer. Most utilities using horizontal collector wells have backup wells or other means to continue water service during maintenance. This maintenance can usually be accomplished over a period of about 1 month, if necessary. In comparison, the maintenance cycles for horizontal collector wells and vertical wells are 7 and 15 years, respectively, at Düsseldorf, Germany, although this trend is reversed in the United States, where vertical wells tend to require maintenance more frequently than collector wells. Collector wells have other applications, including the development of groundwater for artificial recharge and aquifer storage and recovery programs. In addition, seawater collector wells can be used to produce high-quality saline or brackish water for cooling or desalination using the same hydraulic principles used for RBF.

References

Furrer, R., M. Fleig und H.J. Brauch (2000). *Wasserförderung und -aufbereitung im Rheineinzugsgebiet*. Internationale Arbeitsgemeinschaft der Wasserwerke im Rheineinzugsgebiet (IAWR), Amsterdam, The Netherlands.

Hunt, H.C. (1984). "The horizontal radial collector as a recharge well." Poster Presentation, 29th Annual Midwest Ground Water Conference, Lawrence, Kansas.

Hunt, H.C. (1996). "Filtered seawater supplies – Naturally." *International Desalination and Water Reuse Quarterly*, 6(2): 32-37.

Part II:
Contaminant Removal

Chapter 6. Removal of Pathogens, Surrogates, Indicators, and Toxins Using Riverbank Filtration

Jack Schijven, Ph.D.
National Institute of Public Health and the Environment
Microbiological Laboratory for Health Protection
Bilthoven, The Netherlands

Philip Berger, Ph.D.
Ijamsville, Maryland, United States

Ilkka Miettinen, Ph.D.
National Public Health Institute
Division of Environmental Health
Kuopio, Finland

1. Introduction

RBF is a water-treatment process that makes use of surface water that has naturally infiltrated into groundwater through the riverbed or bank(s) and is recovered by a pumping well. Infiltration is typically enhanced by the hydraulic gradient imposed by a nearby pumping-water supply or other well(s). Riverbank filtrate is water emanating from a pumping well that originated nearby as surface water and traveled through the subsurface, mixing to some degree with other groundwater. Through RBF, microbial pathogens, fecal indicator organisms, and other surrogates are removed by contact with aquifer materials. The removal process is most efficient when groundwater velocity is slow and when the aquifer is made of granular materials with open pore space for water flow around the grains. In these granular porous aquifers, the flow path is tortuous, thereby providing ample opportunity for organisms to come into contact with and attach to a grain surface. If detachment does occur, it will typically occur at a very slow rate. Organisms typically remain attached to a grain for long periods. When groundwater velocity is exceptionally slow or when little or no detachment occurs, the organism will become inactivated before it can enter a well. Thus, RBF relies on attachment to the soil and inactivation to remove microorganisms from infiltrating surface water.

The efficiency of RBF to remove microorganisms depends on:

- The efficiency of the various removal processes, of which attachment of the microorganisms to the soil and inactivation are most important.
- The climatic/hydrologic conditions.
- The geometry of the well vis-à-vis the surface-water body.
- The character of the bank material and stream/lake bed.
- The groundwater flow field.

The level of microorganism removal from infiltrating surface water by RBF depends upon the concentration of microorganisms in surface water and the maximum allowable level in the

73

C. Ray et al. (eds.), Riverbank Filtration, 73–116.
© 2002 *Kluwer Academic Publishers. Printed in the Netherlands.*

receiving water, as dictated by legislation or regulation. The concentrations in surface water depend on:

- The number and character of sources that contribute microbiota to the environment.
- Climatic and hydrologic conditions.
- In situ inactivation rates.
- Raw-water source concentration.

In this section, those properties and processes specific to the natural filtration of bacteria, viruses, protozoa, and algae (and algal toxins) are discussed and illustrated, as data allows, with case studies.

2. Why RBF for Microbial Pathogens?

There is a need to use surface water for drinking water; however, surface water is often polluted with pathogens. RBF may be an effective way to remove pathogens from surface water that is used for drinking water. The efficiency of RBF in removing microbial pathogens is the main issue of this chapter.

In The Netherlands, about 67 percent (794×10^6 m³/yr) of all drinking water is delivered from groundwater and 33 percent (399×10^6 m³/yr) from surface water (VEWIN, 1998). Due to government policy to limit desiccation (which affects agricultural production and, especially, nature reserves), groundwater withdrawal is not allowed to increase from the year 2000 onwards. To meet possible future increases in the demand for drinking water, surface water is becoming more important, especially in combination with treatment by soil and aquifer passage (Mülschlegel and Kragt, 1998).

The United States Environmental Protection Agency is developing the proposed Long Term 2 Enhanced Surface Water Treatment Rule (LT2ESWTR) (U.S. Environmental Protection Agency, 2000). The proposed LT2ESWTR applies to public water-supply systems that use either surface water or groundwater under the direct influence of surface water as their raw-water source. By definition, "groundwater under the direct influence of surface water" is groundwater closely associated with surface water. The proposed LT2ESWTR protects public health by identifying those systems that employ conventional filtration, which may need to provide additional protection against the protozoa, Cryptosporidium. The results of site-specific source-water Cryptosporidium monitoring shall result in assigning systems into differing "bin" categories. For each system, additional treatment requirements depend on the assigned bin. Systems will choose technologies to comply with the additional treatment requirements from a "toolbox" of options. One microbial toolbox component is RBF, which is assigned a potential 1.0-log credit for Cryptosporidium removal, where used by a system as a pretreatment method followed by conventional filtration. In the proposed rule, systems with Cryptosporidium concentrations greater than 0.075 oocysts per liter (oocysts/L) must achieve additional treatment.

3. Pathogen Occurrence in Surface Water

Enteric viruses and the pathogenic protozoa Cryptosporidium and Giardia are ubiquitously present in Dutch surface waters (Hoogenboezem et al., 2000; Medema et al., 1996; Schijven et al., 1999a; Theunissen et al., 1998). Surface water is contaminated with pathogenic microorganisms, mainly due to discharges of wastewater and by manure runoff from agricultural land. The highest load of microorganisms into Dutch surface waters is due to import from abroad by the major rivers, the Rhine and the Meuse. From January until March, concentrations are at their maximum, but during the summer, when there is less flow of water and higher temperatures, concentrations are at their minimum. This type of seasonality is similar for pathogenic protozoa, viruses, and fecal indicator bacteria as well. The emission and distribution of these pathogens to Dutch surface waters has been modeled on a national scale (Medema et al., 1997; Schijven et al., 1995, 1996). A high potential for the waterborne transmission of microbial pathogens exists in The Netherlands where surface water is used as the source for drinking-water production; adequate treatment must be guaranteed under all circumstances.

Natural waters are often contaminated by pathogenic bacteria excreted by humans, cattle, and various domestic and wild animals; however, the main source for the pathogenic bacteria entering surface or groundwater is sewage. The concentration of total coliforms, E. coli, or enterococci in sewage may range from 10,000 to more than 10,000,000 cells per 100 mL (Matthess and Pekdeger, 1981; Olson, 1993; Geldreich, 1996). Municipal sewage is a conduit for pathogens like Salmonella to enter surface waters (Geldreich, 1996). Wet-weather flows may result in peak bacterial concentrations due to sewer overflow. For example, Geldreich reported that fecal coliform density was about 2 logs greater in sewer overflow water than in uncontaminated stormwater. Stormwater shock loads need to be accounted for in evaluating RBF efficiency.

Sources for Cryptosporidium oocysts are mammals, including humans. Seventy-nine mammal species are susceptible to Cryptosporidium infection (Fayer et al., 1997). Raw and treated sewage contain significant levels of Cryptosporidium (Robertson et al., 2000). Even when diluted with stormwater in combined sewer overflows (e.g., Bruesch et al., 1999), concentrations may still be sufficient to breakthrough into RBF wells.

Episodic precipitation, recharge, flooding, and scour may amplify the variability of environmental protozoan concentrations. For example, Atherholt et al. (1998) report an association between rainfall and increased oocyst concentration as well as an increase in turbidity associated with both factors. In contrast, watershed modeling by Walker and Stedinger (1999) considered the effects of unusual hydrologic events on surface-water oocyst concentrations. They concluded that daily loading rates showed little variation. These data suggest that only short-duration precipitation periods, and not seasonality, might affect source concentrations.

In the United States, a nationwide survey of pathogenic protozoa, enterovirus, and bacterial-indicator contamination of surface water used as drinking water was recently completed (U.S. Environmental Protection Agency, 2000a). The survey consisted of two parts: the ICR and the Supplemental Survey. The ICR sampled 347 sites (drinking-water supply systems that serve more than 100,000 people) monthly for 18 months. Enterovirus sampling was not required at all sites. The Supplemental Survey sampled 88 sites (both large- and medium-sized systems) biweekly for 1 year for parasitic protozoa using an improved laboratory method. The oocyst analyses were conducted using the ICR method, which showed a 12-percent mean oocyst recovery from samples spiked with a known concentration of oocysts (Messner and Wolpert, 2000). The results of the ICR sampling are reported in Messner and Wolpert (2000) and Rosen and Ellis (2000).

The ICR results show that there was a significant difference between oocyst concentrations in flowing streams and in reservoirs and lakes (Messner and Wolpert, 2000). Based on the measurements of 130 sites that use flowing streams, the oocyst concentration was about a factor of 10 higher in flowing streams than in lakes and reservoirs; however, some reservoirs and lakes have concentrations as high or higher than flowing streams. At measured concentrations of 1 oocyst/L, 25 percent of the flowing stream sites exceeded that value. By comparison, at that same concentration level, 7 percent of the sites using reservoirs and lakes exceeded the 1-oocyst/L concentration level. Monthly sampling in the ICR study (Messner and Wolpert, 2000) also found no statistically significant monthly difference or seasonality for oocyst concentration.

The ICR collected oocyst data from four large RBF systems in the United States that each serve a population greater than 100,000. These sites are:

- Cedar Rapids, Iowa
- Sonoma County, California
- Dayton, Ohio
- Lincoln, Nebraska

The ICR microbial occurrence data for these four sites are shown in Table 6-1 (U.S. Environmental Protection Agency, 2000a).

Table 6-1. ICR Sampling Results for Large RBF Sites

Site	Oocyst-Positive Samples	Cyst-Positive Samples	Enterovirus-Positive Samples	Maximum Enterovirus Concentration (most probable number/100 liters)
Cedar Rapids, IA	0/17	1/17	Not Done	
Sonoma County, CA	0/18	0/18	2/17	7 MPN/100 L
Dayton, OH	0/17	0/17	Not Done	
Lincoln, NE	0/13	0/13	1/13	1 MPN/100 L

Please note that cysts are of *Giardia* origin and oocysts are of *Cryptosporidium* origin.

Schubert (2000) reports a concentration of 6.2 oocysts/100 L in samples from the Rhine River adjacent to the RBF site in Düsseldorf, Germany, but no oocysts were recovered from the RBF wells. An RBF site (Bolton Well Field) in Cincinnati, Ohio, also reported negative oocyst occurrences in 11 samples, each collected in the period after a major precipitation event (Cossins et al., 1997). The vertical wells in the Bolton Well Field range from 30 to 120 m (average 59 m) from the river and are screened at depths of between 9 and 46 m.

4. Health Effects

Most of the waterborne viral, bacterial, and protozoan pathogens are of fecal origin and are transmissible by a (fecal-oral) water route of exposure. These pathogens can cause gastrointestinal illness as well as other more severe illnesses (e.g., hemolytic uremic syndrome). The impact of contaminated water on public health may range from asymptomatic infections to a few days of mild diarrhea, to severe disease requiring a physician's care or hospitalization, to death (Gerba et al., 1996); however, acute gastroenteric illness is most common.

Certain individuals may be at greater risk of serious illness than the general population. Depending on the pathogen, individuals who are at increased risk of developing more severe outcomes from waterborne microorganisms are the very young, the elderly, pregnant women, the immunocompromised (e.g., organ transplants, cancer patients, AIDS patients), those predisposed with other illnesses (e.g., diabetes), and those with a chemical dependency (e.g., alcoholism) (Gerba et al., 1996). For individuals suffering from the disease cryptosporidiosis, those who are immunocompetent will usually recover from illness within 2 weeks. Immunocompromised individuals are more likely to suffer from a chronic and debilitating illness. In these individuals, *Cryptosporidium* may contribute to premature death (U.S. Centers for Disease Control and Prevention, 1997).

Particularly in developing countries, the microbiological contamination of drinking water — either groundwater or riverbank-filtered groundwater — may have profound and severe implications for public health. Contaminated drinking water can contribute to high morbidity and mortality rates from diarrheal diseases and, sometimes, lead to epidemics. The disposal of excreta using land-based systems is a key issue in groundwater quality and public-health protection. The use of inappropriate water supply and sanitation technologies in peri-urban areas leads to severe and long-term public-health risks. The use of poorly constructed sewage treatment works and land application of raw sewage can lead to groundwater contamination close to water-supply sources (Pedley and Howard, 1997).

5. Outbreaks Related to the Use of Riverbank-Filtrated Drinking Water

Despite improved sewage treatment, protected water sources, and improved water purification technology, waterborne epidemics still occur, not only in developing countries, but also in highly industrialized countries (Furtado et al., 1998; Craun et al, 1998; Lack, 1999; Morris and Foster, 2000). Although waterborne disease has largely been controlled in North America, outbreaks continue to occur. To be considered a waterborne outbreak, the disease must cause an acute illness that affects at least two people and must be epidemiologically associated with the ingestion of water (Craun and Calderon, 1996). Most recently, groundwater outbreaks occurred in Brushy Creek, Texas, in the United States (1,300 to 1,500 cases; Bergmire-Sweat et al., 1999), and Walkerton, Ontario, in Canada (2,300 cases; Anonymous, 2000). The number of individuals reported ill from these outbreaks is generally an underestimation of the actual levels of microbial diseases associated with drinking water because endemic levels are not described and the reporting of disease outbreaks is poor (Frost et al., 1996).

According to the World Health Organization's classification (1986, 1998), the bacteria most commonly responsible for waterborne outbreaks belong to the species:
- *Salmonella typhi* (and other salmonellae)
- *Escherichia coli* (E. coli)
- *Vibrio cholerae*
- *Campylobacter spp.*
- *Yersinia enterocolitica*
- *Shigella spp.*

The threat from *Legionella, Pseudomonas aeruginosa, Aeromonas spp.*, and *Mycobacterium* (atypical) is considered moderate, although these bacteria are capable of multiplying in water-supply storage and distribution systems.

The significance of *Campylobacter spp.* as a waterborne pathogen has increased during the last few decades. Several waterborne outbreaks of the disease campylobacteriosis have been reported in the past decades in the United States and elsewhere (Craun, 1986, Van Der Leeden et al.,1990). From 1998 to 2000, four out of a total of 18 reported waterborne outbreaks in Finland were caused by *Campylobacter jejuni* contamination (Miettinen et al., 2000). The significance of *Campylobacters* as a waterborne pathogen is increasing because it can survive for several months in natural waters at low temperatures (Korhonen and Martikainen, 1991). The infectious dose of *Campylobacter* is rather low (about 500 bacteria cells) (Robinson, 1981; Black et al., 1988), enhancing the likelihood of infection.

Table 6-2 identifies cryptosporidiosis outbreaks associated with RBF systems and the available information on possible causes. Cryptosporidiosis outbreaks associated with the failure of an RBF system have been, to date, poorly characterized, if indeed such an outbreak has occurred at all.

Table 6-2. Cryptosporidiosis Outbreaks Associated with RBF in Alluvial Aquifers

Outbreak	Year	Aquifer	Other Surface Water	Collection Device	Cases (laboratory confirmed: estimated)	Comment and/or References
Torbay, Devon, United Kingdom (South West Water)	1992	Littlehempston River Gravel	Dart and Tamar River Blended Supply; Horizontal well water unfiltered	Horizontal Well	108: Unknown	Morris and Foster, 2000; Craun et al., 1998; Vincent et al., 1997
Torbay, Devon, United Kingdom (South West Water)	1995	Littlehempston River Gravel	Dart and Tamar River Blended Supply; Horizontal well water subject to rapid gravity filtration without flocculation	Horizontal Well	557: Unknown	Morris and Foster, 2000; Craun et al., 1998; Waite, 1997.
Unknown	1997	River Gravel	Unknown	Horizontal Well	Unknown: Unknown	Possibly Flood-Associated; Morris and Foster, 2000.
Kitchener-Waterloo, Ontario, Canada	1993	Grand River Sand and Gravel Aquifer	Blended Supply; Grand River Oocyst Concentrations: Range: 77 to 2,075/100 L. Mean: 319/100 L	Vertical Wells, 25- to 35-m Deep, ~10-m Setback.	193: 23,900	Possibly Presumptive Oocysts Recovered from Woolner Well K81 During the Outbreak; Several Months after the Outbreak, Possibly Presumptive Oocysts Were Recovered From Ontario River Well 2; Craun, et al., 1998; Dillon Consulting Ltd., 1997; Welker et al., 1994.
Talent, Oregon	1996	Bear Creek alluvial aquifer		Infiltration gallery	31	Leland, et al., 1993
Ogose, Saitama Prefecture, Japan	1996	River Sediments	Oppe River	Infiltration Gallery Depth = 2 m	125: 8,705	Oocysts Recovered from Source and Tap Water (12 Oocysts/L); Hirata and Hashimoto, 1997; Yamamoto, et al., 2000

The problems are several:

- First, RBF systems are not defined solely as such. A system may use a blend of RBF water and surface water (e.g., outbreaks in 1992 and 1995 at Torbay in the United Kingdom).
- Second, RBF sites may be prone to flooding, and flooding may have adulterated the riverbank filtrate (e.g., an outbreak in 1997 at an unknown site in the United Kingdom).
- Third, oocysts have not been unequivocally recovered in riverbank filtrate during the outbreak to point to the contamination source (e.g., an outbreak in 1993 at Kitchener in Waterloo, Canada).
- Fourth, the outbreak occurred in an infiltration gallery rather than a well (e.g., an outbreak in Ogose, Japan).

Unequivocal outbreaks have occurred due to oocyst-contaminated groundwater supplied by wells, but only in hydrogeologic settings characterized by non-porous media. These outbreaks have occurred in limestone (karst and cavernous karst aquifers) and chalk (fractured [with minor karst] aquifer). These hydrogeologic settings are not typically used as RBF sites; however, it is possible that some might seek to obtain protozoa removal credits for sites in a variety of hydrogeologic settings by claiming that the process is RBF; therefore, Table 6-3 lists available information on the aquifer type for localities where outbreaks have occurred in non-porous media aquifers. More groundwater-associated outbreaks have occurred (e.g., Morris and Foster, 2000), but few data are available so these other outbreaks are not listed Table 6-3.

An outbreak associated with well water in Yakima, Washington, in the United States (15 confirmed; 86 cases) was ascribed to the flow of treated wastewater along the outside of the casing (Dworkin et al., 1996). Other outbreaks associated with production from large diameter Chalk aquifer wells have occurred in the United Kingdom (Morris and Foster, 2000).

Table 6-3. Cryptosporidiosis Outbreaks Associated with the Failure of RBF in Non-Porous Media Aquifers

Outbreak	Year	Aquifer	Other Surface Water	Collection Device	Cases (Laboratory confirmed: estimated)	Comment and/or References
Braun Station, Texas, United States	1984	Edwards Aquifer; Confined Cavernous Karst Limestone	None	Vertical Well	117: 2,006	Possible Cross-Connection; D'Antonio et al., 1985.
Reading, Pennsylvania (Berks County), United States	1991	Unknown; Karst Limestone	None	Vertical Well	Unknown: 551	Moore et al., 1993.
North Thames, United Kingdom (Three Valleys Water)	1997	Chalk Aquifer; Fractured with Minor Karst	Blended with No More than 10-Percent Surface Water	Vertical Well	345: 22 Percent of Potential Controls Excluded Because of Self Reported Gastro-intestinal Illness; 354,000 People Received over 90 Percent of Their Water from the Contaminated Source	No Estimate Made of the Total Number of Illnesses; Willocks et al., 1999; Gray, 1998.
Brushy Creek, Texas, United States	1998	Edwards Aquifer; Cavernous Karst Limestone	None	30+ m Deep Vertical Wells Located More than 400 m from Brushy Creek	89: 1,300 to 1,500	Source Was a Sewer Line Overflow into Brushy Creek; Bergmire-Sweat et al., 1999.

For many decades, no data on waterborne disease outbreaks have been associated with the fecal contamination of (artificial) groundwater in The Netherlands, although enteric viruses and the pathogenic protozoa *Cryptosporidium* and *Giardia* are ubiquitously present in Dutch surface waters (Hoogenboezem et al., 2000; Theunissen et al., 1998).

6. Required Treatment of Surface Water for Drinking-Water Production in the United States, Finland, and The Netherlands: Implications for RBF Treatment

RBF is not, as of yet, an explicitly specified treatment technique in United States drinking-water regulations; however, it is likely that future rulemaking will specify RBF as one of several pretreatment options available to drinking-water systems that currently use conventional filtration, but that have high oocyst (or indicator, if an appropriate indicator is determined) concentrations in the river, lake, or reservoirs.

Under current regulations (U.S. Environmental Protection Agency, 1989), drinking-water systems using RBF may be classified as either:

- Groundwater.
- Groundwater under the direct influence of surface water.
- Surface water.

The determination as to which classification should be applied is made by the primacy agency, typically the state, using guidance (U.S. Environmental Protection Agency, 1991).

If a system in the United States using RBF is determined to be surface water, the system must achieve the following requirements using conventional filtration and disinfection, RBF, or a combination of both:

- 3-log removal and/or inactivation of *Giardia*.
- 4-log removal and/or inactivation of viruses.
- 2-log removal of *Cryptosporidium*.

Currently, inactivation credit for *Cyptosporidium* is not available. Alternatively, an RBF system (determined to be either surface water or groundwater under the direct influence of surface water) may seek to meet the more stringent watershed protection requirements that allow classification as an unfiltered system.

Currently, only two water utilities have received credit for virus and *Giardia* inactivation and/or removal by RBF from their primacy agencies. The City of Kearney in Nebraska was granted credit of 2-log removal of *Giardia* and 1-log removal of viruses. Sonoma County, California, was granted credit of 2.5-log removal of *Giardia* and 1-log removal of viruses for Collector Well 5. Neither site is yet required to meet the *Cryptosporidium* removal requirement because there is a lag period between the publication of regulations and their implementation by the primacy agency. For further details, see the Introduction of this book.

The RBF credits are based upon the putative removal of potential oocyst surrogates or indicator organisms. No standard protocol or guidance exists for evaluating natural filtration credit, so each decision was made independently based on the analysis using available data. Gollnitz et al. (1997) have suggested a protocol and provided an example from Casper, Wyoming, but, to date, no RBF credit has been given using that protocol at that site.

In Finland, all artificially recharged groundwater, including induced surface water (riverbank-filtered water), is classified as groundwater. Unlike in the United States, such classification does not affect drinking-water quality regulatory requirements. All drinking water must meet the same regulatory standards regardless of the treatment procedure. According to European Union regulations, the quality of drinking water is based mainly on monitoring and the occurrence of indicator organisms (enterococci and *E. coli*). Additional analyses of *Clostridium perfringens* are required if surface water is used directly or indirectly (artificial recharge or riverbank filtrate) as raw water (European Union, 1998). Analyses of viruses and protozoa are required if there is reason to suspect a potential danger to human health. Allowable concentrations of enterococci, E. coli, and Clostridium perfringens are zero organisms per 100 mL. Sampling frequency is proportional to the volume of produced drinking water (i.e., an increase in production volume increases the sampling frequency). The microbial quality of drinking water in waterworks producing 100 to 1,000 m^3/d has to be monitored at least four times per year. In waterworks where production is below 100 m^3, each European Union state can independently decide the sampling frequency (for example, Finland is once per year). In Finland, the minimum sample volume for indicator organism analyses is 250 mL.

In The Netherlands, about 39 percent (186×10^6 m^3/yr) of surface water that is used for drinking-water production is treated by soil and aquifer passage (artificial groundwater), either in RBF and/or in dune recharge with pretreated surface water (VEWIN, 1998). According to current guidelines (CBW, 1980), a travel time of 60 days is required at wellhead protection areas and RBF sites. For RBF, the travel time varies between 0.5 and 30 years and, for artificial recharge of pretreated surface water, between 35 and 135 days (Stuyfzand and Lüers, 1996). A 60-day travel time is assumed to be adequate to inactivate pathogenic bacteria to the degree that no health risk exists (Knorr, 1937; CBW, 1980). A similar approach is followed in Germany (Dizer et al., 1984; Matthess et al., 1988).

However, for decades, viruses and protozoa have been recognized as pathogens of major health concern (e.g., Craun and Calderon, 1996; D'Antonio, 1985; Moore et al., 1969; Rose, 1988). Due to their persistence in the environment and their infectivity, enteric viruses and the pathogenic protozoa Cryptosporidium and Giardia may be considered as the most critical waterborne pathogens for drinking-water production in The Netherlands (Medema and Havelaar, 1994). Because of this environmental persistence, a travel time of 60 days may be too short for the sufficient inactivation of viruses and pathogenic protozoa. On the other hand, attachment to the aquifer grains during subsurface transport may contribute significantly to virus and protozoa removal.

The documented persistence and subsurface mobility of viruses, protozoa, and, perhaps, of some bacteria, raises the question to what extent the hygienic quality of drinking water is guaranteed. A new policy for protecting groundwater and for treating surface water by soil and aquifer passage in The Netherlands was proposed (Medema and Havelaar, 1994; VROM, 1995) and was incorporated into legislation in the beginning of 2001. This approach is based on a maximum acceptable infection risk of one per 10,000 persons per year associated with drinking-water consumption and dose-response relationships for pathogens, and has resulted in using maximum allowable concentrations (Regli et al., 1991). These maximum allowable concentrations are given in Table 6-4 for Cryptosporidium, Giardia, and viruses together with the average concentrations observed in Dutch surface waters.

Table 6-4. Concentration of Pathogenic Surface Water in The Netherlands and Required Treatment

	Average Concentration in Surface Water (number/liter)	Maximum Allowable Concentration in Drinking Water (number/liter)	Required Log Removal
Cryptosporidium[a]	1.4 to 87	2.6×10^{-5}	6.4 to 7.5
Giardia[a]	1.5 to 95	5.8×10^{-6}	6.7 to 8
Enteroviruses[b]	0.02 to 8	1.8×10^{-7}	5 to 7

[a]Hoogenboezem et al. (2000).   [b]Schijven et al. (1995).

From the difference between the average concentration in surface water and the maximum allowable concentration in drinking water, the required log removal to produce safe drinking water can be deduced. Depending on the microorganism and the location where surface water is taken in for treatment, concentrations in surface water need to be reduced by 5 to 8 logs by RBF. The World Health Organization has decided to base the 2003 edition of the *Guidelines for Drinking Water Quality* on a similar approach (World Health Organization website, www.who.int, accessed June 24, 2002).

Compliance with these maximum allowable concentrations can only be assessed by the analysis of very large volumes of drinking water (i.e., on the order of 10^5 to 10^7 L). Such precision is considered to be impracticable and another approach for determining compliance must be followed. Pathogenic microorganism concentration in treated water can be calculated from concentrations in source water and the effectiveness of water treatment. In the case of aquifer passage as a water treatment, a computational method for predicting RBF efficiency is needed to estimate the fate and transport of the pathogenic microorganisms during aquifer passage.

7. Hydrology and Hydrogeology

Well Type and Location

Currently, three different well types are typically used to extract water from alluvial aquifers:
- Vertical wells.
- Horizontal collector wells
- Infiltration galleries, which are specialized horizontal wells.

Horizontal collector wells typically consist of a large-diameter vertical caisson with horizontal laterals extending out, usually in all directions, from the bottom of the caisson. At many sites, the laterals may directly underlie the river. In Germany, horizontal filter wells used for drinking water may not extend beyond the edge of the river. Infiltration galleries are typically installed in the riverbank and are used to collect downward infiltrating river water that is pumped or channeled onto the riverbank. Infiltration galleries are not considered further in this discussion.

Horizontal collector wells and vertical wells represent fundamentally differing natural filtration scenarios. Vertical wells, depending on the proximity to the river edge, may capture largely horizontal groundwater flow with lesser vertical flow components. Horizontal collector wells, again depending on the proximity to the river edge, may induce and capture primarily vertical groundwater flow with lesser horizontal flow components. In either case, the vertical component of flow is likely to be significant; however, because the flow fields to the two well types can be roughly dissimilar, in this discussion, horizontal collector wells and vertical wells are analyzed separately.

The proximity of the vertical well and riverbank is herein termed the "horizontal setback distance." For horizontal collector wells, the distance from the bottom of the stream bed (under normal flow conditions) and the lateral under the stream bed is the "vertical setback distance." In general, setback distances are surrogate measures of RBF efficiency in porous media. The greater the setback distance, the greater the filtration efficiency. If average groundwater velocities are measured or calculated, then groundwater travel times can be used interchangeably or in place of setback distances. The analysis of horizontal collector well data will be emphasized herein because these wells typically have the smallest horizontal or vertical setback distance and the shortest travel times and, thus, are more likely to have breakthrough of microorganisms.

Groundwater Flow Field and Alluvial Aquifer Properties

In general, even in humid climates, streams may lose — as well as gain — water along part of their reach. The effect of a pumping well can be to enhance the natural flow (if the stream is losing water) or to reverse the natural flow (if the stream is gaining water). The induced infiltration effects caused by vertical pumping wells are described elsewhere (e.g., Wilson, 1993; Wilson and Linderfelt, 1991; Conrad and Beljin, 1996; Hunt, 1999).

For the case of a (finite length) horizontal collector well underlying surface water, the downward flow from surface water and the lateral inflow to the collector caisson require a three-dimensional approach for a complete solution to the problem. Zhan and Cao (2000) have developed analytical solutions for capture time to a horizontal collector well lateral in an aquifer under a surface-water body; however, the solution assumes a horizontal well of infinite extent. Nevertheless, the solution allows for an efficient computation of travel times from surface water to a horizontal collector well. For example, Zhan and Cao show that the travel time from the surface-water body to the well is about 10 days for a well located near the bottom of a 21-m thick sand aquifer (at a pumping rate of 15-m^3/d/m of screen length).

Riverbed Properties

In an RBF system, the ideal riverbed retards microbial pathogen transport, but provides optimal water recharge to the subsurface. The physical characteristics and hydraulic conditions under which such duality can occur is not currently known. In fact, relatively little is known, either theoretically or through field investigation, about the *in situ* characteristics of the riverbed adjacent to an RBF well and its role in governing flow and transport under normal flow conditions. Studies conducted along the Rhine River constitute the most complete set of field and theoretical investigation, albeit with most results published only in German. The impetus to conduct such studies was concern about the decreasing capacity (clogging) at RBF sites in Düsseldorf, Germany (Wilderer et al., 1985). The Rhine riverbed was inspected in 1964 using a diving bell. It was found that, except for the shipping channel in the middle of the river, the sediment consisted of a relatively impermeable 10-cm thick layer comprised of mineral oil, hydrocarbons, iron, manganese, copper, zinc, and lead (Wilderer et al., 1985).

Yager (1986) conducted a simulation of infiltration to vertical wells in the Susquehanna River alluvial aquifer in Broome County, New York. Yager characterizes the riverbed as made up of a 0.6-m thick layer that is heavily armored with cobbles and boulders and an underlying layer of silt and organic material. The riverbed hydrologic properties were investigated using four drive-point wells. Slug tests indicate that the horizontal hydraulic conductivity ranged from 0.3 to 2 m/d. The wells were placed in riffles where the current is the strongest. In a sensitivity analysis conducted by simulation, the vertical hydraulic conductivity was allowed to vary from 0.0015 to 0.6 m/d. Yager finds that riverbed infiltration is significant in the range of 0.006 to 0.6 m/d vertical hydraulic conductivities. Below the lowest value, little infiltration takes place (<15 percent of the well yield) and the size of the capture zone is governed by lateral flow through the aquifer. Above 0.6 m/d, the induced river infiltration amounts are large (>74 percent of the well yield) and the size of the capture zone remains unchanged as the river supplies the necessary yield. Finally, Yager reports that transient simulations indicate that the simulations are highly sensitive to the value of the vertical (or radial) hydraulic conductivity and uncertainty in this parameter value is the limiting factor in interpreting the simulation results.

As part of an RBF study of heavy metals at the Ansereuilles Well Field near Lille, France, Bourg et al. (1989) collected riverbed sediment core samples to a depth of 80 cm. The lower 60 cm were consolidated sands (55 to 70 percent) with decreasing water content with depth. Iron oxides ranged from 2 to 5 percent, and organic matter was almost 9 percent in the one sample. Clay content increased with depth from 5 to 18 percent, as did silt and gravel. Bourg et al. (1989) found that the sediments retain heavy metals, but remobilization can occur.

Larkin and Sharp (1992) have compiled average riverbed hydraulic conductivity for nine differing river reaches. Values reported ranged from 0.035 to 10 m/d. Conrad and Beljin (1996) report riverbed hydraulic conductivity data from at least four other sites, two with slightly lower conductivity. Both gravel dredging and flood scour have the capability to change the riverbed properties. Gravel dredging is reported to adversely affect water quality in the Surany Well Field (20 horizontal collector wells), which supplies drinking water for Budapest, Hungary (Laszlo and Szekely, 1989). The effects of riverbed hydraulic conductivity must be evaluated using numerical groundwater flow models. Conrad and Beljin (1996) suggest that riverbed hydraulic conductivity can be safely neglected (allowing the use of semi-analytical flow models) if the riverbed hydraulic conductivity differs by no more than a factor of 10 from the aquifer hydraulic conductivity.

According to Gollnitz et al. (1997a), maximum induced infiltration occurs during periods of high stream flow because:

• Streambed thickness decreases and (bulk) permeability increases due to scour.
• The streambed has a larger wetted area.
• Head differential increases with stream stage.

The effects are magnified if the high-river stage event occurs after a period of high pumpage (drawdown) and during warm periods that decrease water viscosity.

8. Microorganism Removal by RBF: Processes

During soil passage, microorganisms may be removed from the aqueous phase primarily by straining, inactivation, and attachment to the aquifer grains (in combination with inactivation). Other removal processes of uncertain significance are sedimentation in connected pores and trapping in dead-end pores. The contributions of these processes to removal and how they are modeled are discussed below. Attachment and hydrophobicity are included within the colloidal filtration paradigm.

Inactivation

a. Virus Inactivation During Saturated Subsurface Transport

Viruses lose their ability to infect host cells with time by inactivation. Viruses are inactivated because of the disruption of coat proteins and the degradation of nucleic acids (Gerba, 1984). Yates et al. (1987) have reviewed the factors that influence the inactivation of viruses and mentioned three reports on inactivation rates for viruses in groundwater (Keswick et al., 1982; Bitton et al., 1983; Yates et al., 1985). Since then, Schijven and Hassanizadeh (2000) have summarized a number of new studies that have been carried out.

Inactivation is usually regarded as a first-order process. The most important factors that influence virus inactivation rates during saturated subsurface transport are temperature, adsorption to particulate matter, and soil microbial activity. Temperature is the most important factor that influences virus inactivation (Hurst et al.,1980; Yates et al., 1985, 1987). Inactivation rates increase with temperature (Hurst et al., 1980; Yates et al., 1985; Jansons et al., 1989; Nasser et al., 1993; Yahya et al., 1993; Blanc and Nasser, 1996).

Schijven and Hassanizadeh (2000) compared the dependence of the inactivation rate coefficient μ_1 on temperature for a number of bacteriophages, using data from several studies. The temperature sensitivity of μ_1 appears to depend on the type of virus. For instance:

- The inactivation of poliovirus 1 is much less sensitive to temperature than that of MS-2 bacteriophage or Echovirus 1.
- Hepatitis A virus may be regarded as relatively insensitive to changes in temperature.
- MS-2 and other F-specific RNA bacteriophages are very sensitive to changes in temperature.

Nevertheless, in a deep-well injection study by Schijven et al. (2000) (i.e., under environmental conditions), a very low value (0.039/day) for μ_1 of MS-2 was measured at the monitoring well at 8 m, near the outer limits of the oxic zone. The temperature was 12°C. Similar μ_1-values were reported in sterilized and non-sterilized groundwater at 10°C by Matthess et al. (1988) for Poliovirus 1 (0.01 to 0.11/day), Coxsackievirus A9 (0.019 to 0.027/day), Coxsackievirus B1 (0.012 to 0.019/day), and Echovirus 7 (0.019 to 0.032/day). Most recently, Norwalk virus survival was estimated in soil-amended groundwater and was not found to differ significantly from the survival of poliovirus and MS-2 bacteriophage (Meschke and Sobsey, 1999).

b. Survival of Bacteria

The number of infective enteric pathogens will decrease with time and will, eventually, decrease to zero or near zero by natural processes. Pathogen persistency depends on how quickly it will perish outside the host. Various abiotic and biotic environmental factors, as well as the properties of the microbe itself, will determine the elimination rate. Temperature, humidity, pH, the amounts of organic matter in soil and aquifer material, rainfall, sunlight, or competitive microorganisms will all affect the survival of a pathogen in water, soil, other unconsolidated material, and within aquifer matrices (Gerba and Bitton, 1984; Crook, 1985). Predation by protozoa and invertebrates also affects the number of bacteria (Hutchinson and Ridgway, 1977).

Bacteria have their own distinct optimum growth temperatures:

- Psychrophilic bacteria have maximum growth below 20°C (Henis, 1987; Vestal and Hobbie, 1988).
- Mesophilic bacteria have optimum growth between 20 to 30°C (Davis et al., 1970; Henis, 1987).
- Thermophilic bacteria grow above 40°C (to 55 or 60°C) (Davis et al., 1970; Henis, 1987; Vestal and Hobbie, 1988).

A decrease in temperature usually prolongs the persistency of microorganisms in soil and aquifer materials (Mirzoev, 1968; Gerba, 1985) and water (Matthess and Pekdeger, 1981; Bitton et al., 1987; Korhonen and Martikainen, 1991; Terzieva and McFeters, 1991).

At typical groundwater temperatures (8 to 25°C), most bacterial pathogens and indicators can remain infectious for weeks or more in groundwater. The indicator bacterium E. coli is detectable in water for a period up to 3 months (Edberg et al, 2000; Maule, 2000). For indicator bacteria, inactivation rates (natural log per day) at groundwater temperatures (in unamended groundwater samples) have been measured as low as:

- 0.53/day (fecal streptococcus; Keswick et al., 1982).
- 0.82/day (fecal coliform; Keswick et al., 1982).
- 0.04/day (estimated) to 0.73/day (E. coli; Nasser and Oman, 1999; Keswick et al., 1982).

For pathogens, inactivation rates have been measured as low as:
- 1.42/day (*Shigella* sp.; McFeters et al., 1974).
- 0.51/day (*Salmonella* sp.; Keswick et al., 1982).
- 0.33/day (*E. coli* O157:H7; Rice et al., 1992).

In a deep-well injection study by Schijven et al. (2000), an inactivation rate coefficient for *E. coli* strain WR-1 was measured in water from the injection well (12°C) and found to be 0.083/day.

One problem associated with detecting pathogenic bacteria is that the bacteria may become dormant in the environment. In this state, they are "viable, but non-culturable," meaning that the organisms have metabolizing activity even though they cannot grow on traditional media (Olson, 1993). As an important example, the pathogen *E. coli* O157:H7 appears to be able to enter a viable, but non-culturable, state in water (Wang and Doyle, 1998). The dormant state prolongs the pathogen survival, thereby increasing the likelihood of a host infection and illness.

Bacteria usually survive better in high pH than in low pH. For example, *Salmonella typhi* survives best in a slightly alkaline environment, but it is inactivated quickly if the pH is between 3 and 4 (McGinnis and DeWalle, 1983). Coliforms survive best in pH 5.5 to 7.5 (McFeters and Stuart, 1972).

Some bacteria can survive by producing resistant spores. Certain gram-positive rods, like *Clostridium* sp. and *Bacillus* sp., can form spores. In unfavorable conditions (dry environment), dormant spores of can survive for years (Davis et al., 1970). The occurrence of aerobic bacterial spores has been studied as possible surrogates for protozoa (Nieminski et al., 2000). According to European Union regulations, *Clostridium perfringens* (spores) have to be monitored if surface water is used as raw water (European Union, 1998).

c. Inactivation of *Cryptosporidium* Oocysts

Oocyst in vitro viability has been measured repeatedly using a surface-water matrix and in vitro excystation and/or dye exclusion assay (e.g., Heisz et al., 1997; Medema et al., 1997; Robertson et al., 1992; Chauret et al., 1995). These data consistently show oocyst viability for 1 to 2 months or longer. For example, Heisz et al. (1997) report 0.12/day (30°C) and that a proportion of the oocysts remained viable after 50 days, regardless of the experimental conditions. Medema et al. (1997) found inactivation rates of 0.023 to 0.056/day (5 to 15°C) in river water and no differences in inactivation rates between 5 and 15°C. From the data of Robertson et al. (1992) and of Chauret et al. (1995), inactivation rates of 0.0051 to 0.0062/day were calculated. Chauret et al. found that the inactivation rate was independent of water temperature up to 20°C. From these studies, a conservative, temperature-independent inactivation rate of 0.007/day was deduced. Unfortunately, in vitro excystation has been shown to overestimate oocyst infectivity, and oocysts that do not excyst are still capable of causing infection (Neumann et al., 2000).

The U.S. National Academy of Science (National Research Council, 2000) presents 1992 data reporting 1-log inactivation at 100 and 180 days, respectively (corresponding to an inactivation-rate coefficient of 0.023 and 0.013/day, respectively), for two *Cryptosporidium* strains examined; however, the National Research Council recognized the overestimate resulting from the use of the excystation or vital dyes. Nevertheless, the National Research Council questions whether a 60-day travel time is sufficiently protective and suggests that a zone of 180 days or more might be required if inactivation is the primary protective barrier. Walker et al. (1998) summarized the oocyst survival literature, including the effects of desiccation, as well as the literature on oocyst transport via overland flow.

Typical groundwater temperatures are lower than river temperatures. For example, initial temperatures at wells in Louisville, Kentucky, are 15 to 17°C and approach river temperature

(26.5°C) only after hours of pumping (Wang et al., 2000). Thus, viability may be longer in groundwater than is measured for surface water. It may be the case that, similar to the viruses, expressions that describe inactivation in surface water are not applicable to groundwater (Hurst et al., 1997). These results indicate that the inactivation measurements applicable to groundwater must be conducted in groundwater rather than in surface water.

Straining

Straining is a purely physical removal process governed by the size of pore throats and microbial particles. McDowell-Boyer et al. (1986) and Harvey (1991) summarized the published and unpublished literature. McDowell-Boyer et al. (1986) report experiments that relate the diameter of the media to the diameter of the particle as a simple ratio. Where the ratio is greater than 20, straining is insignificant. In the 10- to 20-ratio range, straining removal is significant and, below a ratio of 10, no particle penetration through porous media occurs. Thus, for a *Cryptosporidium* oocyst with a 5-micrometer (μm) diameter, straining may be significant for particle sizes below 100 μm (fine sand). Novarino et al. (1997) conclude that, for the Cape Cod (Massachusetts) aquifer, straining would preclude advective movement of protozoa larger than 20 μm.

Herzig et al. (1970) provide a geometric expression for the removal efficiency by straining. Corapcioglu and Haridas (1984) conclude that these results are applicable to bacteria. They suggest that straining should be included in the theory of bacterial transport because bacteria would undergo 3-percent removal for a 1-μm diameter population advecting through silt with a mean grain diameter of 10 μm.

Harvey (1991) and Harvey et al. (1993) use the straining effectiveness criteria for the hetero-geneous media of Matthess and Pekdeger (1986) to identify conditions when straining becomes significant. Harvey (1991) concludes that bacterial straining could not be predicted even for coarse silt (20- to 60-μm diameter). Field measurements (Harvey et al., 1993) were conducted with 0.74-μm (diameter) microspheres in a sandy (~0.5-mm median grain size) aquifer. Column studies were conducted with 4.8-μm (diameter) microspheres and straining seemed to be an important mechanism, but as the results may be dependent on column packing, the measured straining effects on larger particles is uncertain. Straining was not identified as an important removal mechanism for the microspheres in field experiments. Brush et al. (1999) suggest that oocyst straining could explain their experiments on oocyst transport in sand columns because sorption was not significant.

Based on this literature survey, the data suggest that straining may not prevent breakthrough for the pathogenic waterborne protozoa identified above in wells tapping coarse-grained alluvial aquifers when fine-grained riverbed sediments are absent.

Because of their smaller size, the straining of bacteria will be less important than for protozoa and may be negligible in the case of bacterial spores. The straining of viruses should not occur where clogging is insignificant.

Sedimentation in Pores

Oocysts of *Cryptosporidium* have an almost spherical shape and are only slightly heavier than water with a geometric mean density of 1.045 grams per cubic centimeter (g/cm^3) (n=20) (Medema et al., 1998). Stoke's settling velocity in a solution designed to create free oocysts is measured to be 0.35 micrometer per second (μm/s) (Medema et al., 1998). The tenfold higher value for the oocyst concentration in flowing streams, as opposed to lakes and reservoirs in the ICR data, suggests that settling is an important parameter governing occurrence in surface water, although the effects of aggregation and attachment to particles may be profound.

Because the sedimentation velocity is of the same magnitude as groundwater flow velocity under natural gradients in a porous medium (sand), Corapciouglu and Haridas (1984) conclude that for bacteria with a density of 1.02 g/cm^3, pore settling is significant. In contrast, Harvey et al. (1997) conclude that endemic groundwater bacteria (with densities less than 1.019 g/cm^3) seem to be subject to rates of sedimentation that are negligible in comparison to their velocity in the direction of flow. Harvey et al. predict that organisms larger than the bacteria with concomitant greater densities (in the range of oocyst densities) would have a greater sedimentation velocity because the velocity increases with the square of the diameter. For 2.5-μm size flagellates, Harvey et al. measured the apparent settling velocities of 6.4×10^{-8} m/s. This velocity, however, is measured after passage through a porous medium rather than within a single pore; therefore, it represents a settling velocity through a tortuous rather than a simple flow path. Brush et al. (1999) suggest that oocyst settling velocities (on the order of 10^{-4} centimeters per second [cm/s]) should not be significant in pores.

Settling is more likely to occur where groundwater velocities are lowest, such as in the finer grained riverbed material. These fine-grained materials are removed during flooding and, thus, pore sedimentation may be more significant during quiescent periods. More speculatively, transient pressure waves due to on/off pump cycling may retard pore sedimentation. No information is available on trapping in dead-end pores.

Viruses in the environment are often associated with particulate matter or other surfaces, and this has a major effect on their inactivation and transport in the environment (Gerba, 1984). Hejkal et al. (1981) indicated that, in treated wastewater, the majority of enteric viruses are free or attached to particles smaller than 0.3 μm. Metcalf et al. (1984) also showed that enteroviruses and rotaviruses in estuarine water adsorb preferentially to particles smaller than 0.3 μm in diameter. Particles less than 0.3 μm include clays, cell fragments, waste products, and other miscellaneous debris (Levine et al., 1985). Furthermore, Payment et al. (1988) showed that, in river water, 77 percent of indigenous enteric viruses and 66 percent of coliphages were probably free or associated with particles with a diameter of less than 0.25 μm. Thus, a substantial fraction of viruses is attached to wastewater effluent solids and other colloidal particles with a size of less than 0.3 μm.

Colloidal Filtration

Colloidal filtration appears to be the most successful construct suitable for predicting the transport of microorganisms in porous media (Harvey and Garabedian, 1991; Ryan and Elimelech, 1996; Schijven and Hassanizadeh, 2000; Murphy and Ginn, 2000). Colloid filtration theory excludes the effects of flow and diffusion by expressing the attachment rate of microorganisms to soil grains in terms of single collector efficiency η and collision efficiency α. According to this theory, a suspended particle may come into contact with a particle of the solid medium, the collector, either by interception, sedimentation, or diffusion (Yao et al., 1971). The collision efficiency, α, represents the fraction of the particles colliding with the solid grains that remain attached to the collector (Martin et al., 1992). The collision efficiency reflects the net effect of repulsive and attractive forces between the surfaces of the particles and the collector, and depends on the surface characteristics of the virus and soil particles; therefore, the collision efficiency depends on pH, organic carbon content, and ionic strength.

Colloid filtration was applied by Schijven et al. (1998) for calculating collision efficiencies from removal data by artificial dune recharge of total coliforms and thermotolerant coliforms, fecal streptococci, spores of sulphite-reducing clostridia, and F-specific RNA bacteriophages. It was assumed that concentrations in the infiltrating water did not change within a time-scale of a few days and, thus, a steady-state model was applied. The dispersion of the transported microorganisms was neglected because the monitoring wells were located at only 2 and 4 m from the infiltration canal and the soil consisted of fine dune sand. In the case of bacteriophages (F-specific bacteriophage MS-2, somatic *salmonella* phage PRD-1, somatic coliphage M-1) and viruses (poliovirus), column (Bales et al., 1991, 1993; Kinoshita et al., 1993) and field studies (Bales et al., 1997; Pieper et al., 1997; Schijven et al., 1999, 2000) have shown that adsorption is reversible and kinetically limited with detachment rates being much slower than attachment rates. Several column studies (Hornberger et al., 1992; Scholl and Harvey, 1992; Camper et al., 1993; McCaulou et al., 1994, 1995; Rijnaarts et al., 1995; Hendry et al., 1997) and a field study (Harvey and Garabedian, 1991) showed that the attachment of bacteria mainly determines the level of breakthrough.

Based on these studies, one may reasonably assume that detachment is negligible; however, note that in the case of very persistent microorganisms, like *Clostridium* spores or oocysts of *Cryptosporidium*, this assumption may not be valid. Under steady-state conditions, such microorganisms will eventually breakthrough, and actual removal rates depend on the time period of contamination.

As a worst-case approach, field studies by Schijven et al. (1998, 1999) were conducted in wintertime when pore water temperature is low and, hence, inactivation rates are very low. In the study by Schijven et al. (1998), travel times to the monitoring wells were only 1 to 2 days, justifying the neglect of inactivation of bacteriophages and of all the fecal indicator bacteria that were studied. The growth of fecal indicator bacteria may also be neglected under these conditions.

This leads to the following equation, from which the collision efficiency α may be calculated (Yao et al., 1971):

$$\alpha = -\log\left(\frac{C}{C_0}\right)\frac{2}{3}\frac{2.3d_c}{(1-n)\,\eta L} \tag{1}$$

Where, C is the concentration at monitoring well [N/L]; C_0 is the concentration in infiltrating water [N/L]; d_c is the average diameter of single collector (grain size) [m]; n is the porosity of the soil; η is the single collector efficiency; and L is the travel distance [m].

The single collector efficiency η was calculated using the following relationship due to Martin et al. (1992):

$$\eta = 1.0A_s\,N_{Lo}^{1/8}N_R^{15/8} + 0.00388A_s\,N_G^{1.2}N_R^{0.4} + 4A_s^{1/3}N_{Pe}^{-2/3} \tag{2}$$

Here, $N_R = d_p/d_c$ accounts for interception; $N_G = d_p^2\,(\rho_p - \rho)g\,/(18\mu vn)$ for gravity effects; $N_{Lo} = 4H/(9\mu d_p^2 vn)$ for van der Waals interactions; and $N_{Pe} = d_p vn/D_{BM}$ for diffusion. In these definitions, d_p and d_c represent the microorganism particle sizes and soil grain sizes [m], respectively; g is the gravitational acceleration; ρ and ρ_p are the density of water and the microorganism particle, respectively; $\mu = \rho \times 0.000947/(T + 42.5)^{1.5}$ is the dynamic viscosity [kg/m sec], with T the water temperature [°C]; $H = 6.2 \times 10^{-21}$ is the Hamaker constant [J] for the bacterium-glass-water interface (Rijnaarts et al., 1995); $D_{BM} = K_B(T + 273)/(3\pi d_p \mu)$ is the diffusion coefficient [m²/s], with Boltzmann-constant $K_B = 1.38 \times 10^{-23}$ (J/K); and $A_s = 2(1-\gamma^5)/(2-3\gamma+3\gamma^5-2\gamma^6)$ is Happel's porosity-dependent parameter, with $\gamma = (1-n)^{1/3}$.

Bacteriophages are small in size (d_p= 21 to 30 nm [Havelaar, 1993]), and their transport in the immediate vicinity of the collector surface is dominated by Brownian diffusion. In this case, the single collector efficiency is restricted to the last term in Equation 2, the Smoluchowski-Levich approximation (Penrod et al., 1996).

Total coliforms, fecal coliforms, and fecal streptococci have sizes in the range of 0.3 to 1.1 μm by 0.6 to 6 μm. From these sizes, a geometric mean range of the bacterial diameter d_p= (length × width)$^{1/2}$ (Rijnaarts et al., 1993) was estimated: 0.4 to 2.6 μm. Bacterial density ρ_p was in the range of 1,040 to 1,130 kilograms per cubic meter (kg/m^3), as suggested by Bouwer and Rittmann (1992). Spores of sulphite-reducing clostridia were assumed to have a size of 1 μm (Stegeman et al,. 1980; Lund and Peck, 1994) and a buoyant density of 1,270 kg/m^3 (Tisa et al., 1982).

9. Surrogate Microorganisms and Other Indicators

Surrogate Viruses

The removal of pathogenic viruses under field conditions can be studied only if contamination levels are high enough. Usually, this is not the case. Only in exceptional situations may permission be obtained to seed pathogenic viruses in the field. At a site that is in use for drinking-water production, this will never be allowed; therefore, surrogate viruses that are not pathogenic, but are still representative for the transport behavior of pathogenic viruses, are needed. A surrogate virus is suitable if its inactivation and adsorption are similar to that of pathogenic viruses under given conditions. This implies that it should be possible to predict the removal of pathogenic viruses by passage through aquifer material from the removal of the surrogate virus.

Usually, bacteriophages are used as surrogate viruses. Bacteriophages offer the following advantages:

- Bacteriophages are not pathogenic to humans, but infect a specific host bacterium.
- Bacteriophages can be prepared in large quantities (10^{10} to 10^{12} phages per millimeter), allowing the seeding of high numbers. This makes it possible to show removals of up to 11 logs.
- The assay of bacteriophages is relatively easy, whereas the analysis of pathogenic viruses is much more complex, time consuming, and sometimes not possible at all.

Bacteriophages MS-2, PRD-1, and ϕX-174 — as well as naturally occurring F-specific RNA bacteriophages — have been used extensively to study virus transport under various column and field conditions. Surrogates that best mimic the properties of pathogenic viruses are used as indicators of virus or fecal contamination. These indicator viruses are described in the following sections.

a. Bacteriophage MS-2

MS-2 is an icosahedral phage with a diameter of 27 nm and a low isoelectric point of 3.5. The three-dimensional structure of its capsid is known at the atomic level (Penrod et al., 1996). MS-2 may be considered as a relatively conservative tracer for virus transport in saturated sandy soils at pH 6 to 8 and with a low organic carbon content as, under those conditions, it showed little or no adsorption (Bales et al., 1989; Powelson et al., 1990; Herbold-Paschke et al., 1991; Kinoshita et al., 1993; Jin et al., 1997; Schijven et al., 1999).

In most soils, the attachment of MS-2 is also relatively low compared to most other viruses (Goyal and Gerba, 1979; Herbold-Paschke et al., 1991; Bradford et al., 1993; Farrah and Preston, 1993; Bales et al., 1993; Sobsey et al., 1995; Penrod et al., 1996; Redman et al., 1997; Jin et al., 1997; DeBorde et al., 1999).

With regard to inactivation, MS-2 is less stable than several pathogenic viruses and is inactivated faster at higher temperatures; however, at temperatures lower than 7°C, its inactivation rate is very low and similar to that of PRD-1 (Yates et al., 1985; Yahya et al., 1993; Blanc and Nasser, 1996; Schijven et al., 1999, 2000).

b. Bacteriophage PRD-1

PRD-1 is an icosahedral bacteriophage with a diameter of 62 nm and an inner lipid membrane (Bales et al., 1991, 1995; Caldentey et al., 1990). Its isoelectric point lies between 3 and 4 (Loveland et al., 1996). PRD-1 may be considered as a worst-case model virus because of its low inactivation rate of between 10 to 23°C (Yahya et al., 1993; Blanc and Nasser, 1996). And, because of its larger size, PRD-1 is of interest as a representative of rotaviruses and adenoviruses (Sinton et al., 1997).

With regard to attachment characteristics, PRD-1 seems to behave less conservatively than MS-2 (Bales et al., 1991; Kinoshita et al., 1993; Powelson et al., 1993; Dowd et al., 1998), possibly because it is more hydrophobic than MS-2 (Shields and Farrah, 1987; Bales et al., 1991; Kinoshita et al., 1993; Lytle and Routson, 1995). In the field studies by DeBorde et al. (1999) and Schijven et al. (1999), the removal of PRD-1 was similar to that of MS-2. Apparently, PRD-1 may also be considered as a relatively conservative model virus, similar to MS-2, under field conditions in sandy soils at pH 6 to 8 and with low organic carbon content and a low concentration of multivalent cations. In addition, PRD-1 is more stable at higher temperatures (12 to 23°C).

c. Bacteriophage φX-174

Bacteriophage φX-174 is less hydrophobic than MS-2 (Shield and Farrah, 1987). In studies on the retention of viruses by barrier materials (like membranes, condoms, and testing gloves), φX-174 is regarded as the best model virus because it exhibits the least electrostatic and hydrophobic interaction (Shields and Farrah, 1987; Lytle and Routson, 1995; Fujito and Lytle, 1996). Bacteriophage φX-174 has essentially no charge at neutral pH (isoelectric point = 6.6 to 6.8) and a size of 27 nm (Fujito and Lytle, 1996; Dowd et al., 1998).

Jin et al. (1997) found that the breakthrough of φX-174 in columns with Ottawa sand attached significantly, whilst MS-2 did not. It was suggested that this difference in adsorption behavior was a reflection of the difference in the isoelectric points of the two viruses. In field studies by DeBorde et al. (1998, 1999), φX-174 appeared to be very stable (i.e., its inactivation was negligible over a period of about half a year).

To conclude, φX-174 may be a relatively conservative model virus because of its low hydrophobicity (Shields and Farrah, 1987) and stability (DeBorde et al., 1998, 1999). In aquifers, where hydrophobic interactions would significantly increase virus removal, φX-174 could be a better choice as a model virus than MS-2 or PRD-1; however, the value of pH will strongly determine whether φX-174 will behave conservatively.

d. F-specific RNA Bacteriophages

F-specific RNA bacteriophages have similar physical properties as enteroviruses, especially with respect to size (Bitton, 1980; Havelaar, 1993). MS-2 belongs to Group I of F-specific RNA bacteriophages (Havelaar, 1986). As naturally present model viruses, these bacteriophage are of high interest to represent enteroviruses in various treatment processes of surface water, including aquifer passage. Before entering a treatment (like aquifer passage), enteroviruses and F-specific

RNA bacteriophages have largely followed the same path (i.e., both have passed the sewerage system, followed by sewage treatment, discharge into surface water, and some kind of pretreatment before recharge into an aquifer). It may be reasoned that, along this path from the sewerage system to the point of recharge into an aquifer, viruses that are less stable or that adsorb readily to solid surfaces have already disappeared. This suggests that a selection has taken place leading to the preservation of very stable and poorly adsorbing viruses (i.e., worst-case viruses). This selection has been the same for F-specific RNA bacteriophages and enteroviruses.

In surface water, F-specific RNA bacteriophages occur in numbers of 10^2 to 10^4 times higher than enteroviruses (Havelaar et al., 1993); therefore, it has been possible to show 4- to 6-log units removal of F-specific RNA bacteriophages by RBF (Havelaar et al., 1994).

In an alluvial gravel aquifer where effluent was irrigated, Sinton et al. (1997) studied the transport of somatic coliphages, F-specific RNA bacteriophages, and fecal coliforms in one experiment and of MS-2 and E. coli J6-2 in a second experiment. Concentrations of all these microorganisms were similarly reduced, about 9 logs after 400 m of transport. The concentration of rhodamine WT dye was reduced about 7 logs over this distance. This shows that removal of F-specific RNA bacteriophages and MS-2 (also an F-specific RNA bacteriophage) is similar.

To conclude, F-specific RNA bacteriophages — as a group of naturally occurring viruses — are useful model viruses for the behavior of viruses during subsurface transport. F-specific RNA bacteriophages behave relatively conservatively, like MS-2, and they have been shown to be very persistent. Moreover, naturally present F-specific RNA bacteriophages may consist of stable and poorly adsorbing viruses prior to treatment by aquifer passage.

Surrogates and Indicators for Parasitic Protozoa

a. Surrogate Protozoa

Given the difficulties in monitoring for low levels of oocysts and cysts, surrogates and indicators have a key role in assessing whether oocyst breakthrough is possible at an existing or potential RBF site; however, the use of surrogates and indicators are fraught with difficulties, the greatest of which is the lack of sufficient co-occurrence data in oocyst breakthrough samples from RBF sites.

Moulton-Hancock et al. (2000) attempted to address this issue using data from all groundwater settings, including springs, wells, and infiltration galleries. They found that 16 separate genera and higher taxa were significant predictors indicative of *Cryptosporidium* and/or *Giardia* contamination. The two biota with the greatest statistical significance are the diatom genera, *Navicula* and *Synedra*. The data do not suggest that any one organism or taxa is an oocyst indicator; rather, the authors emphasize that a variety of data must go into an evaluation, including microbiota and hydrogeologic data.

One problem with the use of the results reported in Moulton-Hancock et al. (2000) is that springs used as drinking-water sources are typically karst springs, and the biota associated with such sites will likely be very different from the biota found in a porous media site. Unfortunately, Moulton-Hamilton et al. had insufficient data to analyze each type of hydrogeologic site separately.

In the following, it is assumed that diatom genera data, especially *Synedra* occurrence, have some implications, however uncertain, about oocyst occurrence in RBF sites. It must be emphasized that no laboratory sand-column studies have compared the mobility of diatoms with the mobility of oocysts. This lack of experimental mobility data, combined with limited co-occurrence data from RBF sites and, perhaps, the inappropriate combining of karst and porous media microbiota data to determine statistical predictors, all suggest that any conclusions about the use of diatoms as surrogates or indicators for oocysts are premature. In the following discussion, the occurrence of diatoms will be examined to see if further insight can be gained about the use of any indicator or surrogate for a site in which no oocyst breakthrough has been observed. Such detailed diatom occurrence data are not available for sites in which oocyst breakthrough has occurred, as the locations of most of those sites are unknown.

Heinemann et al. (1996) collected data from the Platte River and from five vertical wells in the Kilgore Island Well Field in Kearney, Nebraska, over an 18-week period (June 12 to October 12, 1995) (Table 6-5). Pumping demands varied from 19 million to 53 million liters per day (L/d) and Platte River flows varied from 0 to 475 m³/s. Four of the wells selected for sampling had previous histories of high levels of indicators and one had much lower levels. The wells are 16.7- to 18-m deep, with a well screen in the interval from 10 m to total depth.

Table 6-5. Diatom Occurrence Data (Predominant Genera: Total Diatom Concentration) for River and RBF Wells, Summer, 1995, Kilgore Island Well Field in Kearney, Nebraska (Heinemann et al., 1996)

	June	July	August	September	October
River	Centrales: 800,000/L	Centrales: 1,300,000/L	Not Available	Unknown: 5,300,000/L	Not Available
Well 1	Synedra: 0.3/L	Pennales: 0.008/L	Pennales: 0.018/L	Not Detected	Pennales: 0.015/L
Well 2	Unknown: 12/L	Synedra: 0.17/L	Not Detected	Not Detected	Not Detected
Well 3	Synedra: 190/L	Pennales: 0.005/L	Pennales: 0.005/L	Not Detected	Not Detected
Well 4	Synedra: 0.003/L	Synedra: 0.092/L	Not Detected	Not Detected	Not Detected
Well 5	Synedra: 16.9/L	Fragilaria: 0.003/L	Not Detected	Not Detected	Not Detected

Concentrations represent the number of specific organisms per liter of water.

The available data suggest that the total diatom concentration is increasing from early to late summer, as might be expected, due to increased solar insolation and lowered river discharge supporting a diatom bloom. The predominant diatom in the river in early summer appears to be *Centrales*. In contrast, *Centrales* is not typically found in RBF wells. One might surmise that *Centrales* is not hydrodynamically favored for transport through porous media. The highest concentrations of *Synedra* are most commonly found in RBF wells during the early summer months, suggesting that *Synedra* is more capable of being transported through porous media. Moulton-Hancock et al. (2000) have found a similar result based on the compilation of data from a large number of sites in a variety of hydrogeologic settings, both porous and non-porous media.

The increased breakthrough concentrations in early summer suggest that either:
- *Synedra* concentrations in the river were higher in early summer, or
- Platte River flow conditions were more conducive to early summer breakthrough.

Data from Verstraeten et al. (1999) suggest that the early summer period of high flow may be significant. Verstraeten et al. measured flow and contaminants in the Platte River and RBF wells in Lincoln, Nebraska (about 200-km downriver from Kearney), in early June 1995. They found that the peak atrazine concentration occurred in the wells on June 5, with a 6-day lag period. The velocity of the Platte River is about 2 to 3.5 kilometer per hour (km/h), yielding about a 3-day river travel time from Kearney to Lincoln. Thus, the peak poor-quality flow (as determined by the atrazine concentration) occurred in late May. These data suggest the possibility that the spring runoff season in the Platte River watershed provides conditions that are more favorable for diatom breakthrough in wells, perhaps due to bottom scour or increased hydraulic head.

Because the sites that have been given RBF credit or are under consideration for RBF credit do not have a known history of oocyst breakthrough at the well, surrogate and indicator organisms as well as physical data (such as turbidity, particle counts, and temperature) have been used. Typically, a surrogate suite of organisms is described based on the microscopic particulate analysis method (Vasconcelos and Harris, 1992) and is used in the decision-making process. The microscopic particulate analysis method does not identify biota at their lowest taxonomic levels (except for *Giardia*). Thus, diatoms are only identified as such and are not identified by genus.

b. Microscopic Particulate Analysis as an Indicator of *Cryptosporidium* Oocysts

The microscopic particulate analysis method was developed as a guide to supplement standard methods in making decisions (using all available data) as to which groundwater systems should be classified as systems using groundwater under the direct influence of surface water. The microscopic particulate analysis method was not developed for use in determining RBF credit. Moulton-Hancock et al. (2000) participated in developing the microscopic particulate analysis method and have conducted an analysis of the microscopic particulate analysis scoring method using oocyst and cyst breakthrough as a measure of the method's validity. They conclude that the general categories of high, medium, and low risk are valid measures of potential risk, but that the individual numerical score based on the presence and counts of surrogate organisms was not a better predictor than the general groupings.

Mikels (1992) compared the turbidity, temperature, and approximate microscopic particulate analysis organism count data for two horizontal collector wells, one in Kalama, Washington, and the other in Kennewick, Washington, over a period of 15 months. No *Giardia* or coccidian protozoa (which includes members of the genus *Cryptosporidium*) were recovered from the wells, despite low occurrence levels in the river. These data support the conclusions of Moulton-Hancock et al. (2000) that the microscopic particulate analysis organism count data are not, by themselves, significant predictors.

Variability in the microscopic particulate analysis method was evaluated by Nelson (1996). Over a 2.5-year time period, 30 microscopic particulate analysis samples were collected from a single well in the Willamette Well Field in Springfield, Oregon (well screen depth is 6 m; distance to surface water is 15 m). Microscopic particulate analysis scores ranged from 1 to over 30 and, perhaps, showed seasonal trends. In typical usage, an microscopic particulate analysis score of 20 or over is considered to be a high-risk setting.

Gollnitz et al. (1997) state that the microscopic particulate analysis method is too sensitive to evaluate most groundwater and suggest that RBF credit can be determined based primarily on the removal of diatoms and algae. Thus, 2-log removal of algae implies, based on the method of Gollnitz et al. (1997), a 2-log removal of oocysts. In this scheme, a reduction in algae concentration from 1 million/L in the river to 10,000/L in the well might be considered significant. In Gollnitz et al. (1997a), algae concentrations in wells are plotted against calculated rates of induced infiltration for vertical wells in the Bolton Well Field. Concentrations of algae in well water were at their peak values during the periods of highest-induced infiltration.

Microscopic particulate analysis has the advantage of including organisms that may have the same size range as oocysts. In evaluating RBF efficiency using only algae and diatoms, Clancy and Stendhal (1996) enumerate algae and diatoms into three size ranges (2 to 7 µm, 7 to 15 µm, and

>15 μm); however, oocysts are almost equi-dimensional, being only slightly oblate. Xiao et al. (2000) summarize data for all *Cryptosporidium* and report data from a study of 30 oocysts. *Cryptosporidium parvum* was found to range in length from 4.8 to 5.6 μm, in width from 4.2 to 4.8 μm, and with a length-to-width ratio that ranged from 1.04 to 1.33. Few organisms, including diatoms and other algae, have the same size and shape as oocysts. By analogy with the bacteria, shape factors are important in governing the hydrodynamics of transport in porous media. Weiss et al. (1995) compared the ratio of bacterial cell width to cell length and found that cell shape affected transport. Shorter cell lengths appeared to be favored among those exiting the column as compared with the dimensions of influent organisms.

Total coliform data collected for 1 year (beginning in May 1992) were used by the primacy agency to determine that Collector Well 5 of Sonoma County (California) should be regulated as groundwater under the direct influence of surface water (CH2M Hill, 1993). During that period, 11 of 60 samples (18 percent) were total coliform positive. Coliform data were collected from each lateral; one lateral had 77-percent positive samples and another 55-percent positive samples. Because horizontal collector wells mix relatively clean water from the aquifer with poorer-quality riverbank infiltrate, coliform counts from the laterals are significant. Additional total coliform data were collected for a year, beginning in February 1997 (Price et al., 1999). These data show that high total coliform counts at Collector Well 5 correlate with high levels of river flow. When the Russian River flow is below 85 m^3/s, less than 9 percent of the samples are total coliform positive. When flow exceeds 141 m^3/s, more than 40 percent of the samples are total coliform positive. These (and other) data were successfully used to make the case that Collector Well 5 was only groundwater under the direct influence of surface water during periods when the Russian River exceeded a certain flow rate. See the Introduction of this book for further details.

Total coliform data were also collected at the Kearney, Nebraska, site as part of the study to support RBF credit (Heinemann et al., 1996). At the ninetieth percentile, the concentration of total coliforms in the river is between 220 and 290 MPN/100 mL; at the same percentile, the concentration in vertical wells is between 0.8 and 1.2 MPN/100 mL.

The use of particle-count data as a surrogate is an area of ongoing research. Particle counters are unable to differentiate between organic and inorganic particles. Also, there are particles in the well water, such as iron and manganese precipitates, that are not present in river water, resulting, perhaps, in low estimates of natural filtration efficiency (Dillon Consulting, Ltd., 1997).

Nelson (1996) examined the decrease with distance of particles in the 4- to 15-μm range at the Willamette and McKenzie Well Fields. For vertical wells ranging up to 91 m from the river, the initial particle count above 100,000 was decreased 2 logs or more with distance, but with some variability, including at least one sample at a distance of 30 m with <1-log decrease. Both Sonoma County, California, and Kearney, Nebraska, made successful use of particle-count data in their application for RBF credit. At Sonoma County, particle counts in the river and Collector Well 5 indicated better than a 2.5-log removal of 4- to 10-μm range particles in more than 95 percent of the samples (Price et al., 1999). At Kearney, particles greater than 2 μm were measured in the river and in five vertical wells. Average log removals are reported to range from 2.05 to 2.27 (Heinemann et al., 1996). At nine vertical RBF wells in Kitchener in Waterloo, Ontario, in Canada (Woolner, Pompeii, and Forwell Well Fields), particle (5 to15 μm) log removals ranged from 1.3 to 2.6 log. For smaller particles (2 to 5 μm), removals ranged from 0.9 to 2.2 log.

It is not clear what the limited particle and microbial removal data imply about oocyst removal. To date, there exists insufficient data to evaluate any surrogate or indicator organism, suite of organisms, or physical parameter (e.g., turbidity) as measures of RBF efficiency. More research is needed in this area.

The column studies by Harter et al. (2000) suggest that oocysts can be sufficiently mobile in a coarse-grained alluvial aquifer under high hydraulic head differential to reach a horizontal collector well lateral underlying a river or, perhaps, a shallow vertical well adjacent to the river. Removal of the fine-grain riverbed material during flood scour may be sufficient to allow a contaminant slug to enter the aquifer. As the high-water stage and heads decline, oocysts can become remobilized from their initial attachment points and continue to be transported with the groundwater flow toward the well or lateral. If the aquifer is poor in fine-grained material, organic particles or oxide grain coatings, oocyst mobility is enhanced, albeit at lower oocyst concentrations. The data reporting oocyst breakthrough to four horizontal collector wells by Moulton-Hancock et al. (2000) suggest that this scenario can and does occur.

No one oocyst surrogate is, given available data, a good predictor of oocyst breakthrough; however, as field data are collected using the lowest taxonomic levels and as column studies compare the mobility of oocysts and surrogates, it is possible that some organisms, such as the diatom *Synedra*, may become suitable indicators of RBF efficiency. Existing data using various indicators and surrogates do not clearly indicate the absence of a potentially significant public health risk.

10. Removal by RBF and Artificial Infiltration

Removal of Viruses

Table 6-6 lists the removal data obtained from the field studies on RBF and dune recharge in The Netherlands.

These removal data show effective removal of bacteriophages. The same can probably be said for viruses, considering the high removal of the bacteriophages. The removal of the viruses could only be measured at shorter travel distances. Similarly, the removal of bacteriophages MS-2 and PRD-1 was effective in the field studies on dune recharge (Schijven et al., 1999; Figure 6-1) and deep well injection (Schijven et al., 2000; Figure 6-2). By dune passage, concentrations of both bacteriophages were reduced about 3-logs within the first 2.4 m and another 5 logs in a linear fashion within the following 27 m (see Figure 6-1).

Table 6-6. Log Removal of Microorganisms from Surface Water by RBF and from Pretreated Surface Water by Dune Recharge in The Netherlands

	Riverbank Filtration			Dune Recharge
	Rhine at Remmerden[a]	Meuse at Zwijndrecht[a]	Meuse at Roosteren[b]	Heemskerk[c]
Travel Distance (m)	30	25	13 25 150	2 4
Travel Time (days)	15	63	7 18 43	1 2
FRNAPH[d]	6.2		3.9 5.1 7.3	3.1 4
SOMCPH[e]			4.0 5.9 6.7	
Enteroviruses	≥ 2.6	≥ 2.7	1.7	
Reoviruses	≥ 4.8	≥ 4.7	2.8	
TOTCOL[f]	≥ 5.0	≥ 5.0		0.85
THCOL[g]	≥ 4.1	≥ 4.1	4.1 4.5 6.2	0.86
SSRC[h]	≥ 3.1	≥ 3.6	3.3 3.9 5.0	1.9
FSTREP[i]	≥ 3.2	≥ 3.5		

[a]Havelaar et al. (1995). [b]Medema et al. (2000). [c]Schijven et al. (1998). [d]F-specific RNA bacteriophages. [e]Somatic coliphages. [f]Total coliforms. [g]Thermotolerant coli. [h]Spores of sulphite-reducing clostridia. [i]Fecal streptococci.

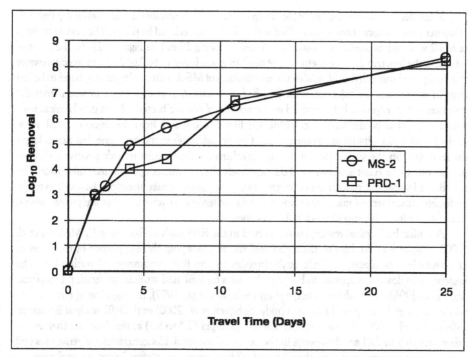

Figure 6-1. Removal of bacteriophages MS-2 and PRD-1 by dune passage (Schijven et al, 1999).

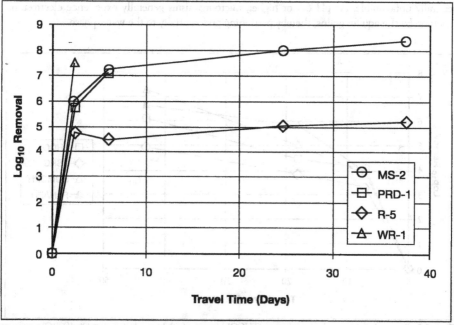

Figure 6-2. Removal of bacteriophages MS-2 and PRD-1, spores of *Clostridium biferementans* R-5, and *E. coli* WR-1 by deep well injection (Schijven et al, 2000).

In the deep-well injection study (Schijven et al., 2000), it was found that, within the first 8 m of aquifer passage, concentrations of MS-2 and PRD-1 were reduced by 6 logs. The concentrations of MS-2 were reduced only by about 2 logs in the following 30 m (see Figure 6-2). At the point of injection, the inactivation rate coefficient of free MS-2 was found to be 0.010/day. In injection water that had passed through 8 m of aquifer, the inactivation of MS-2 bacteriophages was found to be less than in water from the injection well. The higher inactivation rate of MS-2 in water from the injection well may probably be ascribed to the activity of aerobic bacteria. From geochemical mass balances, it could be deduced that, within the first 8-m distance from the injection well, ferric oxyhydroxides precipitated as a consequence of pyrite oxidation in the oxic zone, but not at larger distances (i.e., in the anoxic part of the aquifer). Ferric oxyhydroxides provide positively charged patches on to which fast attachment of the negatively charged microorganisms may take place. The non-linear logarithmic reduction of concentrations with distance may, therefore, be ascribed to the preferable attachment of microorganisms to patches of ferric oxyhydroxides that are present within 8-m distance from the injection point, but not thereafter.

A similar high-initial removal was observed in the RBF study at Roosteren by Medema et al. (2000) (Figure 6-3). At this site, the high initial removal may possibly be explained by the presence of preferable attachment to ferric oxyhydroxides in the first centimeters of infiltration. The presence of redox zones appears to be general at sites for RBF and artificial infiltration (Stuyfzand and Lüers, 1996). In the dune recharge study (Schijven et al., 1999), the high initial removal rate was less than in the deep-well injection study (Schijven et al., 2000) or the RBF study at Roosteren (Medema et al., 2000). This is probably due to higher pH (7.3 to 8.3) at the dune site than at the other sites (6.5 to 7.0) and less pyrite in the aeolian dune sand. Concentrations of reoviruses and enteroviruses were reduced more than 2.6 and 1.7 logs, respectively, after 7 days of travel time.

The collision efficiencies calculated from the removal data from Schijven et al. (1998, 1999, 2000) are very low in all cases (Table 6-7), meaning that the conditions for attachment are generally unfavorable. At pH 6.5 or higher, microorganisms generally experience electrostatic repulsion by the aquifer grains, thereby partioning preferentially to the water phase.

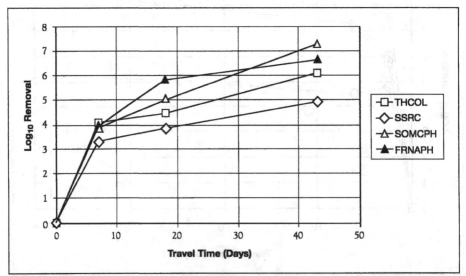

Figure 6-3. Removal of thermotolerant coliforms (THCOL), spores of sulphite-reducing clostridia (SSRC), somatic coliphages (SOMCPH), and F-specific RNA bacteriophages (FRNAPH) by RBF (Medema et al, 2000).

Table 6-7. Collision Efficiencies of Microorganisms from Studies on the Removal of Microorganisms by Dune Recharge and Deep Well Injection in The Netherlands

| | Dune Recharge | | | | Deep Well Injection | |
	Heemskerk[a]		Castricum[b]		Someren (Pilot Site)[c]	
Travel Distance (m)	2	4	2.4	30	8	38
Travel Time (days)	1	2	1.7	25	2.4	38
FRNAPH[d]	0.0020	0.00078				
TOTCOL[e]	0.0068					
THCOL[f]	0.0077					
SSRC[g]	0.0097					
MS-2[h]			0.0014	0.00027	0.0014	0.00020
PRD-1[h]			0.0024	0.00043		
R-5[i]					0.0080	0.0010

[a]Schijven et al., 1998. [b]Schijven et al., 1999. [c]Medema et al., 2000. [d]F-specific RNA bacteriophages. [e]Total coliforms. [f]Thermotolerant coliforms. [g]Spores of sulphite-reducing clostridia. [h]Bacteriophages MS-2 and PRD-1. [i]Spores of Clostridium bifermentans R-5.

Removal of Bacteria

Bacterial transport in soil and other unconsolidated or consolidated material depends strongly on material properties. In fine sand, the migration of bacteria is limited and most of the bacteria are removed at the beginning of infiltration, many even within the first 0.5 m (Bellamy et al., 1985; Preuss, 1996; Albrechtsen et al., 1998). In sandy gravel, Pang et al. (1998) report the detection of Bacillus subtilus endospores after transport of 90 m from an injection well. In a gravel aquifer, Sinton et al. (1997) report E. coli J6-2 was recovered 401 m from an injection well. These data suggest that some coarse-grained aquifers may provide high water yields, but may have low efficiency in removing bacteria by RBF, if fine-grained riverbed sediments are absent.

A decrease in temperature decreases the removal of coliforms and heterotrophic bacteria during artificial sand filtration (Bellamy et al., 1985), and water salinity may also have an effect (Gannon et al., 1991).

In Dutch studies, it was found that fecal indicator bacteria are removed to a similar extent as bacteriophages (see Table 6-6); however, clostridia spores seem to be removed less in the RBF study at Roosteren (Medema et al., 2000). A similar observation was made at the deep-well injection study (Schijven et al., 2000), where spores of Clostridium bifermentans R-5 were removed less than bacteriophage MS-2. Within the first 8 m of aquifer passage, concentrations of R-5 spores were reduced by 5 logs, but were reduced negligibly in the following 30 m. The high initial removal of the spores may also be ascribed to attachment to preferable sites of ferric oxyhydroxides that are present in the oxic zone, but not in the anoxic zone. In the RBF study by Medema et al. (2000), the high initial removal of the bacteria may be due to similar preferable sites of ferric oxyhydroxides, but in only in the first few centimeters (see Figure 6-3).

Collision efficiencies for fecal indicator bacteria (see Table 6-7) are also low but, generally, a bit higher compared to those for bacteriophages. On the other hand, single collector efficiencies for fecal indicator bacteria are generally lower due to the greater size of the bacteria.

The capability of a horizontal collector well to remove total coliform bacteria was studied by Wang et al. (2000). The well is located about 30 m from the river with laterals at 25-m below ground level. The laterals extend under the river. Daily (for 7 months) total coliform concentrations in the river ranged from 9 to 33,040 MPN/100 mL. Well coliform occurrence (combined data, not from a single lateral) were only occasionally present and only once exceeded 100 MPN/100 mL. For another horizontal collector well, Arora et al. (2000) measured total coliform in 15 monthly samples from the Wabash River, but in only 2 monthly samples from a well with a lateral 20-m below the river bottom. Also at that site, *Clostridium* and bacteriophage were frequently detected in river water, but only twice in well samples. Bacterial removal data from two other RBF sites are discussed in the following section on protozoa removal as the bacterial data were used to estimate protozoa removal efficiencies.

Miettinen et al. (1997) studied how the number and metabolic activity of heterotrophic bacteria changes during RBF of humus-rich (12.1 ± 1.4 mg/L) lake water (Figure 6-4 and Table 6-8).

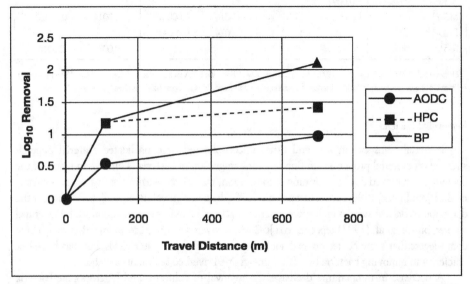

Figure 6-4. The mean $\log_{10}$ removals of microbial parameters measured as acridine orange direct counts (AODC), heterotrophic plate counts (HPC), and bacterial production activity (BP) in water samples obtained from Hietasalo Bank Filtration Plant. Note: Figure 6-4 and Table 6-8 display the same data.

Table 6-8. Microbial Activity (Mean ± Standard Deviation) Measured as Acridine Orange Direct Counts (AODC), Heterotrophic Plate Counts (HPC), and Bacterial Production Activity (BP) in Water Samples Obtained from Hietasalo Bank Filtration Plant

Sample Location	Lake	110 m	700 m	Abstraction Wells (Mixed Water)
Travel Time		1 Week	3 to 4 Weeks	2/3 of Water: 150 m and 1/3 from 750 m (Time: 1 Week/3 to 4 Weeks)
AODC (cells/mL)	1,640,000 ± 805,000	448,000 ± 288,000 (0.56)	170,000 ± 49,000 (0.98)	239,000 ± 113,000 (0.84)
HPC (cfu/mL)	11,900 ± 8,700	740 ± 840(1.2)	450 ± 390 (1.4)	700 ± 1,000 (1.2)
BP (ngC /L × h)	390 ± 670	24 ± 24 (1.2)	3 ± 3 (2.1)	10 ± 5 (1.6)

Note: Figure 6-4 and Table 6-8 display the same data.   ngC = Nanograms of carbon.

The studies were carried out at an RBF water plant located on an esker island, Hietasalo, where 90 percent of all produced drinking water is riverbank filtrated. An esker is a sinuous deposit of sand and gravel formed by flowing water in a tunnel at the base of a continental glacier. The study showed that, after RBF, there was a more than tenfold reduction in heterotrophic microbial numbers. HPCs, total microbial counts, and microbial biomass decreased rapidly after infiltration from the lake. The microbial parameters decreased by 1 log during the filtration (see Table 6-8). The removal rate of heterotrophic bacteria was lower than in previously presented studies concerning the removal of coliforms. The most likely reason for decreased removal is indigenous heterotrophic microbial growth/activity in the subsurface; however, in this study, it was also found that coliform bacteria were predominantly removed during RBF and were detected only occasionally in filtered water (Miettinen et al., 1997).

Breakthrough of Cryptosporidium *Oocysts at RBF Sites*

The protozoans of concern (or potential concern), include *Cryptosporidium parvum*, *Giardia lamblia*, *Cyclospora* sp., and members of the Microsprididea ("Microsporidia") class, seven genera (10 species) of which have been recovered in humans (Mota et al., 2000). *Cryptosporidium parvum* is the primary protozoan pathogen of concern because of its:
- Small size.
- Ubiquity and survival in the environment.
- Resistance to disinfectants.
- Ability to amplify in a variety of mammalian hosts.
- Potential for health consequences.

This discussion will focus on *Cryptosporidium parvum*, but will address other protozoa, as the topic warrants.

Cryptosporidium and *Giardia* removal by RBF is an area of current research. On the one hand, the relatively larger size of protozoan oocysts (*Cryptosporidium*) and cysts (*Giardia*) suggests that, in the absence of bacterial or viral aggregation or attachment to larger particles, protozoan removal may be more efficient than for the smaller bacteria and viruses. On the other hand, the available data show that cysts and oocysts do occur in well water from alluvial aquifers, albeit perhaps not in sufficient concentration to cause observable health effects in a single individual or a small population. As implicit in the Dutch regulations and other epidemiology literature, health effects of small (below detectable levels) microbial concentrations that large populations are exposed to are not easily observable. The size ranges for the pathogenic protozoa of concern are shown in Table 6-9.

Table 6-9. Size of Pathogenic Protozoa

Pathogenic Protozoa	Size Range (μm)
"Microsporidia"	1.5 to 5
Cryptosporidium Parvum	4 to 6
Giardia Lamblia	8 to 18
Cyclospora sp.	8 to 10

Dowd et al. (1998a) identified one of the pathogenic microsporidia from a single groundwater sample (using polymerase chain reaction). A suspected waterborne outbreak of microsporidiosis in 1995 was retrospectively identified in Lyon, France (Cotte et al., 1999). The water system uses riverbank-filtered water from 114 wells along the Rhone River in France with subsequent chlorination and, for one distribution area, lake water that is also filtered. The outbreak appeared to be clustered primarily within the distribution area that used both lake water and riverbank filtrate. Ooi et al. (1995) reported (in New Hampshire, United States) at least one case of gastrointestinal illness caused by *Cyclospora* that may be associated with drinking contaminated well water or swimming in a pool filled with well water. While suggestive, these data do not demonstrate that RBF systems should be concerned with pathogenic protozoa other than *Cryptosporidium* and *Giardia*. Some indicator or pathogenic bacteria can, at the extreme end of their size range, approach the size of oocysts:

- *E. coli* (1 to 6 μm).
- *Clostridium perfringens* (3 to 9 μm) (vegetative cells).
- *Klebsiella* sp. (0.6 to 6 μm).
- *Streptococcus faecalis* (0.5 to 10 μm).

The enhancing effect of flooding and scour on contaminant breakthrough in alluvial wells has been investigated for agricultural chemicals (Ray et al., 1998; Verstraeten et al., 1999), but not yet been studied in cases of *Cryptosporidium* contamination. In principal, as with agricultural chemicals, the potential for *Cryptosporidium* breakthrough should be enhanced because flooding increases the potentiometric head and, likely, flushes oocysts from land sources (e.g., combined sewer overflows) into surface water. Scour typically removes any fine-grained material in the riverbed that has the greatest potential to filter *Cryptosporidium*, and it decreases the transport distance through the alluvium. Gollnitz et al. (1997a, 1997) have investigated the transient hydraulic properties of the river/aquifer system at the Bolton Well Field in Cincinnati, Ohio, using a similar working hypothesis. That is, the vertical wells are at greatest risk for oocyst breakthrough during periods of high-river stage when, for that system, the greatest induced infiltration rates occur. Aerobic endospores (e.g. *Bacillus subtilus*), about one micron in diameter, are beginning to be widely used as oocyst surrogates in field investigations to predict *Cryptosporidium* removal (Berger, 2002).

a. Sorption and Hydrophobicity

For oocysts, Brush et al. (1998) summarize and compare previous oocyst isoelectric point measurements. They report that those adhesion properties governed by the electrophoretic mobility can be altered by the purification method. Brush et al. found no clear isoelectric point. They also measured oocyst hydrophobicity and found that the hydrophobicity of the oocyst surface changes as the oocysts age after they are excreted.

The mobility of oocysts in saturated columns was investigated by Brush et al. (1999) using the convection-dispersion equation with equilibrium (instantaneous, reversible) sorption. In sand, Brush et al. found that the oocysts had a retardation value of 1.0, indicating that the oocysts were transported with the same velocity as pore water without measurable sorption. In contrast, under dynamic conditions of recharge and maximum water holding conditions, Mawdsley et al. (1996) found only small amounts of oocyst leaching through intact silt loam soils and no oocyst leaching through intact loamy sand soil.

Harter et al. (2000) conducted column studies with concomitant colloidal filtration calculations for oocyst transport in sand. In coarse sand columns, oocyst pore velocities were measured to be 16 m/d. Other colloidal filtration column data were collected by Adin et al. (2000). Harter et al. (2000) conclude, based on column studies, that the colloidal filtration theory is adequate to predict gross behavior in coarse-grained porous media; however, reversible detachment, which is not considered in colloidal filtration theory, appears significant and can represent a continued oocyst source after passage of a contamination slug. Nevertheless, actual breakthrough is highly dependent on the inactivation rate of oocysts.

Oocyst breakthrough to wells has occurred in the absence of an identified outbreak. Hancock et al. (1998) report finding oocysts in seven of 149 vertical wells and five of 11 horizontal collector wells, based on samples submitted by public-water supplies from locations throughout the United States. On average, two samples were necessary to detect oocysts in vertical wells and three samples were required to detect oocysts in horizontal collector wells. *Giardia* cysts were found in two of 149 vertical wells and in four of 11 horizontal collector wells. Table 6-10 provides unpublished detail about those wells and samples with oocyst breakthrough summarized in Hancock et al. (1998), for which some additional information is available.

The limited diatom data available for RBF sites (and samples) with oocyst breakthrough are shown in Tables 6-10 and 6-11. These data show that diatoms typically, but not always, co-occur with oocyst breakthrough in RBF sites, sometimes at high concentrations. The available data do not identify the diatom species, such as *Navicula* or *Synedra*, which are statistically significant predictors of oocyst co-occurrence (Moulton-Hancock et al., 2000). Presumably, both diatom species are well represented as the data shown are from the same raw data set used by Moulton-Hancock et al., albeit compiled differently.

Table 6-10. Wells with Oocyst or Cyst Breakthrough and with Ancillary Data

Well	Sample	Crypto/ 100 L	Distance to Surface Water (m)	Well Depth (m); Type	Giardia/ 100 L	Diatoms/ 100 L	MPA Score
A	1	5.3			2.1	38.8	84
	2	0.5			ND	5.0	41
B	1	2.1			ND	0	30
	2	3.4			1.6	25.3	80
C	1	0.5			ND	29.3	37
	2	2.6			1.8	5.0	66
	3	5.0			ND	0.3	41
D	1	0.5	60	6; Vert.	ND	0.5	33
E	1	15.0		137; Vert.	120.0	78.7	108
F	1	4.5	15	26; Horiz.	8.9	237.7	103
	2	0.8			ND	234.3	51
G	1	527.8			ND	0	35
H	1	7.9	8,000	33; Vert.	ND	6,427	82
I	1	383.8	15	30; Vert.	ND	0	53
J	1	18.5		45; Vert.	ND	Not Taken	
K	1	0.3		27; Vert.	ND	Not Taken	
L	1	8.5	250	13; Horiz.	ND	Not Taken	
M	1	ND	115	9	0.5	0	21

Unpublished data provided by C. Moulton-Hancock, CDH Diagnostic, Inc., Loveland, Colorado.
Vert. = Vertical well.   Horiz. = Horizontal collector well.   ND = Not detected.   MPA = Microscopic particulate analysis.

Table 6-11. Oocyst, Cyst, and Diatom Breakthrough in Well F

Month	Cryptosporidium per 100 L	Giardia per 100 L	Diatoms per 100 L
March	Not Detected	Not Detected	0.26
March	Not Detected	Not Detected	3.43
April	0.8	Not Detected	176.0
June	Not Detected	Not Detected	6,396.0
December	4.5	9.0	237.0
December	Not Detected	Not Detected	3.17

Unpublished data provided by C. Hancock, CDH Diagnostic, Inc., Loveland, Colorado.

Well F is particularly interesting because it is a horizontal collector well of typical construction (27 m deep; 15 m from surface water) that was sampled six times, two of which show oocyst and cyst breakthrough. Table 6-11 shows the available sampling results from Well F.

While not definitive or statistically significant, oocyst breakthrough appears to be associated with higher diatom concentrations.

Although the positive sites are unidentified, it is likely that many or most were public water-supply utilities. The sample analyses were conducted by CDH Diagnostic, Inc., a commercial laboratory in Colorado that obtained samples in the normal course of business, much of which is to analyze public water-supply well samples for protozoa and indicator organisms, as suggested by the Surface Water Treatment Rule guidance document (U.S. Environmental Protection Agency, 1989, 1991).

The available data, both published and unpublished, unequivocally show oocyst breakthrough in drinking-water wells. No additional information is available that documents the contamination source; however, given the proximity to surface water and the co-occurrence of cysts, diatoms, and other surrogates (evident by a microscopic particulate analysis score above 20), surface water is the likely source. Where surface water is nearby and pumping rates are large (data unavailable), induced infiltration from surface water is likely. Presumably, the four horizontal collector wells are located in alluvial aquifers because their design is specifically targeted for such a location; however, Collector Well L is reported to be located 240 m from surface water, so that well might be designed to operate as an infiltration gallery and, thus, is excluded here from further consideration. Horizontal collector wells or vertical wells in close proximity to surface water are likely RBF wells. By these criteria, five wells are RBF wells (three collector wells [including Well F] and two vertical wells [Wells D and I]). Thus, oocyst breakthrough may be surmised for five RBF wells.

It is not known how many, if any, of the vertical wells with oocyst breakthrough reported by Hancock et al. (1998) are located in alluvial aquifers; however, oocysts (0.3/100 L, based on a composite of seven samples) were recovered from a vertical well (Jeffersonville, Indiana, Well 9) located (screened interval 13- to 27-m below ground surface) about 60 m from the Ohio River (Arora et al., 2000). The adjacent Wabash River oocyst concentrations were below detection limits. Male-specific coliphage (MS-2) were detected in both the river and Well 9 (0.09 plaque forming unit (pfu)/100 mL).

Rosen et al. (1996) report oocysts from seven of 17 utilities using groundwater that is sufficiently connected with surface water so as to be regulated as a surface-water utility. Because the locations for these wells are not known, some (or all) of these sites may be included among those reported by Hancock et al. (1998). Archer et al. (1995) report oocyst recovery in a non-community well in Door County, Wisconsin, at a concentration of 1 oocyst/1,175 L; however, groundwater in Door County is produced from a highly vulnerable fractured dolomite, so the oocyst occurrence may not be the result of induced infiltration from surface water. Oocyst breakthrough in wells in the United States and internationally is reported (Rose et al., 1991; Lisle and Rose, 1995), but again, information is unavailable to determine whether induced infiltration to a public well has occurred. Presumptive oocysts were identified in other wells, including some located in alluvial aquifers, but a lack of information about laboratory methods preclude further conclusions.

Gollnitz et al. (1997) proposed a specific methodology for the use of surrogate indicators to identify RBF wells that do not require further treatment to protect public health. The major rationale proposed for the use of surrogates was the absence of any "scientifically reported" occurrence of oocysts or cysts in groundwater. The results of Hancock et al. (1998) revise the statement of Gollnitz et al. (1997) that the occurrence of *Giardia* or *Cryptosporidium* in groundwater has not yet been scientifically reported.

b. *Cryptosporidium* Risk Characterization

Because of the natural filtration capability of porous media, oocyst breakthrough is unlikely to occur in large contamination slugs, as might be the case for a non-porous media. First arrival concentrations will likely be low, reflecting the probability that only a small oocyst subset has found the fastest flowpaths (among the large number of possibilities) and, thus, arrived first at the well.

The assumption used herein is that cryptosporidiosis risk is below regulatory concern during normal water levels and operations and only becomes significant during periods of high-water stage. A large outbreak requires a massive oocyst contamination slug, which is an unlikely occurrence given the natural filtration properties of porous media. The assumption is supported by the absence of any unequivocally recognized outbreak occurring at an RBF site. Outbreaks are recognized sometimes by chance, but more often when large numbers of people become ill. For example, Eisenberg et al. (1998) simulated conditions occurring during the 1993 Milwaukee, Wisconsin, cryptosporidiosis outbreak and concluded that a smaller, unrecognized outbreak occurred prior to the larger outbreak

During high water, scour can remove the protective fine-grained riverbed sediment, and any high intensity precipitation can increase oocyst runoff to surface water. Thus, the risk characterization herein will focus on the risks associated with the expected low oocyst concentrations that might breakthrough to a well during seasonal high water or flooding. Later arrivals may not be capable of causing infection; however, peak oocyst plume concentrations may govern the overall risk characterization if:

- The surface water has high oocyst levels at all times.
- There are very short travel times.
- There is insufficient natural filtration.

This scenario is not addressed herein because it is less likely than the scenario where the risk is governed by the first arrival concentrations.

In any microbial risk analysis, a key — and sometimes measurable — parameter is the dose required to cause infection in 50 percent of the test subjects (ID_{50} dose). For *Cryptosporidium parvum*, ID_{50} dose data are available for three Genotype 2 strains. Table 6-12 shows the ID_{50} dose for these three organisms (Okhuysen et al., 1999).

Table 6-12. ID$_{50}$ Dose for Three Genotype 2 *Cryptosporidium Parvum* Strains (Okhuysen, 1999)

Strain	Number of Test Subjects (people)	ID$_{50}$ Dose (oocysts)
Iowa	29	87
UCP	17	1,042
TAMU	14	9

Most drinking-water risk analyses assume that consumption is, on average, about 1 to 2 L of water per day. First arrival oocyst concentrations are likely to be low in an RBF site, but it is conceivable that the oocyst concentration can be as high as nine oocysts. Thus, low first arrival oocyst concentrations, as low as about 4 to 5 oocysts/L, are capable of infecting significant numbers of people. The number of people who, once infected, become ill is variable, but is 39 percent in one study with the Iowa strain (Okhuysen, 1999).

Messner et al. (2001) considered the possibility that the three test strains are a subset of a larger population of strains with variable ID$_{50}$ doses. Considering two possibilities, a universe of three strains or more than three strains, Messner et al. extrapolated existing high-dose data into the low dose range (using an exponential model). They found that the daily risk of infection to an individual from ingesting one oocyst is either 2 or 3 percent. Based on the ICR data, about 25 percent of flowing streams have oocyst concentrations greater than 1 oocyst/L. Thus, low first arrival doses are still capable of causing significant numbers of infections, especially if the consuming population is large.

11. Cyanobacteria (Blue-Green Algae)

Cyanobacteria grow in fresh, brackish, and seawater. Only few cyanobacterial species can form massive growths (e.g., algal blooms or blue-green scum) (Sivonen et al., 1990). Massive occurrence of cyanobacteria (blooms) leads to rapid deterioration in the quality of water because some species produce hepatotoxins or neurotoxins (Gorham and Carmichael, 1980; Carmichael, 1992; Lahti et al., 1995). A study from Finland shows that 44 percent of cyanobacterial blooms contained toxic species. Usually, hepatotoxic blooms are more common than neurotoxic blooms (Carmichael et al., 1985; Sivonen et al., 1990).

The most common hepatotoxins are microcystins and nodularin. Microcystins are cyclic heptapeptides and nodularin is a cyclic pentapeptide (Carmichael et al., 1988; Carmichael, 1992). There are three main types of neurotoxins:

- Anatoxin-a.
- Anatoxin-a(s).
- Saxitoxins.

Structurally, anatoxin-a is a secondary amine: 2-acetyl-9-azabicyclo(4.2.1) non-2-ene. Anatoxin-a(s) is an organophosphate: N-hydroxyguanidine methyl phosphate ester. And, saxitoxins are a group of carbamate alkaloid neurotoxins that are found in some dinoflagellates (Carmichael, 1992).

Toxic cyanobacterial blooms have been associated with cattle poisoning (Sivonen et al., 1990). Also, domestic pets (such as dogs) and wildlife have become ill or died after ingesting water containing toxic cyanobacteria (Persson et al., 1984; Edler et al., 1985). Massive growth of toxic cyanobacteria in surface waters causes problem for human recreational use (Pilotto et al., 1997). There are also reports of human illness associated with blooms of toxic cyanobacteria in raw water reservoirs (Kuiper-Goodman et al., 1999); therefore, the occurrence of cyanobacterial cells and their toxins in surface water creates a serious public-health problem for drinking-water production.

The World Health Organization has presented a normative limit value (1 μg/L) for the concentration of microcystin-LR in drinking water (World Health Organization, 1998).

Although traditional drinking-water treatment using chemical coagulation can remove cyanobacterial cells efficiently (Lepistö et al., 1994; Zabel and Melbourne, 1980), it may be inefficient for the removal of cyanobacterial neurotoxins or hepatotoxins (Keijola et al., 1988; Lepistö et al., 1994). More effective treatment processes that eliminate cyanobacterial toxins are activated carbon filtration, slow sand filtration, and ozonation (Keijola et al., 1988; Hrudey et al., 1999).

The removal of cyanobacterial cells and their toxins have been studied by sand column filtration experiments (Lahti et al., 1996; Lahti et al., 1998; Vaitomaa, 1998). Microcystin removal was evaluated using columns filled with lake bottom sediments (taken from a riverbank-filtering waterworks) and with glacially-derived sediments taken from an esker deposit. These studies showed a high removal of blue-green cyanobacteria cells. Cyanobacterial biomass removal was 98 to 99 percent. The removal of (hepatotoxins) microcystins were not as effective as the removal of cyanobacteria cells (biomass). The column experiments also showed that the removal of hepatotoxins was higher in lake bottom sediment columns than in columns filled with esker deposit material. In these experiments, biodegradation was estimated to account for 50 to 70 percent of the total removal of hepatotoxins (Lahti et al., 1998; Vaitomaa, 1998).

There are few published reports documenting the removal of cyanobacterial cells and their toxins in full-scale RBF treatment plants. In Finland, data from artificially recharging groundwater works have shown that cyanobacteria cells and their toxins are not usually found in filtered water, even during massive growth of cyanobacteria and algae in the raw water (Lahti et al., 1993; Hult et al., 1997; Lahti et al., 1998); however, in another study, Lahti et al. (1998) have also shown that although the majority of cyanobacteria and microcystins were removed during RBF, traces of microcystins (<0.1 μg/L) and single cells of cyanobacteria were found in riverbank-filtrated water even after as much as a 100-m subsurface-filtration path length. Anaerobic conditions seemed to favor the persistence of microcystins in water.

12. Conclusions

RBF for the removal of microbial contaminants is, in principle, an efficient system. Given sufficient flow-path length and time, microbial contaminants will be removed or inactivated to levels protective of public health. The field experiments of Schijven et al. (1998, 1999, 2000) serve as a benchmark for field removal under relatively homogenous and steady-state conditions in a saturated sand aquifer. Under optimal conditions (canal infiltration into dune sand), RBF can achieve up to 8-log virus removal over a distance of 30 m in about 25 days. Greater removal efficiency may be expected for bacteria, protozoa, and algae under the same conditions. These high removal efficiencies can be expected to protect public health to minimize risk levels, given appropriate flow-path lengths and retention times.

It is likely that RBF efficiency will be diminished by short path lengths, high heterogeneity, coarse matrices, high gradients, and accompanying high velocities — features common to many riverbank-filtered water-supply systems. The available data for viruses, bacteria, protozoa, algae, and toxins suggest that some RBF sites may, on occasion, not achieve sufficient removal efficiency. This deficiency is well known and, therefore, many water-treatment systems rely on additional treatment barriers, especially disinfection, but also pre-filtration treatment and/or conventional filtration. *Cryptosporidium* and algal toxins are little affected by disinfection and, therefore, require greater attention if disinfection is the only supplemental barrier. *Cryptosporidium* breakthrough to RBF wells has occurred and public-health protection may have been compromised, although the data are incomplete and, as a result, the health risk is uncertain. For bacteria, viruses, and algae, the available data suggest little public-health risk because supplemental treatment has provided sufficient protection when removal efficiency is poor.

For water-supply systems with degraded removal efficiency and no supplemental treatment, the available data suggest that high vigilance is required. The active attenuation processes in riverbank-filtered porous media, even in a sub-optimal setting or time, will mitigate the public-health impact; therefore, the health effect, while significant, may be unrecognizable under current public-health surveillance conditions. This feature is both a benefit and a liability of RBF. The benefit is that RBF processes are always working to minimize contaminant breakthrough concentrations; the liability is that an RBF failure to completely remove microbial or toxin contaminants will most likely result in difficult-to-recognize, short periods with modest contaminant concentrations. Although RBF has the capability to achieve high removals, there will be times when a concentrated microbial or toxin contaminant slug will be insufficiently attenuated and a public-health risk occurs.

References

Adin, A, P. Biswas, G. Dutair, V. Sethi, and C. Patterson (2000). "Characterizing *Cryptosporidium parvum* as a waterborne particle in granular filtration." *Proceedings, International Symposium on Emerging Pathogens*, American Water Works Association, Denver, Colorado.

Albrechtsen H.J., R. Boe-Hansen, M. Henze, and P.S. Mikkelsen (1998). "Microbial growth and clogging in sand column experiments simulating artificial recharge of groundwater." *Artificial Recharge of Groundwater*, Peters et al., eds., A.A. Balkema, Rotterdam, The Netherlands.

Anonymous (2000). "Waterborne outbreak of gastroenteritis associated with a contaminated municipal supply, Walkerton, Ontario, May-June 2000." *Canada Communicable Disease Report*, 26-20: 170-173.

Archer, J.R., J.R. Ball, J.H. Standridge, S.R. Greb, P.W. Rasmussen, J.P. Masterson, and L. Boushon (1995). "*Cryptosporidium spp.* Oocyst and *Giardia spp.* cyst occurrence, concentrations, and distribution in Wisconsin waters." *Publication WR420-95*, Wisconsin Department of Natural Resources Management, Madison, Wisconsin.

Arora, H., M. LeChevallier, R. Aboytes, E. Bouwer, C. O'Melia, W. Ball, W. Weiss, and T. Speth (2000). "Full-scale evaluation of riverbank filtration at three midwest water treatment plants." *Proceedings, Water Quality Technology Conference*, American Water Works Association, Denver, Colorado.

Atherholt, T.B., M.W. LeChevallier, W.D. Norton, and J.S. Rosen (1998). "Effect of rainfall on *Giardia* and *Crypto.*" *Journal American Water Works Association*, 90(9): 66-80.

Bales, R.C., C.P. Gerba, G.H. Grondin, and S.L. Jensen (1989). "Bacteriophage transport in sandy soil and fractured tuff." *Applied and Environmental Microbiology*, 55: 2061-2067.

Bales, R.C., S.R. Hinkle, T.W. Kroeger, and K. Stocking (1991). "Bacteriophage adsorption during transport through porous media: Chemical perturbations and reversibility." *Environmental Science and Technology*, 25: 2088-2095.

Bales, R.C., S. Li, K.M. Maguire, M.T. Yahya, and C.P. Gerba (1993). "MS-2 and poliovirus transport in porous media: Hydrophobic effects and chemical perturbations." *Water Resources Research*, 29: 957-963.

Bales, R.C., S. Li, K.M. Maguire, M.T. Yahya, C.P. Gerba, and R.W. Harvey (1995). "Virus and bacteria transport in a sandy aquifer, Cape Cod, MA." *Ground Water*, 33: 653-661.

Bales, R.C., S. Li, T.C.J. Yeh, M.E. Lenczewski, and C.P. Gerba (1997). "Bacteriophage and microsphere transport in saturated porous media: Forced-gradient experiment at Borden, Ontario." *Water Resources Research*, 33: 639-648.

Bellamy, W.D., D.W. Hendricks, and G.S. Logsdon (1985). "Slow sand filtration: Influences of selected process variables." *Journal American Water Works Association*, 77(12): 62-66.

Berger, P. (2002). "Removal of *Cryptosporidium* using bank filtration." *Riverbank Filtration: Understanding Contaminant Biogeochemistry and Pathogen Removal*, C. Ray ed., NATO Science Series, Kluwer Academic Publishers, Dordrecht, The Netherlands, 85-121.

Bergmire-Sweat, D., K. Wilson, L. Marengo, Y.M. Lee, W.R. MacKenzie, J. Morgan, K. Von Alt, T. Bennett, V.C.W. Tsang, and B. Furness (1999). "Cryptosporidiosis in Brush Creek: Describing the epidemiology and causes of a large outbreak in Texas, 1998." *Proceedings, International Conference on Emerging Infectious Diseases*, American Water Works Association, Denver, Colorado.

Bitton, G. (1980). *Introduction to environmental virology*. John Wiley & Sons, New York, New York.

Bitton, G., S.R. Farrah, R.H. Ruskin, J. Butner, and Y.J. Chou (1983). "Survival of pathogenic and indicator organisms in groundwater." *Ground Water*, 21: 405-410.

Bitton, G., J.E. Maruniak, and F.W. Zettler (1987). "Virus survival in natural ecosystems." *Survival and dormancy of microorganisms*, Y. Henis, ed., John Wiley & Sons, New York.

Black, R.E., M.M. Levine, M.L. Clements, T.P. Hughes, and M.J. Blaser (1988). "Experimental *Campylobacter jejuni* infection in humans." *Journal of Infectious Diseases*, 157(3): 472-479.

Blanc, R., and A. Nasser (1996). "Effect of effluent quality and temperature on the persistence of viruses in soil." *Water Science and Technology*, 33: 237-242.

Bourg, A.C.M., D. Darmendrail, and J. Ricour (1989). "Geochemical filtration of riverbank and migration of heavy metals between the Deule River and the Ansereuilles Alluvion-Chalk Aquifer (Nord, France)." *Geoderma*, 44: 229-244.

Bouwer, E.J., and B.E. Rittmann (1992). "Comment on 'Use of colloid filtration theory in modeling movement of bacteria through a contaminated aquifer.'" *Environmental Science and Technology*, 26: 400-401.

Bradford, S.M., A.W. Bradford, and C.P. Gerba (1993). "Virus transport through saturated soils." *Weiner Mitteilungen Wien*, 12: 143-147.

Bruesch, M.E., P. Biedrzycki, D. Gieryn, A. Singh, M.S. Gradus, K. Blair, and J. MacDonald (1999). "Baseline occurrence and distribution of *Cryptosporidium* and *Giardia* in the Milwaukee River watershed." *Proceedings, Water Quality Technology Conference*, American Water Works Association, Denver, Colorado.

Brush, C.F., M.F. Walter, J.L. Anguish, and W.C. Ghiorse (1998). "Influence of pretreatment and experimental conditions on electrophoretic mobility and hydrophobicity of *Cryptosporidium parvum* oocysts." *Applied and Environmental Microbiology*, 64(11): 4439-4445.

Brush, C.F., W.C. Ghiorse, L.J. Anguish, J. Parlange, and H.G. Grimes (1999). "Transport of *Cryptosporidium parvum* oocysts through saturated columns." *Journal of Environmental Quality*, 28: 809-815.

Caldentey, J., J.K.H. Bamford, and D.H. Banford (1990). "Structure and assembly of bacteriophage PRD-1, an *Escherichia coli* virus with a membrane." *Journal of Structural Biology*, 104: 44-51.

Camper, A.K., J.T. Hayes, P.J. Sturman, W.L. Jones, and A.B. Cunningham (1993). "Effects of motility and adsorption rate coefficients on transport of bacteria through saturated porous media." *Applied and Environmental Microbiology*, 59: 3455-3462.

Carmichael, W.W., C.L.A. Jones, N.A. Mahmood, and W.C. Theiss (1985). "Algal toxins and water-based diseases." *CRC Critical Reviews in Environmental Control*, 15: 275-313.

Carmichael, W.W., V.R. Beasley, D.L. Bunner, J.N. Eloff, I. Falconer, P. Gorham, K.I. Harada, T. Krisnamurthy, M.J. Yu, R.E. Moore, K. Rinehart, M. Runnegar, O.M. Skulberg, and M. Watanabe (1988). "Naming of cyclic heptapeptide toxins of cyanobacteria (blue-gree algae)." *Toxicon*, 26: 971-973.

Carmichael, W.W. (1992). "Cyanobacteria secondary metabolites — The cyanotoxins." *Journal of Applied Bacteriology*, 72: 445-459.

Commissie Bescherming Waterwingebieden (CBW) (1980). *Richtlijnen en aanbevelingen voor de bescherming van waterwingebieden*. Vereniging Van Exploitanten van Waterleidingbedrijven in Nederland (VEWIN)-Rijksinstituut voor drinkwatervoorziening (RID), Rijswijk, The Netherlands (In Dutch).

CH2M Hill (1993). *Final Report on the Russian River Demonstration Study*. Report available from the Sonoma County Water Agency, Santa Rosa, California.

Chauret, C., P. Chen, S. Springthorpe, and S. Sattar (1995). "Effect of environmental stressors on the survival of *Cryptosporidium* oocysts." *Proceedings, Water Quality Technology Conference*, American Water Works Association, Denver, Colorado.

Clancy, J.L., and D. Stendahl (1996). "Ground water or surface water — Microscopic evaluation of an Ontario River well system." *Proceedings, Water Quality Technology Conference*, American Water Works Association, Denver, Colorado.

Conrad, L.P., and M.S. Beljin (1996). "Evaluation of an induced infiltration model as applied to glacial aquifer systems." *Journal American Water Resources Association*, 32(6): 1209-1220.

Corapcioglu, M.Y., and A. Haridas (1984). "Transport and fate of microorganisms in porous media: A theoretical investigation." *Journal of Hydrology*, 72: 149-169.

Cossins, F.A., W.D. Gollnitz, J. DeMarco, D.J. Hartman, D.H. Metz, and J. Swertfeger (1997). "The Cincinnati Water Works' development of a Groundwater Parasite Monitoring Program and analytical results." *Proceedings, International Symposium on Waterborne Cryptosporidium*, American Water Works Association, Denver, Colorado, 195-199.

Cotte, L., M. Rabodonirina, F. Chapuis, F. Bailly, F. Bisseul, C. Raynal, P. Gelas, F. Persat, M. Piens, and C. Trepo (1999). "Waterborne outbreak of intestinal microsporidiosis in persons with and without human immunodeficiency virus infection." *Journal of Infectious Diseases*, 180: 2003-2008.

Craun, G.F. (1986). *Waterborne diseases in the United States*. CRC Press, Boca Raton, Florida.

Craun, G.F., and R.L. Calderon (1996). "Microbial risks in ground water systems: Epidemiology of waterborne outbreaks." *Proceedings, The Ground Water Foundations' 12th Annual Fall Symposium*, American Water Works Association, Denver, Colorado, 9-15.

Craun, G.F., S.A Hubbs, F. Frost, R.L. Calderon, and S.H. Via (1998). "Waterborne outbreaks of Cryptosporidiosis." *Journal American Water Works Association*, 90(9): 81-91.

Crook, J. (1985). "Water reuse in California." *Journal American Water Works Association*, 77(7): 60-71.

D'Antonio, R.G., R.E. Winn, and J.P. Taylor (1985). "A Waterborne outbreak of Cryptosporidiosis in normal hosts." *Annals of International Medicine*, 103(6-1): 886-888.

Davis, B.D., R. Dulbecco, H.N. Eisen, H.S. Ginsberg, and W.B. Wood (1970). *Microbiology*, (Second Edition). A Harper International, New York, New York.

DeBorde, D.C., W.W. Woessner, B. Lauerman, and P.N. Ball (1998). "Virus occurrence in a school septic system and unconfined aquifer." *Ground Water*, 36: 925-834.

DeBorde, D.C., W.W. Woessner, Q.T. Kiley, and P. N. Ball (1999). "Rapid transport of viruses in a floodplain aquifer." *Water Resources*, 33: 2229-2238.

Dillon Consulting Ltd. (1997). *Final Draft of the Regional Municipality of Waterloo Grand Reservoir and Pumping Station Project (95-2669)*. Unpublished report to the Regional Municipality of Waterloo, Waterloo, Canada.

Dizer, H., A. Nasser, and J.M. Lopez (1984). "Penetration of different human pathogenic viruses into sand columns percolated with distilled water, groundwater, or wastewater." *Applied and Environmental Microbiology*, 47: 409-415.

Dowd, S.E., S.D. Pillai, S. Wang, and M.Y. Corapcioglu (1998). "Delineating the specific influence of virus isoelectric point and size on virus adsorption and transport through sandy soils." *Applied and Environmental Microbiology*, 64: 405-410.

Dowd, S.E., C.P. Gerba, and I. Pepper (1998a). "Confirmation of the human-pathogenic microsporidia *Enterocytozoon bieneusi*, *Encephalitozoon intestinalis*, and *Vittaforma corneae* in water." *Applied and Environmental Microbiology*, 64(9): 3332-3335.

Dworkin, M.S., D.P. Goldman, T.G. Wells, J.M. Kobyashi, and B. Herwaldt (1996). "Cryptosporidiosis in Washington State: An outbreak associated with well water." *Journal of Infectious Diseases*, 174: 1372-1376.

Edberg, S.C., E.W. Rice, R.J. Karlin, and M.J. Allen (2000). "*Escherichia coli*: The best biological drinking water indicator for public health protection." *Journal of Applied Microbiology*, S6: 106S-116S.

Edler, L., S. Fernö, M.G. Lind, R. Lundberg, and P.O. Nilsson (1985). "Mortality of dogs associated with blooms of cyanobacterium *Nodularia spumigena* in the Baltic Sea." *Ophelia*, 24: 103-109.

Eisenberg, J.N.S., E.Y.W. Seto, J.M. Colford, Jr., A. Olivieri, and R.C. Spear (1998). "An analysis of the Milwaukee Cryptosporidiosis outbreak based on a dynamic model of the infection process." *Epidemiology*, 9(3): 255-263.

European Union (1998). Council Directive 98/83/EC on the quality of water intended for human consumption.

Farrah, S.R., and D.R. Preston (1993). "Adsorptions of viruses to sand modified by *in situ* precipitation of metallic salts." *Wiener Mitteilungen Wein*, 12: 25-29.

Fayer, R., J.M. Trout, T.K. Graczyk, C.A. Farley, and E.J. Lewis (1997). "The potential role of oysters and waterfowl in the complex epidemiology of *Cryptosporidium parvum*." *Proceedings, International Symposium on Waterborne Cryptosporidium*, American Water Works Association, Denver, Colorado, 153-158.

Frost, F.J., G.F. Craun, and R.L. Calderon (1996). "Waterborne disease surveillance." *Journal American Water Works Association*, 88: 66-75.

Fujito, B.T., and C.D. Lytle (1996). "Elution of viruses by ionic and nonionic surfactants." *Applied and Environmental Microbiology* 62: 3470-3473.

Furtado, C., G.K. Adak, J.M. Stuart, P.G. Wall, H.S. Evans, and D.P. Casemore (1998). "Outbreaks of waterborne infectious intestinal disease in England and Wales, 1992-5." *Epidemiology and Infection* (121): 109-119.

Gannon, J., Y. Tan, P. Baveye, and M. Alexander (1991). "Effects of sodium chloride on transport of bacteria in saturated aquifer material." *Applied and Environmental Microbiology*, 57: 2497-2501.

Geldreich, E.E. (1996). "Pathogenic agents in freshwater resources." *Hydrological Processes*, 10: 315-333.

Gerba, C.P. (1984). "Applied and theoretical aspects of virus adsorption to surfaces." *Advances in Applied Microbiology*, 30: 133-168.

Gerba, C.P. (1985). "Microbial contamination of the subsurface." *Ground water quality*, C.H. Ward, W. Giger, and P.L. McCarty, eds., John Wiley & Sons, New York, New York.

Gerba, C.P., and G. Bitton (1984). "Microbial pollutants: Their survival and transport pattern to groundwater." *Groundwater pollution microbiology*, G. Bitton and C.P. Gerba, eds., John Wiley & Sons, New York, New York, 66-88.

Gerba, C.P., J.B. Rose, and C.N. Hass (1996). "Sensitive populations: Who is at the greatest risk?" *International Journal Food Microbiology*, 30: 113.

Gollnitz, W.D., J.L. Clancy, and S.C. Garner (1997). "Reduction of microscopic particulates by aquifers." *Journal American Water Works Association*, 89(11): 84-93.

Gollnitz, W.D., F. Cossins, D. Hartman, and J. DeMarco (1997a). "Impact of induced infiltration on microbial transport in an alluvial aquifer." *Proceedings, Water Quality Technology Conference*, American Water Works Association, Denver, Colorado.

Gorham, P.R., and W.W. Carmichael (1980). "Toxic substances from freshwater algae." *Progress in Water Technology*, 12: 189-198.

Goyal S.M., and C.P. Gerba (1979). "Comparative adsorption of human enteroviruses, simian rotavirus, and selected bacteriophages to soils." *Applied and Environmental Microbiology*, 38: 241-247.

Gray, M.J. (1998). *Assessment of water supply and associated matters in relation to the incidence of Cryptosporidiosis in West Herts and North London in February and March 1997*. United Kingdom Drinking Water Inspectorate, London, England.

Hancock, C.M., J.B. Rose, and M. Callahan (1998). "Crypto and *Giardia* in U.S. groundwater." *Journal American Water Works Association*, 90(3): 58-61.

Harter, T., S. Wagner, and E.R. Atwill (2000). "Colloid transport and filtration of *Cryptosporidium parvum* in sandy soils and aquifer sediments." *Environmental Science and Technology*, 34: 62-70.

Harvey, R.W. (1991). "Parameters involved in modeling movement of bacteria in groundwater." *Modeling the environmental fate of microorganisms*, C.J. Hurst, ed., American Society for Microbiology, Washington, D.C., 89-114.

Harvey, R.W., and S.P. Garabedian (1991). "Use of colloid filtration theory in modeling movement of bacteria through a contaminated sandy aquifer." *Environmental Science and Technology*, 25(1): 178-185.

Harvey, R.W., N.E. Kinner, D. MacDonald, D.W. Metge, and A. Bunn (1993). "Role of physical heterogeneity in the interpretation of small-scale laboratory and field observation of bacteria, microbial-sized microsphere, and bromide transport through aquifer sediments." *Water Resources Research*, 29(8): 2713-2721.

Harvey, R.W., D.W. Metge, N. Kinner, and N. Mayberry (1997). "Physiological considerations in applying laboratory-determined buoyant densities to predictions of bacterial and protozoan transport in groundwater: Results of in situ and laboratory tests." Environmental Science and Technology, 31(1): 289-295.

Havelaar, A.H. (1986). "F-specific bacteriophages as model viruses in water treatment processes." Ph.D. thesis, University of Utrecht, Utrecht, The Netherlands, 240 p.

Havelaar, A.H., M. Van Olphen, and J.F. Schijven (1994). "Removal and inactivation of viruses by drinking-water treatment processes under full-scale conditions." Proceedings, 17th Biannual International Conference, International Association of Water Quality (now International Water Association), London, England, 233-240.

Havelaar, A.H. (1993). "Bacteriophages as models of enteric viruses in the environment." ASM News, 59: 614-619.

Havelaar, A.H., M. Van Olphen, and Y.C. Drost (1993). "F-specific RNA bacteriophages are adequate model organisms for enteric viruses in fresh water." Applied and Environmental Microbiology, 59: 2956-2962.

Havelaar, A.H., M. Van Olphen, and J.F. Schijven (1995). "Removal and inactivation of viruses by drinking-water treatment processes under full-scale conditions." Water Science and Technology, 31: 55-62.

Hejkal, T.W., F.M. Wellings, A.L. Lewis, and P.A. LaRock (1981). "Distribution of viruses associated with particles in wastewater." Applied and Environmental Microbiology, 41: 628-634.

Heinemann, T.J., W.D. Bellamy, K.W. Stocker, M.A. Bowman, and J.C. Fischer (1996). Pursuing alternative treatment credit for a groundwater under the direct influence of surface water: Case study for Kearney, Nebraska. Unpublished CH2M Hill report for the City of Kearney, available from the Nebraska Department of Health, Lincoln, Nebraska.

Heisz, M., C. Chauret, P. Chen, S. Springthorpe, and S.A. Sattar (1997). "In vitro survival of Cryptosporidium oocysts in natural waters." Proceedings, International Symposium on Waterborne Cryptosporidium, American Water Works Association, Denver, Colorado, 71-175.

Hendry, M.J., J.R. Lawrence, and P. Maloszewski (1997). "The role of sorption in the transport of Klebsiella oxytoca through saturated silica sand." Ground Water, 35: 575-584.

Henis, Y. (1987). Survival and dormancy of bacteria in survival and dormancy of microorganisms, Y. Henis, ed., John Wiley & Sons, New York, New York.

Herbold-Paschke, K., U. Straub, T. Hahn, G. Teutsch, and K.Botzenhart (1991). "Behaviour of pathogenic bacteria, phages, and viruses in groundwater during transport and adsorption." Water Science and Technology, 24: 301-304.

Herzig, J.P., D.M. Leclerc, and P. Le Goff (1970). "Flow of suspensions through porous media-application to deep filtration." Industrial and Engineering Chemistry, 62: 8-35.

Hirata, T., and A. Hashimoto (1997). "A field survey on occurrence of Giardia cysts and Cryptosporidium oocysts in sewage treatment plants." Proceedings, International Symposium on Waterborne Cryptosporidium, American Water Works Association, Denver, Colorado, 183-193.

Hoogenboezem, W., H.A.M. Ketelaars, G.J. Medema, G.B.J. Rijs, and J.F. Schijven (2000). Cryptosporidium en Giardia: Vookomen in rioolwater, mest, en oppervlaktewater met Zwem-em drinkwaterfunctie (Cryptosporidium and Giardia: Presence in wastewater, manure, and in surface water for recreational purposes and for drinking water production). Association of River Waterworks (RIWA)/National Institute of Public Health and the Environment (RIVM)/Institute for Inland Water Management and Waste Water Treatment (RIZA), Bilthoven, The Netherlands (in Dutch).

Hornberger G.M., A.L. Mills, and J.S. Herman (1992). "Bacterial transport in porous media: Evaluation of a model using laboratory observations." Water Resources Research, 28: 915-938.

Hrudey, S., M. Burch, M. Drikas, and R. Gregory (1999). "Remedial measures." Toxic cyanobacteria in water: A guide to their public health consequences, monitoring and management, I. Chorus and J. Bartram, eds., The World Health Organization, E & FN Spon, London, England, 275-312.

Hult, A., U. Beckman-Sund, T. Möller, E. Willen, and B. Erlandsson (1997). Algtoxiner i sjö- och dricksvatten (Toxins of algae in lake and drinking water. Report 19/97). National Food Administration, Uppsala, Sweden (in Swedish).

Hunt, B. (1999). "Unsteady stream depletion from ground water pumping." Ground Water, 37(1): 98-102.

Hurst, C.J., C.P. Gerba, and I. Cech (1980). "Effects of environmental variables and soils characteristics on virus survival in soil." Applied and Environmental Microbiology, 40: 1067-1079.

Hurst, C.J., J. Mosher, and M.V. Yates (1997). "Modeling the compatibility of viral datasets." Proceedings, Water Quality Technology Conference, American Water Works Association, Denver, Colorado.

Hutchinson, M., and J.W. Ridgway (1977). "Microbiological aspects of drinking-water supplies." Aquatic Microbiology, F.A. Skinner and J.M. Shewan, eds., Academic Press, London, England.

Jansons, J., L.W. Edmonds, B. Speight, and M.R. Bucens (1989). "Survival of viruses in groundwater." Water Research, 23: 301-306.

Jin, Y., M.V. Yates, S.S. Thompson, and W.A. Jury (1997). "Sorption of viruses during flow through saturated sand columns." Environmental Science and Technology, 31: 548-55.

Keijola, A.M., K. Himberg, A.L. Esala, K. Sivonen, and L. Hiisvirta (1988). "Removal of cyanobacterial toxins in water treatment processes: Laboratory and pilot-plant experiments." Toxicity Assessment, 3(5): 643-656.

Keswick, B.H., C.P. Gerba, S.L. Secor, and I. Cech (1982). "Survival of enteric viruses and indicator bacteria in groundwater." Journal of Environmental Science and Health, A12: 903-912.

Kinoshita, T., R.C. Bales, K.M. Maguire, and C.P. Gerba (1993). "Effect of pH on bacteriophage transport through sandy soils." Journal of Contaminant Hydrology, 14: 55-70.

Knorr, N. (1937). "Die schutzzonenfrage in der trinkwasser-hygiene." Das Gas- und Wasserfach, 80: 330-355.

Korhonen, L.K., and P.J. Martikainen (1991). "Survival of Escherichia coli and Campylobacter jejuni in untreated and filtered lake water." Journal of Applied Bacteriology, 17: 379-382

Kuiper-Goodman T., I. Falconer, and J. Fitzgerald (1999). "Human health aspects." *Toxic cyanobacteria in water: A guide to their public health consequences, monitoring, and management*, I. Chorus and J. Bartram, eds., The World Health Organization, E&FN Spon, London, England, 113-153.

Lack, T. (1999). "Water and health in Europe: An overview." *British Medical Journal*, 318: 1678-1682.

Lahti, K., L. Lepistö, J. Niemi, and M. Färdig (1993). "Removal efficiency of algae and especially cyanobacteria in different waterworks." *Publications of the Water and Environmental Administration - Series A143*, National Board of Waters and the Environment, Helsinki, Finland (Abstract in English).

Lahti, K., J. Ahtiainen, J. Rapala, K. Sivonen, and S.I. Niemelä (1995). "Assessment of rapid bioassays for detecting cyanobacterial toxicity." *Letters in Applied Microbiology*, 21: 109-114.

Lahti, K., J. Silvonen, A. L. Kivimäki, and K. Erkomaa (1996). "Removal of cyanobacteria and their hepatotoxins from raw water in soil and sediments columns." *Proceedings of an International Symposium on Artificial Recharge of Groundwater*, Finnish Environment Institute, Helsinki, Finland, 187-195.

Lahti, K, J. Vaitomaa, A.L. Kivimäki, and K. Sivonen (1998). "Fate of cyanobacterial hepatotoxins in artificial recharge of groundwater and in bank filtration." *Artificial recharge of groundwater*, J.H. Peters et al., eds., A.A. Balkema, Rotterdam, The Netherlands.

Larkin, R.G., and J.M. Sharp, Jr. (1992). "On the relationship between river-basin geomorphology, aquifer hydraulics, and ground-water flow direction in alluvial aquifers." *Geological Society of America Bulletin*, 104: 1608-1620.

Laszlo, F., and F. Szekely (1989). "Modeling of groundwater flow and quality changes around bank filtration well fields." *Contaminant Transport in Groundwater*, Kobus and Kinzelbach, eds., A.A. Balkema, Rotterdam, The Netherlands.

Leland, D., J. McAnulty, W. Keene, and G. Stevens (1993). "A cryptosporidiosis outbreak in a filtered water supply." *Journal American Water Works Association*, 85:34-42.

Lepistö, L., K. Lahti, and J. Niemi (1994). "Removal of cyanobacteria and other phytoplankton in four Finnish waterworks." *Algological Studies*, 75: 167-181.

Levine, A.D., G. Tchobanoglous, and T. Asano (1985). "Characterization of the size distribution of contaminants in wastewater: Treatment and reused implications." *Journal Water Pollution Control Federation*, 57: 805-816.

Lisle, J.T., and J.B. Rose (1995). "*Cryptosporidium* contamination of water in the USA and UK: A Mini-review." *Journal of Water Supply, Research, and Technology – Aqua*, 44(3): 103-117.

Loveland, J.P., J.N. Ryan, G.L. Amy, and R.W. Harvey (1996). "The reversibility of virus attachment to mineral surfaces." *Colloids Surfaces A: Physicochemical and Engineering Aspects*, 107: 205-221.

Lund, B.M., and M.W. Peck (1994). "Heat resistance and recovery of spores of non-proteolytic *Clostridium botulinum* in relation to refrigerated, processed foods with an extended shelf-life." *Journal of Applied Bacteriology Symposium Supplement*, 76: 115S-128S.

Lytle, C.D., and L.B. Routson (1995). "Minimized virus binding for test barrier materials." *Applied and Environmental Microbiology*, 61: 643-649.

Martin, M.J., B.E. Logan, W.P. Johnson, D.G. Jewett, and R.G. Arnold (1996). "Scaling bacterial filtration rates in different sized porous media." *Journal of Environmental Engineering*, 122 :407-415.

Martin, R.E., E.J. Bouwer, and L.M. Hanna (1992). "Application of clean-bed filtration theory to bacteria deposition in porous media." *Environmental Science and Technology*, 26: 1053-1058.

Matthess, G., and A. Pekdeger (1981). "Concepts of a survival and transport model of pathogenic bacteria and viruses in groundwater." *Science of the Total Environment*, 21: 149-159.

Matthess, G., and A. Pekdeger (1985). "Survival and transport of pathogenic bacteria and viruses in ground water." *Water Quality*, C.H. Ward and P. McCarty, eds., John Wiley & Sons, Inc., New York, New York, 472-482.

Matthess, G., A. Pekdeger, and J. Schroeter (1988). "Persistence and transport of bacteria and viruses in groundwater – A conceptual evaluation." *Journal of Contaminant Hydrology*, 2: 171-188.

Maule, A. (2000). "Survival of verocytotoxigenic *Escherichia coli* O157 in soil, water, and on surfaces." *Journal of Applied Microbiology Symposium Supplement*, 88: 71S-78S.

Mawdsley, J., A.E. Brooks, and R.J. Merry (1996). "Movement of the protozoan pathogen *Cryptosporidium parvum* through three contrasting soil types." *Biology and Fertility of Soils*, 21: 30-36.

McCaulou, D.R., R.C. Bales, and J.F. McCarthy (1994). "Use of short pulse experiments to study bacteria transport through porous media." *Journal Contaminant Hydrology*, 15: 1-14.

McCaulou, D.R., R.C. Bales, and R.G. Arnold (1995). "Effect of temperature-controlled transport of bacteria and microspheres through saturated sediment." *Water Resources Research*, 31: 271-280.

McDowell-Boyer, L.M., J.R. Hunt, and N. Sitar (1986). "Particle transport through porous media." *Water Resources Research*, 22: 1901-1921.

McFeters, G.A., and D.G. Stuart (1972). "Survival of coliform bacteria in natural waters: Field and laboratory studies with membrane-filter chambers." *Applied Microbiology*, 25(5): 805-811.

McFeters, G.A., G.K. Bissonnette, J.J. Jewelski, C.A. Thomson, and D.G. Stuart (1974). "Comparative survival of indicator bacteria and enteric pathogens in well water." *Applied and Environmental Microbiology*, 27(5): 823-829.

McGinnis, J.A., and F. DeWalle (1983). "The movement of typhoid organisms in saturated, permeable soil." *Journal American Water Works Association*, 75(6): 266-271.

Medema, G.J., M. Bahar, and F.M. Schets (1997). "Survival of *Cryptosporidium parvum*, *Escherichia coli*, fecal enterococci, and *Clostridium perfringens* in river water: Influence of temperature and autochthonous microorganisms." *Water Science and Technology*, 35(11-12): 249-252.

Medema, G.J., P.M. Schets, P.F. M.Teunis, and A.H. Havelaar (1998). "Sedimentation of free and attached *Cryptosporidium* oocysts and *Giardia* cysts in water." *Applied and Environmental Microbiology*, 64(11): 4460-4466.

Medema, G.J., M.H.A. Juhasz-Hoterman, and J.A. Luitjen (2000). "Removal of microorganisms by bank filtration in a gravel-sand soil." *Proceedings, International Riverbank Filtration Conference*, Internationale Arbeitsgemeinschaft der Wasserwerke im Rheineinzugsgebiet (IAWR), Amsterdam, The Netherlands.

Medema, G.J., and A.H. Havelaar (1994). "Microorganisms in water: A health risk." *Report #289202002*, National Institute of Public Health and the Environment (RIVM), Bilthoven, The Netherlands (in Dutch).

Medema, G.J., H.A.M. Ketelaars, and W. Hoogenboezem (1996). "*Cryptosporidium* and *Giardia* in the Rhine and the Meuse." *Report 289202 015*, National Institute of Public Health and the Environment (RIVM), Bilthoven, The Netherlands (in Dutch).

Medema, G.J., J.F. Schijven, A.C.M. de Nijs, and J.G. Elzenga (1997). "Modeling of the discharge of *Cryptosporidium* and *Giardia* by domestic sewage and their dispersion in surface water." *Proceedings, International Symposium on Waterborne Cryptosporidium*, American Water Works Association, Denver, Colorado.

Messner, M.J., and R.L. Wolpert (2000). "Occurrence of *Cryptosporidium* in the nation's drinking water sources - ICR data analysis." *Proceedings, Water Quality Technology Conference*, American Water Works Association, Denver, Colorado.

Messner, M.J., C.L. Chappell, and P.C. Okhuysen (2001). "Risk assessment for *Cryptosporidium*: A hierarchial Bayesian analysis of human dose response data." *Water Research*, 35: 3934-3940.

Meschke, J.S., and Sobsey, M.D. (1999). "Comparative survival of Norwalk virus, poliovirus, and F+RNA coliphage MS-2 to soils suspended in groundwater." *Proceedings, Water Quality Technology Conference*, American Water Works Association, Denver, Colorado.

Metcalf, T.G., V.C. Rao, and J.L. Melnick (1984). "Solids associated viruses in a polluted estuary." *Monographs in Virology*, 15: 97-110.

Miettinen, I.T., T.K. Vartiainen, and P.J. Martikainen (1997). "Changes in bacterial biomass and production in ground water during bank filtration of lake water." *Canadian Journal of Microbiology*, 43(12): 1126-1132.

Miettinen, I.T., O. Zacheus, C.H. Von Bonsdorff, and T. Vartiainen (2000). "Waterborne epidemics in Finland 1998-99." *Proceedings, First World Water Congress of the International Water Association, The Symposium on Health Related Water Microbiology*, International Water Association, London, England.

Mikels, M.S. (1992). "Characterizing the influence of surface water on water produced by collector wells." *Journal American Water Works Association*, 84(9): 77-84.

Mirzoev, G.G. (1968). "Extent of survival of Dysentery Bacilli at low temperatures and self-disinfection of soil and water in the far north." *Gigiena I Sanitariya*, 33:437-439 (Summary in English).

Moore, A.C., B.L. Herwaldt, G.F. Craun, R.L. Calderon, A.K. Highsmith, and D.D. Juranek (1993). "Surveillance for waterborne disease outbreaks — United States, 1991-1992." *Morbidity and Mortality Weekly Report*, 42, *Surveillance Summary SS-5*, U.S Centers for Disease Control and Prevention, Atlanta, Georgia

Morris, B.L., and S.S.D. Foster (2000). "*Cryptosporidium* contamination hazard assessment and risk management for British groundwater sources." *Water Science and Technology*, 41(7): 67-77.

Mota, P., C.A. Rauch, and S.C. Edberg (2000). "*Microsporidia* and *Cyclospora*: Epidemiology and assessment of risk from the environment." *Critical Reviews of Microbiology*, 26(2): 69-90.

Moulton-Hancock, C., J.B. Rose, G.J. Vasconcelos, S.I. Harris, P.T. Klonicki, and G.D. Sturbaum (2000). "*Giardia* and *Cryptosporidium* occurrence in groundwater." *Journal American Water Works Association*, 92: 117-123.

Murphy, E.M., and T.R. Ginn (2000). "Modeling microbial processes in porous media." *Hydrogeology Journal*, 8: 142-158.

Mülschlegel, J.H.C., and F.J. Kragt, (1998). "Waterwinning en waterverbruik bij Doelgroepen." *Report #703717003*, National Institute of Public Health and the Environment (RIVM), Bilthoven, The Netherlands (in Dutch).

Nasser, A.M., Y. Tchorch, and B. Fattal (1993). "Comparative survival of *E. coli*, F+bacteriophages, HAV, and poliovirus 1 in wastewater and groundwater." *Water Science and Technology*, 27: 401-407.

Nasser, A.M., and S.D. Oman (1999). "Quantitative assessment of the inactivation of pathogens and indicator viruses in natural water sources." *Water Research*, 33: 1748-1752.

National Research Council (2000). "*Watershed management for potable water supply: Assessing the New York City strategy.*" National Academy Press, Washington, D.C.

Nelson, D.O. (1996). "Determination of groundwater under the direct influence of surface water: Observations regarding hydraulic connection, MPAs, and particle counting." Text and slides of a presentation at the American Water Works Association Conference in Bellevue, Washington, May 3, 1996, available from Dr. Dennis Nelson, Oregon Department of Human Resources, Portland, Oregon.

Neumann, N.F., L.L. Gyurek, L. Gammie, G.Finch, and M. Belosevic (2000). "Comparison of animal infectivity and nucleic acid staining for assessment of *Cryptosporidium parvum* viability in water." *Applied and Environmental Microbiology*, 66(1): 406-412.

Nieminski, E.C., W.D. Bellamy, and L.R. Moss (2000). "Using surrogates to improve plant performance." *Journal American Water Works Association*, 92(3): 67-78.

Novarino, G., A. Warren, A. Butler, G. Lambourne, A. Boxshall, J. Bateman, N.E. Kinner, R.W. Harvey, R.A. Mosse, and B. Teltsch (1997). "Protistan communities in aquifers: A review." *FEMS Microbiology Reviews*, 20: 261-275.

Okhuysen, P.C., C.L. Chappell, J.H. Crabb, C.R. Sterling, and H.L. DuPont (1999). "Virulence of three distinct *Cryptosporidium parvum* isolates for healthy adults." *Journal Infectious Disease*, 180: 1275-1281.

Olson, B.H. (1993). *Pathogen occurrence in source waters: factors affecting survival and growth in safety of water disinfection: Balancing chemical and microbial risks*, G.F. Craun, ed., ILSI Press, Washington D.C.

Ooi, W.W., S.K. Zimmerman, and C.A. Needham (1995). "Cyclospora species as a gastrointestinal pathogen in immunocompetent hosts." *Journal of Clinical Microbiology*, 33(5): 1267-1269.

Pang, L., M. Close, and M. Noonan (1998). "Rhodamine WT and *Bacillus subtilis* transport through an alluvial gravel aquifer." *Ground Water*, 36: 112-122.

Payment, P., E. Morin, and M. Trudel (1988). "Coliphages and enteric viruses in the particulate phase of river water." *Canadian Journal of Microbiology*, 34: 907-910.

Pedley, S., and G. Howard (1997). "The public health implications of microbiological contamination of groundwater." *Quarterly Journal Engineering Geology*, 30: 179-188.

Penrod, S.L., T.M. Olson, and S.B. Grant (1996). "Deposition kinetic of two viruses in packed beds of quartz granular media." *Langmuir*, 12: 5576-5587.

Persson, P., K. Sivonen, K. Keto, K. Kononen, M. Niemi, and H. Viljamaa (1984). "Potentially toxic blue-green algae (cyanobacteria) in Finnish natural waters." *Aqua Fennica*, 14(2): 147-154.

Pieper, A.P., J.N. Ryan, R.W. Harvey, G.L Amy, T.H. Illangasekare, and D.W. Metge (1997). "Transport and recovery of bacteriophage PRD-1 in a sand and gravel aquifer: Effect of sewage-derived organic matter." *Environment Science and Technology*, 31: 1163-1170.

Pilotto, L.S., R.M. Douglas, M.D. Burch, S. Cameron, M. Kirk, C.T. Cowie, S. Hardiman, C. Moore, and R.G. Attewell (1997). "Health effects of recreational exposure to cyanobacteria (blue-green algae) during recreational water-related activities." *Australian New Zealand Journal of Public Health*, 21: 562-566.

Powelson, D.K., J.R. Simpson, and C.P. Gerba (1990). "Virus transport and survival in saturated and unsaturated flow through soil columns." *Journal of Environmental Quality*, 19: 396-401.

Powelson, D.K., C.P. Gerba, and M. Yahya (1993). "Virus transport and removal in wastewater during aquifer recharge." *Water Research*, 27: 583-590.

Preuss, G. (1996). "Succession of microbial communities during bank filtration and artificial groundwater recharge." *Proceedings of an International Symposium on Artificial Recharge of Groundwater*, Finnish Environment Institute, Helsinki, Finland, 215-221.

Price, M.L., J. Flugum, P. Jeanne, and L. Tribbet-Peelen (1999). "Sonoma county finds groundwater under the direct influence of surface water depends on river conditions." *Abstracts, International Riverbank Filtration Conference*, National Water Research Institute, Fountain Valley, California, 25-27.

Ray, C., T.W.D. Soong, G.S. Roadcap, and D.K. Borah (1998). "Agricultural chemicals: Effects on wells during floods." *Journal American Water Works Association*, 90(7): 90-100.

Redman, J.A., S.B. Grant, T.M. Olson, M.E. Hardy, and M.K. Estes (1997). "Filtration of recombinant Norwalk virus particles and bacteriophage MS-2 in quartz sand: Importance of electrostatic interactions." *Environmental Science and Technology*, 31: 3378-3383.

Regli, S., J.B. Rose, C.N. Haas, and C.P. Gerba (1991). "Modeling the risk from *Giardia* and viruses in drinking water." *Journal American Water Works Association*, 213: 76-84.

Rice, E.W., C.H. Johnson, D.K. Wild, and D.J. Reasoner (1992). "Survival of *Escherichia coli* O157:H7 in drinking water associated with a waterborne disease outbreak of hemorrhagic colitis." *Letters in Applied Microbiology*, 13: 38-40.

Rijnaarts, H.H.M., W. Norde, E.J. Bouwer, J. Lyklema, and A.J.B. Zehnder (1993). "Bacterial adhesion under static and dynamic conditions." *Applied and Environmental Microbiology*, 59: 3255-3265.

Rijnaarts, H.H.M., W. Norde, E.J. Bouwer, J.L. Lyklema, and A.J.B. Zehnder (1995). "Reversibility and mechanism of bacterial adhesion." *Colloids and Surfaces B: Biointerfaces*, 4: 5-22.

Robertson, L.J., A.T. Campbell, and H.V. Smith (1992). "Survival of *Cryptosporidium parvum* oocysts under various environmental pressures." *Applied and Environmental Microbiology*, 58: 3494-3500.

Robertson, L.J., C.A. Paton, A.T. Campbell, P.G. Smith, M.H. Jackson, R.A. Gilmour, S.E. Black, D.A. Stevenson, and H.V. Smith (2000). "*Giardia* cysts and *Cryptosporidium* oocysts at sewage treatment works in Scotland, U.K." *Water Research*, 34(8): 2310-2322.

Robinson, D.A. (1981). "Infective dose of *Campylobacter jejuni* in milk." *British Medical Journal*, 282: 1584.

Rose, J.B., C.P. Gerba, and W. Jakubowski (1991). "Survey of potable water supplies for *Cryptosporidium* and *Giardia*." *Environmental Science and Technology*, 25(8): 1393-1400.

Rosen, J.S., M.S. LeChevallier, and A. Roberson (1996). "Development and analysis of a national protozoa database." *Proceedings, Water Quality Technology Conference*, American Water Works Association, Denver, Colorado.

Rosen, J.S., and B. Ellis (2000). "The bottom line on the ICR microbial data." *Proceedings, Water Quality Technology Conference*, American Water Works Association, Denver, Colorado.

Ryan, J.N., and M. Elimelech (1996). "Colloid mobilization and transport in groundwater." *Colloids Surfaces A: Physicochemical Engineering Aspects*, 107: 1-56.

Schijven, J.F., W. Hoogenboezem, P.J. Nobel, G.J. Medema, and A. Stakelbeek (1998). "Reduction of FRNA-bacteriophages and faecal indicator bacteria by dune infiltration and estimation of sticking efficiencies." *Water Science and Technology*, 38: 127-131.

Schijven, J.F., W. Hoogenboezem, S.M. Hassanizadeh, and J.H. Peters (1999). "Modeling removal of bacteriophages MS-2 and PRD-1 by dune recharge at Castricum, The Netherlands." *Water Resources Research*, 35: 1101-1111.

Schijven, J.F., G.J. Medema, A.J. Vogelaar, and S.M. Hassanizadeh (2000). "Removal of microorganisms by deep well injection." *Journal of Contaminant Hydrology*, 44: 301-327.

Schijven, J.F., and S.M. Hassanizadeh (2000). "Removal of viruses by soil passage: Overview of modeling, processes, and parameters: Critical reviews." *Environmental Science and Technology*, 30: 49-127.

Schijven, J.F., J.A. Annema, A.C.M. de Nijs, J.J.H. Theunissen, and G.J. Medema (1995). "Enteroviruses in surface waters in The Netherlands - Emission and distribution calculated with PROMISE and WATNAT: A pilot study." *Report #289202006*, National Institute of Public Health and the Environment (RIVM), Bilthoven, The Netherlands (in Dutch).

Schijven, J.F., G.J. Medema, A.C.M. de Nijs, and J.G. Elzenga (1996). "Emission and distribution of *Cryptosporidium*, *Giardia*, and enteroviruses via domestic wastewater." *Report #289202 014*, National Institute of Public Health and the Environment (RIVM), Bilthoven, The Netherlands (in Dutch).

Schijven J.F., H.A.M. de Bruin, G.B. Engels, and E.J.T.M. Leenen (1999a). "Emission of *Cryptosporidium* and *Giardia* by domestic farm animals." *Report #289202 023*, National Institute of Public Health and the Environment (RIVM), Bilthoven, The Netherlands (in Dutch).

Scholl, M.A., and R.W. Harvey (1992). "Laboratory investigations on the role of sediment surface and groundwater chemistry in transport of bacteria through a contaminated sandy aquifer." *Environment, Science and Technology*, 26: 1410-1417.

Schubert, J. (2000). "Entfernung von Schwebstoffen und Mikroorganismen sowie Verminderung der Mutagenitat bei der Uferfiltration." *Gas- und Wasserfach Wasser/Abwasser*, 141(4): 218-225 (in German).

Shields, P.A., and S.R. Farrah (1987). "Determination of the electrostatic and hydrophobic character of enteroviruses and bacteriophages." *Program Abstracts, 87th Annual Meeting of the American Society for Microbiology*, American Society for Microbiology, Washington D.C.

Sinton, L.W., R.K. Finlay, L. Pang, and D.M. Scott (1997). "Transport of bacteria and bacteriophages in irrigated effluent into and through an alluvial gravel aquifer." *Water Air Soil Pollution*, 98: 17-42.

Sivonen, K., S.I. Niemelä, R.M. Niemi, L. Lepistö, T.H. Luoma, and L.A. Räsänen (1990). "Toxic cyanobacteria (blue-green algae) in Finnish fresh and coastal waters." *Hydrobiologica*, 190: 267-275.

Sobsey, M.D., R.M. Hall, and R.L. Hazard (1995). "Comparative reductions of hepatitis A virus, enteroviruses, and coliphage MS-2 in miniature soil columns." *Water Science and Technology*, 31: 203-209.

Stegeman, H., H.M.C. Put, and L.P.M. Langeveld (1980). *Sporevormende bacterien in voedingsmiddelen*. Centrum voor landbouwpublicaties en landbouwdocumentatie, Wageningen, The Netherlands.

Stuyfzand, P.J., and F. Lüers (1996). *Gedrag van milieugevaarlijke stoffen bij oeverfiltratie en kunstmatige infiltratie. Effecten van bodempassage gemeten langs stroombanen*. Mededelina 125 (Behavior of environmental pollutants during bank filtration and artificial recharge. Effects of soil passage measured along flow paths [Report 125]), Kiwa, Niewegein, The Netherlands (in Dutch).

Theunissen, J.J.H., P.J. Nobel, R. Van de Heide, H.A.M. de Bruin, D. Van Veendendaal, W.J. Lodder, J.F. Schijven, G.J. Medema, and D. Van de Kooij (1998). "Enterovirus concentrations at intake points for drinking-water production." *Report #289202 013*, National Institute of Public Health and the Environment (RIVM), Bilthoven, The Netherlands (in Dutch).

Tisa, L.S., T. Koshikawa, and P. Gerhardt (1982). "Wet and dry bacterial spore densities determined by buoyant sedimentation." *Applied and Environmental Microbiology*, 43: 1307-1310.

Terzieva, S.I., and G.A McFeters (1991). "Survival and injury of *Escherichia coli*, *Campylobacter jejuni*, and *Yersinia enterocolitica* in stream water." *Canadian Journal of Microbiology*, 37: 785-790.

U.S. Centers for Disease Control and Prevention (1997). *Cryptosporidium and water: A public health handbook*, Centers for Disease Control and Prevention, U.S. Department of Health and Human Services, Atlanta, Georgia.

U.S. Environmental Protection Agency (1989). "National primary drinking water regulations: Disinfection, turbidity, *Giardia lamblia*, viruses, legionella, and heterotrophic bacteria: Final rule." *Federal Register*, 54: 27544.

U.S. Environmental Protection Agency (1991). *Guidance manual for compliance with the filtration and disinfection requirements for public water supplies using surface water sources*. American Water Works Association, Denver, Colorado.

U.S. Environmental Protection Agency (2000). "Stage 2 Microbial and Disinfection Byproducts Federal Advisory Committee agreement in principle." *Federal Register*, 65: 83015.

U.S. Environmental Protection Agency (2000a). "ICR Auxiliary 1 Database, Version 5.0." *EPA 815-C-00-002*, U.S. Environmental Protection Agency, Washington, D.C.

Vaitomaa, J. (1998). "Fate of blue-green algae and their hepatotoxins during infiltration — Experiments with soil and sediment columns." *The Finnish Environment, 174*, The Finnish Environmental Institute, Helsinki, Finland (Abstract in English).

Van der Leeden, F., F.L. Troise, and D.K. Todd (1990). *The water encyclopedia*, (Second Edition). Lewis Publishers, Chelsea, Michigan.

Vasconcelos, J., and S. Harris (1992). "Consensus method for determining groundwaters under the direct influence of surface water using Microscopic Particulate Analysis (MPA)." *EPA 910/9-92-029*, U.S. Environmental Protection Agency, Washington, D.C.

Verstraeten, I.M., J.D. Carr, G.V. Steele, E.M. Thurman, K.C. Bastian, and D.F. Dormedy (1999). "Surface water-ground water interaction: Herbicide transport into municipal collector wells." *Journal of Environmental Quality*, 28: 1396-1405.

Vestal, J.R., and J.E. Hobbie (1988). "Microbial adaptations to extreme environments." *Microorganisms in action: Concepts and applications in microbial ecology*, J.M. Lynch and J.E. Hobbie, eds., Blackwell Scientific Publications, Oxford, United Kingdom.

VEWIN (1998). *Waterleidingstatistiek (1997) (Drinking Water Statistics [1997]).* Vereniging van Exploitanten van Waterleidingbedrijven (VEWIN), Rüswijk, The Netherlands (in Dutch).

Vincent, R.J. (1997). *Assessment of water supply and associated matters in relation to the incidence of cryptosporidiosis in Torbay in 1992.* United Kingdom Drinking Water Inspectorate, London, England.

VROM (1995). *Infectierisico van virussen en parasitaire protozoa via drinkwater* (Risk of infection by viruses and parasitic protozoa in drinking water). Note in preparation of a policy. Draft March 17, 1995. Directie DWL, VROM, The Hague, The Netherlands (in Dutch).

Waite, W.M. (1997). *Assessment of water supply and associated matters in relation to the incidence of cryptosporidiosis in Torbay in August and September 1995.* United Kingdom Drinking Water Inspectorate, London, England.

Walker, F.J., C.D. Montemagno, and M.B. Jenkins (1998). "Source water assessment and nonpoint sources of acute toxic contaminants: A review of research related to survival and transport of *Cryptosporidium parvum.*" *Water Resources Research,* 34(12): 3383-3392.

Walker, F.R., and J.R. Stedinger, (1999). "Fate and transport model of *Cryptosporidium.*" *Journal of Environmental Engineering,* 125: 325-333.

Wang, G., and M.P. Doyle, (1998). "Survival of enterhemorrhagic *Escherichia coli* O157:H7 in water." *Journal of Food Protection,* 61: 662-667.

Wang, J.Z., R. Song, and S.A. Hubbs (2000). "Particle removal through riverbank filtration process." *Proceedings, Water Quality Technology Conference,* American Water Works Association, Denver, Colorado.

Weiss, T.H., A.L. Mills, G.M. Hornberger, and J.S. Herman (1995)." Effect of bacterial cell shape on transport of bacteria in porous media." *Environmental Science and Technology,* 29(7): 1737-1740.

Welker, R., R. Porter, W.B. Pett, M.R. Provart, and M. Schwartz (1994). "Cryptosporidiosis outbreak in Kitchener-Waterloo: Assessment and future prevention." *Proceedings, American Water Works Association Annual Conference,* American Water Works Association, Denver, Colorado, 55-101.

Wilderer, P.A., U. Forstner, and O.R. Kuntschik (1985). "The role of riverbank filtration along the Rhine River for municipal and industrial water supply." *Artificial recharge of groundwater,* T. Asano, ed., Butterworth Publishers, Boston, Massachusetts, 509-528.

Willocks, L., A. Crampin, L. Milne, C. Seng, M. Susman, R. Gair, S. Shafi, R. Wall, R. Wiggins, and N. Lightfoot (1999). "A large outbreak of Cryptosporidiosis associated with a public-water supply from a deep chalk borehole." *Communicable Disease and Public Health,* 1: 239-243.

Wilson, J.L. (1993). "Induced infiltration in aquifers with ambient flow." *Water Resources Research,* 29(10): 3503-3512.

Wilson J.L., and W.L. Linderfelt (1991). "Groundwater quality in pumping wells located near surface water bodies." *Tech Completion Report 261,* New Mexico Water Resource Research Institute, Las Cruces, New Mexico.

World Health Organization (1986). *Guidelines for drinking-water quality,* Volume 2, Health criteria and other supporting information, World Health Organization, Geneva, Switzerland.

World Health Organization (1998). *Guidelines for drinking-water quality,* Second Edition. Addendum to Volume 2, Health Criteria and Other Supporting Information, World Health Organization, Geneva, Switzerland.

Xiao, L., U.M. Morgan, R. Fayer, R.C.A. Thompson, and A.A. Lal (2000). "*Cryptosporidium* systematics and implications for public health." *Parasitology Today,* 16(7): 287-292.

Yager, R.M. (1986). "Simulation of ground-water flow and infiltration from the Susquehanna River to a shallow aquifer at Kirkwood and Conklin, Broome County, New York." *Water-resources Investigations Report 86-4123,* U.S. Geological Survey, Denver, Colorado.

Yahya, M.T., L. Galsomies, C.P. Gerba, and R.C. Bales (1993). "Survival of bacteriophages MS-2 and PRD-1 in ground-water." *Water Science and Technology,* 27: 409-412.

Yamamoto, N., K. Urabe, M. Takaoka, K. Nakazawa, A. Gotoh, M. Haga, H. Fuchigami, I. Kimata, M. Iseki (2000). "Outbreak of Cryptosporidiosis after Contamination of the Public Water Supply in Saitama Prefecture, Japan, in 1996." *Kansenshogaku Zasshi (Journal Japanese Association for Infectious Diseases),* 74(6):518-526.

Yao, K.M., M.T. Habibian, and C.R. O'Melia (1971). "Water and wastewater filtration: Concepts and applications." *Environmental Science and Technology,* 5: 1105-1112.

Yates, M.V., S.R.Yates, J. Wagner, and C.P. Gerba (1987). "Modeling virus survival and transport in the subsurface." *Journal of Contaminant Hydrology,* 1: 329-345.

Yates, M.V., C.P. Gerba, and L.M. Kelley (1985). "Virus persistence in groundwater." *Applied and Environmental Microbiology,* 49: 778-781.

Zabel, T.F., and J.D. Melbourne (1980). *Flotation in developments in water treatment-1,* W.M. Lewis, ed., Applied Science Publishers Ltd., London, England.

Zhan, H., and J. Cao (2000). "Analytical and semi-analytical solutions of horizontal well capture times under no-flow and constant-head boundaries." *Advances in Water Resources,* 23: 835-848.

Chapter 7. Riverbank Filtration Case Study at Louisville, Kentucky

Jack Wang, Ph.D.
Louisville Water Company
Louisville, Kentucky, United States

1. Introduction

RBF is the result of a process of collecting water from wells or filtration galleries that are recharged by a river that flows through alluvial valleys in which the wells or filtration galleries are located. It is a natural process that has been used for more than 100 years. RBF has been shown to have a positive effect on many water-quality parameters (Sontheimer, 1980; Hubbs, 1981; Wang et al., 1995; National Water Research Institute, 1999; Kühn and Müller, 2000). It is an effective method for reducing turbidity and microbial contaminants in surface water, as particles and microorganisms may be removed from the aqueous phase by straining and attachment to the aquifer grains during soil passage. Microorganisms may also be inactivated during soil passage due to changes in their living environment.

RBF is also effective for removing NOM and organic contaminants in surface water. Adsorption and biodegradation are the primary mechanisms in RBF that lead to a reduction of NOM, trace organic contaminants, and other contaminants that traditional drinking-water treatment processes target (Kuhlmann and Kaczmarzcyk, 1995). Depending on their biodegradability, anthropogenic contaminants are partly removed from surface water. In addition, groundwater dilution plays a significant role in the reduction of NOM through the RBF process. Literature on the removal of NOM and organic micropollutants through RBF is plentiful. RBF processes greatly reduce TOC, pesticides, and pharmaceuticals in source waters (Sontheimer, 1980; Wilderer et al., 1985; Wang et al., 1995; Kühn and Müller, 2000). RBF is also an effective process for reducing taste- and odor-causing compounds in drinking water, which may not be removed by conventional treatment methods (Juttner, 1995).

The RBF process has been used for more than 100 years in Europe (Kühn and Müller, 2000). The process has also been used in many communities in the United States, although it is not widely recognized. In the United States, the process is often considered as groundwater under the direct influence of surface water because the filtrate of the RBF process often blends with groundwater that is naturally present in the aquifer; however, by closely examining the data, it is clear that many facilities using groundwater under the direct influence of surface water facilities are actually using the RBF process. This chapter summarizes the results of RBF studies conducted at Louisville, Kentucky, in the United States.

2. Previous Studies Conducted at Louisville, Kentucky

Since as early as the 1940s, it has been recognized that there is a direct connection between the alluvial aquifer and the Ohio River at Louisville, Kentucky (Roragaugh, 1948). Under conditions of heavy groundwater pumping near the riverbank, water from the Ohio River is drawn into the aquifer, which results in induced RBF. The Louisville Water Company recognized induced

C. Ray et al. (eds.), Riverbank Filtration, 117–145.
© 2002 *Kluwer Academic Publishers. Printed in the Netherlands.*

RBF as a potentially effective treatment process for removing selected river-borne contaminants and started to investigate the effectiveness of RBF for removing disinfection byproduct precursors in the late 1970s (Hubbs, 1981).

In the late 1970s and early 1980s, Hubbs (1981) studied the reduction of TOC, THM formation potential (FP), and synthetic organic chemicals through the riverbank. In his study, Hubbs used a production well that was located about 82 m from the river with a pump rate ranging from 0.113 to 0.139 m³/s, and four sample wells around the production well to monitor water quality. Of the four sample wells, Sample Well 2 was located between the river and the production well, and was 23 m from the river. He found that water-quality parameters for Sample Well 2 closely followed those for river water, indicating induced RBF. Temperature profiles indicated that the water travel time from the river to Sample Well 2 was about 3 months under the pumping conditions mentioned above. Although Hubbs did not measure the percentage of water in Sample Well 2 that was contributed by induced RBF, the contribution can be estimated from the hardness data that he collected. The data indicated that at least 95 percent of the water in Sample Well 2 was from the Ohio River.

By comparing water-quality data collected from the river and Sample Well 2, Hubbs concluded that the concentration of general organic content in Ohio River water, as indicated by TOC and THM FP, was reduced by about 50 percent as a result of RBF. He speculated that the reduction was primarily associated with the removal of suspended sediment in the river.

Although the study showed the great potential of RBF as a water-treatment process, the company was not ready to pursue this alternative at that time; therefore, this research was discontinued at the Louisville Water Company after the early 1980s. In the early 1990s, however, many waterborne disease outbreaks attributed to *Cryptosporidium* in drinking water had raised serious concerns over the effectiveness of conventional water-treatment processes to produce safe drinking-water supplies; therefore, in 1993, the Louisville Water Company made a decision to reinvestigate the effectiveness of RBF as a treatment process for controlling disinfection byproduct precursors as well as microbial contaminants.

The second study began in April 1994 (Wang et al., 1995). The well field used in this study was the same as that used by Hubbs (1981) during the late 1970s and early 1980s with an exception in that an additional sample well was installed between the river and Sample Well 2. The new sample well (Well 5) was about 9 m from the river.

In the second study, Wang et al. (1995) monitored the effectiveness of RBF for removing pathogens, particles, turbidity, NOM, disinfection byproduct precursors, and pesticides. They also compared RBF with conventional treatment processes currently used by the Louisville Water Company. Wang et al. found that RBF is an effective process for particle and organic matter removal. Up to 2.4 logs of particle removal was achieved through RBF while only 1.5 logs of particle reduction was achieved through conventional treatment processes. They concluded that, under the conditions tested, RBF without any other post-treatment processes could produce water with similar quality (turbidity and total particles) as compared to the quality of the finished water at the Louisville Water Company, which has well-operated conventional treatment plants.

In their study, Wang et al. (1995) found that the RBF process can greatly reduce chlorine demand as well as THM and HAA FPs in river water. Their results indicated that a 50-percent reduction in disinfection byproduct FPs was achieved by RBF. Also, they found that RBF could reduce as much as 60 percent of TOC in river water, while conventional treatment processes removed about 25 percent of TOC. This suggested that the reduction of TOC was not necessarily associated with the removal of suspended sediment in the river. Physical-chemical interaction or biological activities in the riverbank may play an important role in the reduction of TOC.

In addition to NOM and disinfection byproduct precursor removal, the RBF process can effectively remove herbicides present in river water. Wang et al. (1995) found no detection of atrazine and alachlor in the infiltrated water, while the concentrations in river water were sometimes as high as 3.0 μg/L.

The most important finding from the study by Wang et al. (1995) was that water quality in Sample Well 2 and in the new sample well are almost the same, although the new sample well was much closer to the river. This may suggest that beyond a critical depth, which may be shallow, the impact on water quality may not be significant.

In addition to the above studies, the Louisville Water Company also invited U.S. Environmental Protection Agency scientists to conduct a mutagenicity study for Louisville Water Company waters, including riverbank-infiltrated water from Sample Well 5 during the fourth quarter of 1995. In the study, concentrations of TOC, ultraviolet absorbance at 254-nm wavelength, assimilable organic carbon, aldehydes, THMs, HAAs, total organic halogen (TOX), total coliform and HPC, and mutagenicity were assessed for raw (river) water, chlorinated raw water, well water (from Sample Well 5), chlorinated well water, settled and finished water from the treatment plant, and water from the distribution systems. The results confirmed previous findings that a 50- to 60-percent reduction of NOM could be achieved through the RBF process. Additionally, the assimilable organic carbon level in riverbank-filtered water was significantly lower than that in the raw and finished waters (250 μg/L versus 400 to 470 μg/L). Although the concentration of *Pseudomonas fluorescens* (P-17) was similar in all tested waters, riverbank filtrate had a much lower concentration for Spirillum NOX. The study also showed that the mutagenicity of waters correlated well with TOX data.

Based on these studies, the Louisville Water Company decided to implement the RBF process at full-scale. In 1999, the company installed a 0.88-m³/s RBF demonstration well and associated monitoring facilities. Extensive water-quality monitoring was conducted during the first 2 years of operation and the results are presented in the following sections.

3. Description of the RBF Facility

The RBF facility is a horizontal collector well with a design capacity of 0.88 m³/s. This collector well is located on the bank of Ohio River at River Mile 592 (starting at the origin of the river, which is River Mile Zero), about 30 m from the river. It is a 6-m diameter caisson with seven horizontal laterals. The caisson is about 34-m deep from ground level. The length of each lateral is 73 m for the four laterals oriented towards the river and 61 m for the remaining time. The laterals are 30-cm diameter stainless-steel wire-wound screens and are installed about 24-m below ground level. The laterals extending under the riverbed are covered with about 15 m of aquifer material. The schematic of the collector well and the orientation of the laterals are shown in Figure 7-1.

Two types of monitoring were conducted around the demonstration facility: water-quality monitoring and hydraulic monitoring. Water-quality monitoring facilities included four area-wide water-quality sampling wells, the collector well and its individual laterals, and three sampling wells in the riverbed. Hydraulic-monitoring facilities include piezometer wells around the collector well and pressure/temperature transducers installed in the riverbed.

Figure 7-2 shows the locations of four area-wide water-quality sampling wells. The purpose of these wells was to obtain baseline groundwater quality and to assess the impact of groundwater dilution on riverbank-filtered water quality. As shown in the figure, M1 was installed across the river from the collector well to assess the potential impact of groundwater from the other side of the river (about 1,200 m from the collector well). This well was installed into upper portions of the

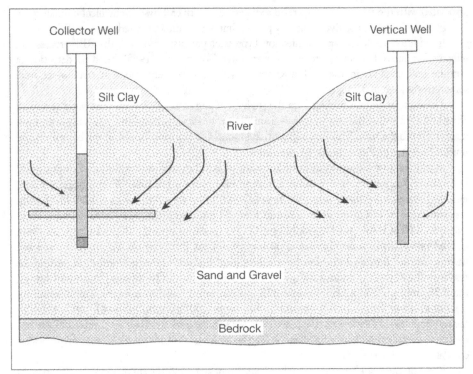

Figure 7-1. Schematic of the RBF process.

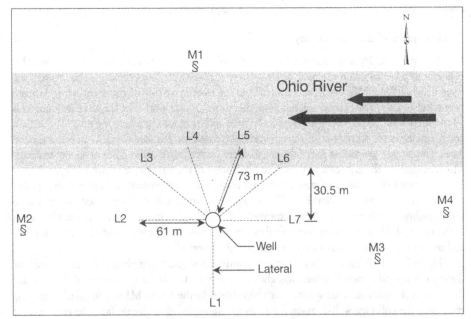

Figure 7-2. Locations of area-wide water-quality sampling wells. L = Laterals; M = Monitoring wells.

weathered bedrock of the area. Water-quality results from this well show the groundwater to be more characteristic of water found in limestone. M2 was located 152-m downstream from the collector well, between the collector well and the sludge lagoons of the water treatment plant. This well was installed to monitor the potential recharge from the lagoons. The well was also equipped with a pressure transducer that records water level and water temperature every 30 minutes. The third well, M3, was installed about 245-m upstream from the collector well, between the collector well and a potential contamination source (a residential septic tank). Both M2 and M3 were inside of the influence zone of the collector. The fourth well, M4, was located farther upstream from the collector well (about 600 m) and was not affected by the pumping of the collector well.

Three sampling wells were installed into the riverbed, about 23 m from the bank, at depths of 0.6 m (W1), 1.5 m (W2), and 2.75 m (W3) below the streambed to capture water-quality changes during the first 3 m of RBF. These sampling wells were installed directly above a lateral that extends about 15-m beneath the riverbed (L4) (see Figure 7-2). The sampling wells have a screen with a diameter of 5 cm and a length of 15 cm. The screen is connected to Tygon tubing, which is buried at the bottom of the river and is extended to the riverbank. A peristaltic pump was used to take water samples out of the sampling wells from the riverbank. In addition, transducers were installed parallel to the sampling wells at these depths. Temperature and hydraulic head data were obtained from the transducers. To collect samples from individual laterals, submersible sampling pumps were installed at the tip of three laterals of the collector well to collect water-quality samples from individual laterals. The laterals equipped with sampling pumps were Laterals 1, 2, and 4 (L1, L2, and L4).

4. Determination of Water Travel Time and Groundwater Dilution

To assess how effectively the RBF process improves water quality, one must determine both the water travel time through the process as well as the extent of groundwater mixing. This step is critical in RBF studies. First, the correct determination of water travel time is the key to appropriate sampling procedures with which the quality of river water and riverbank filtrate can be directly compared. Second, determining the effects of groundwater dilution will assist in calculating the true effectiveness of the RBF process. In the study, water-quality parameters, such as temperature and total hardness, were used as the tracer for determining water travel time and groundwater dilution.

River Water and Groundwater Characteristics

Although the river and aquifer are hydraulically connected (Rorabaugh, 1948, 1949), significant water-quality differences exist between the two water sources. Groundwater in the alluvial aquifer is a mixture of three components: subsurface flow from consolidated rocks along valley walls, groundwater that flows upward from limestone bedrock to the alluvial aquifer, and direct filtration of precipitation through the floodplain. Rorabaugh (1949) estimated the flow from the valley wall to be about 490 $m^3/d/km$ at this project location, and the recharge from rainfall on the flood plain to be about 15 cm/yr. Lyverse et al. (1996) estimated that the precipitation provided about 43 percent of the recharge, while the flow from the valley walls and upward from the limestone bedrock was about 39 percent. As a result of the flow from the bedrock, the groundwater has a much higher hardness than that from the river. Also, the temperature, turbidity, TOC concentration, and total hardness of Ohio River water change significantly during the year, depending on weather conditions and river flow, while those of the groundwater remain nearly constant.

Figure 7-3 shows river-water temperature results during 1999 and 2000. As expected, the river-water temperature changed seasonally, with the highest temperature occurring in the summer seasons and the lowest temperature occurring in the winter. During the 2-year period, the highest river-water temperature at the project site was 31.8°C and the lowest was 0.5°C, with an average 16.7°C. The 2-year river turbidity results are shown in Figure 7-4.

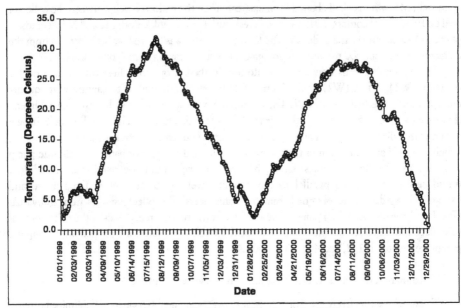

Figure 7-3. Ohio River water temperature profile (January 1999 to December 2000).

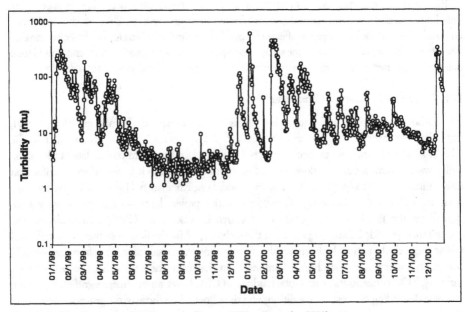

Figure 7-4. Ohio River water turbidity results (January 1999 to December 2000).

The turbidity of river water highly depends on river flow conditions. Normally, the river flows at a velocity of about 0.2 m/s. During the winter and spring seasons, turbidity is usually high due to increased river flows as a result of precipitation. It is common to have events where river turbidity is greater than 100 ntu during this period. The highest river turbidity recorded in the past 5 years was 1,500 ntu in March 1997, when the Ohio River flooded the area. River velocity was about 0.7 m/s at that time. In contrast, river turbidity is very low during the summer and fall seasons, usually below 5 ntu, due to low river flow conditions. During this period, the river is more like a lake due to the pool created by the McAlpine Dam (River Mile 603) and river flow is usually at 0.1 to 0.2 m/s.

The total hardness of river water also fluctuates seasonally, as shown in Figure 7-5, with the lowest hardness occurring during the spring and the highest occurring during October and November. The average total hardness of river water is about 145 mg/L as $CaCO_3$; however, the level can be as low as 90 mg/L as $CaCO_3$ during flooding and as high as 200 mg/L as $CaCO_3$ in November. The higher hardness level in the river during October and November is a result of water discharged from upstream reservoirs for the purpose of flood control and groundwater contribution to the river.

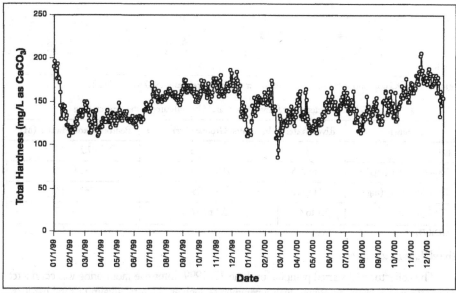

Figure 7-5. Ohio River water total hardness results (January 1999 to December 2000).

The TOC concentration of the Ohio River is usually between 2 to 4 mg/L. Figure 7-6 shows TOC data collected during the project period, including TOC results from the sampling wells in the riverbed and one of the laterals of the collector well.

Many samples were collected from sampling wells in the riverbed (W1, W2, and W3) before the collector well began production. Since the river functions as a gaining stream prior to the pumping of the collector well, the samples represented the conditions of ambient groundwater. Samples were also collected monthly from the four area-wide water-quality sampling wells for 6 months after the collector well had been pumping for 12 months. The samples collected from Wells M1 and M4 represented groundwater conditions as the sampling location was outside the cone of depression of the collector well. In addition, the U.S. Geological Survey maintains several monitoring sites in this area with automatic recording devices to record the groundwater level and temperature. The results show that groundwater in this region has a temperature of 14.5 to 15.5°C. The water-quality comparison between river water and groundwater is summarized in Table 7-1.

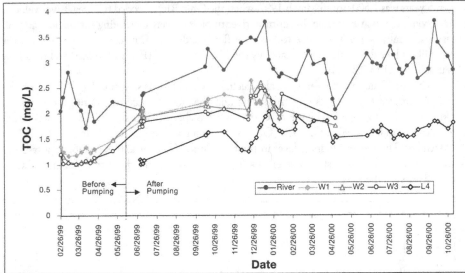

Figure 7-6. TOC concentrations of Ohio River water and RBF (February 1999 to October 2000).

Table 7-1. Characteristics of River Water and Groundwater in the Project Area

Parameters	River Water	Infiltrated Groundwater (M4)	Bedrock Groundwater (M1)
pH	7.7 to 7.9	7.4 to 7.5	7.2 to 7.3
Total Hardness (mg/L)[a]	90 to 205	280 to 290	530 to 582
Total Alkalinity (mg/L)[a]	50 to 110	235 to 250	260 to 280
TOC (mg/L)	2.1 to 4.9	0.3 to 0.6	0.4 to 0.7

[a] As $CaCO_3$.

Travel-Time Determination

The collector well started pumping on June 23, 1999. Intensive monitoring was conducted during the initial pumping. Water-quality parameters, such as pH, total alkalinity, total hardness, conductivity, and temperature, were analyzed to track changes in water quality during the initial pumping period. The change in water-quality characteristics served as a tracer for water traveling from the river to the collector well.

During the first 7 hours of pumping, temperature data were recorded every 5 to 15 minutes at three sampling wells in the riverbed. Figure 7-7 shows the results of temperature profiles of the river and Sampling Wells W1, W2, and W3 during the initial pumping hours. The river temperature was constant at 26.5°C on the day of initial pumping. Before pumping started, the temperature at W1, W2, and W3 was 17.4°C, 15.6°C, and 15.1°C, respectively. The results verified that the river was a gaining stream under no pumping conditions as the temperatures at these wells were very close to the groundwater temperature of this aquifer, which varies between the narrow range of 14.5 and 15.5°C.

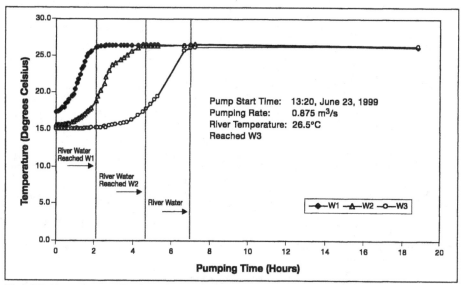

Figure 7-7. Temperature profile of RBF during initial pumping

After pumping started, the temperature in the sampling wells quickly approached river-water temperature, indicating the occurrence of filtration from the river to the aquifer. The temperature of W1 approached river-water temperature after 2 hours of pumping, while the temperatures of W2 and W3 approached river-water temperature after 4.5 and 7.0 hours, respectively.

Water level in the collector well decreased from 128 to 120 m above mean sea level during the first hour of pumping, and then stabilized at 120 m above mean sea level. The river stage was also stable at 128 m above mean sea level during this period; therefore, the hydraulic head of the RBF process can be assumed to be constant during the initial pumping period.

Temperature data can be used to estimate the filtration velocity from the river to the sampling wells. Based on the depths of the sampling locations and the temperature results, it is estimated that the maximum vertical filtration velocity of the RBF process was 0.3 to 0.4 m/h during the initial pumping period. The velocity estimated this way is the actual velocity of flow through the riverbed, which is appreciably greater than the approach velocity by a factor of 1/p, where p is the effective porosity of the riverbed (which might be about 0.4); therefore, the approach velocity was about 0.12 to 0.16 m/h during the initial pumping. This rate is in the same rate range of typical slow sand filters, which is 2.0 to 5.0 m/d per day (Ellis, 1985), suggesting that the analogy between slow sand filtration and RBF is valid.

Other water-quality monitoring results supported the temperature data. Figure 7-8 shows the total hardness results of the river and three sampling wells during the initial pumping period. The results show that the total hardness from W1 and W2 approached those of river water after 2 and 5 hours of pumping, respectively.

Monitoring results from the laterals of the collector well during initial pumping are shown in Figures 7-9 to 7-11. Based on estimates of water-travel velocity, the riverbank filtrate should have reached L4 — the lateral directly below the three sampling wells — after 2 days of pumping; however, water-quality data were not consistent with each other. It is likely that the samples collected from laterals during the initial pumping stage were a mixture of riverbank filtrate and groundwater.

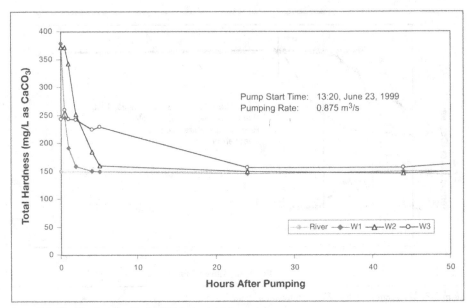

Figure 7-8. Total hardness profile for RBF during initial pumping.

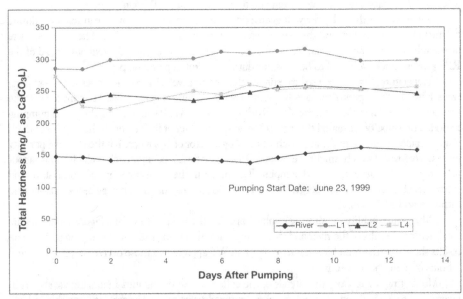

Figure 7-9. Total hardness concentration in collector well laterals.

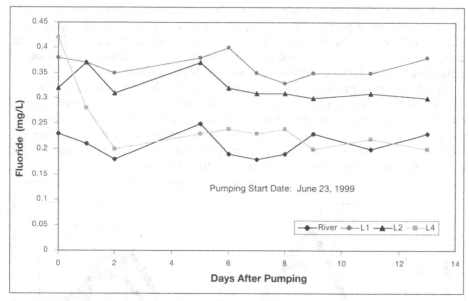

Figure 7-10. Fluoride concentration in collector well laterals.

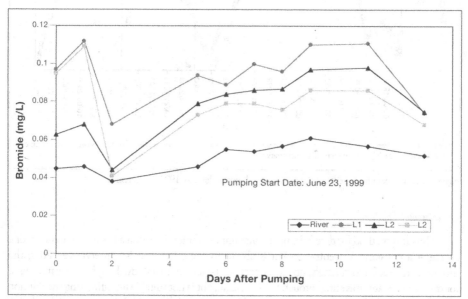

Figure 7-11. Bromide concentration in collector well laterals.

Nevertheless, it is clear that a portion of the riverbank filtrate had reached the collector well laterals after 2 days of pumping. In addition, the results have shown that water quality was different in different laterals, indicating a different extent of groundwater mixing. It is obvious that the laterals closer to the river (L2 and L4) received more riverbank filtrate, while the lateral farther from the river (L1) received the least amount of riverbank filtrate.

It is difficult to determine the overall water travel time from the river to the collector well as there were multiple flow paths between the river and the well. The water entrance velocity from the river to the collector well was much larger at the locations near the collector well and was near zero at the edge of the cone of depression; therefore, no attempt was made to calculate the composite travel time. However, the temperature profiles shown in Figure 7-12 suggest that, overall, it took about 4 weeks for river water to reach the collector well.

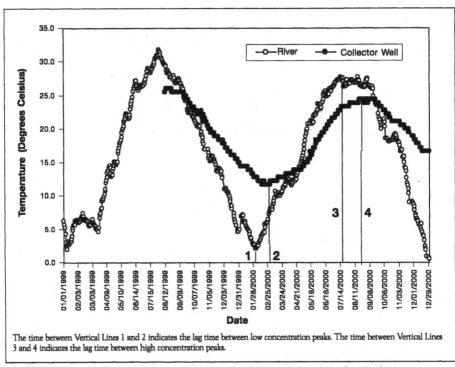

The time between Vertical Lines 1 and 2 indicates the lag time between low concentration peaks. The time between Vertical Lines 3 and 4 indicates the lag time between high concentration peaks.

Figure 7-12. Temperature profiles of the river and well water (January 1999 to December 2000).

Groundwater Dilution

The ideal method for determining the amount of dilution is to find a tracer that exists in one source at a constant concentration but is absent in the other, and is conservative during the filtration process. Unfortunately, no ideal tracer can be used in this study. Temperature can be used for tracing the water traveling through the riverbank, but is not suitable for estimating the dilution effect (because of differing heat capacities for different materials), and it is not conservative during the filtration process. Other water-quality parameters, such as bromide, fluoride, pH, and dissolved oxygen, cannot be used as tracers in this study as their concentrations are either similar in both sources or they are not conservative during the filtration process; however, the parameter of total hardness is suitable for estimating the mixing effect, although it is not an ideal parameter. The total hardness concentration difference between the two sources is large enough. The conservative assumption of total hardness through the RBF process is also logical due to the nature of aquifer materials (sand and gravel).

Groundwater in the aquifer mainly comes from two sources: recharge from precipitation and flow from the valley wall and bedrock. Studies have shown that the contributions from these two sources are almost equal (Lyverse et al., 1996). The portion from the valley wall and bedrock has a total hardness level of 570 mg/L as $CaCO_3$ (the results from M4), and the portion from the precipitation recharge has a total hardness level of 280 mg/L as $CaCO_3$ (the results from M3); therefore, the average groundwater hardness is about 425 mg/L as $CaCO_3$.

Table 7-2 lists the average total hardness levels at each sampling location during different seasons, and Table 7-3 lists the calculated percentage of groundwater dilution at the corresponding sites. The data show that, during initial pumping, the sampling wells in the riverbed virtually received all water from the river (only 0.6- to 4-percent groundwater dilution), while the water pumped from the collector well contained a large percentage of groundwater (40 percent). As expected, the laterals close to the river (L2 and L4) had less dilution from groundwater than those farther from the river (35 to 37 percent versus 56 percent). As pumping continued, the change in flow paths resulted in different mixing dynamics. The laterals close to the river (L2 and L4) received more and more water from the river source, while the lateral on the land side (L1) was influenced by groundwater.

Table 7-2. Average Total Hardness Level at Different Stages of RBF (mg/L as $CaCO_3$)

Sample Time	Sample Location						
	River	W1 (0.6 m)	W2 (1.5 m)	W3 (2.75 m)	L4 (15 m)	L2	L1
Initial pumping	148	150	150	159	250	245	303
4th Quarter, 1999	172	182	181	190	228	220	279
1st Quarter, 2000	141	160	162	158	237	179	300
2nd Quarter, 2000	140	NA	NA	NA	169	174	333
3rd Quarter, 2000	148	NA	NA	NA	203	176	345

NA = Not analyzed.

Table 7-3. Groundwater Dilution at Each Sampling Location

Sample Time	Sample Location					
	W1 (0.6 m)	W2 (1.5 m)	W3 (2.75 m)	L4 (15 m)	L2	L1
Initial pumping	0.6%	0.6%	4.0%	37%	35%	56%
4th Quarter, 1999	4.0%	4.0%	7.0%	22%	19%	42%
1st Quarter, 2000	7.0%	8.0%	6.0%	34%	13%	56%
2nd Quarter, 2000	NA	NA	NA	10%	12%	68%
3rd Quarter, 2000	NA	NA	NA	20%	10%	71%

NA = Not analyzed.

5. NOM and Disinfection Byproduct Precursor Removal

Many studies have shown that RBF is an effective process for removing NOM in source waters because of the biological degradation and physical-chemical processes that occur in the riverbank (or riverbed, in this study). In this study, the effectiveness of the RBF process for NOM and disinfection byproduct formation precursor removal was evaluated using:

- TOC.
- Ultraviolet absorbance at 254 nm.
- Biodegradable organic carbon.
- Disinfection byproduct formation potential analyses.

TOC samples were collected weekly from different locations of the RBF process. Figure 7-13 shows the TOC results as a function of filtration depth. The results indicate that the TOC concentration decreased as the filtration depth increased. The TOC concentration in the river ranged from 2.1 to 4.9 mg/L, with an average of 2.9 mg/L. The average TOC concentration decreased to 2.1 mg/L and the maximum was reduced to 2.5 mg/L after 2.75 m of filtration. Additional filtration distance and some dilution from groundwater further reduced the average TOC concentration to 1.6 mg/L (in L4).

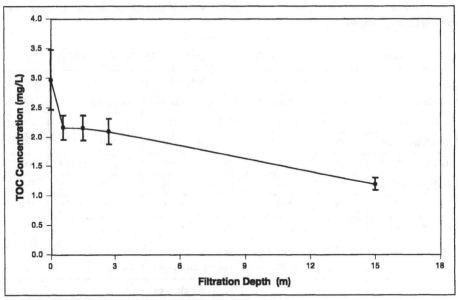

Figure 7-13. Removal of TOC as a function of filtration depth.

The TOC removal percentages were 12, 14, and 23 percent after 0.6, 1.5, and 2.75 m of filtration (W1, W2 and W3), respectively, during the initial 2 weeks of pumping between June 23 and July 7, 1999; however, the removal increased to about 30 percent after 3 months of pumping, as shown in Figure 7-14. The increase may be a result of process acclimation. The results in Figure 7-14 show that there was no significant difference among 0.6 m (W1), 1.5 m (W2), and 2.75 m (W3) of filtration for TOC removal after 3 months of pumping, suggesting that the physical removal that occurred at the surface of the riverbed may have been the dominant mechanism for TOC removal. Further reductions of TOC after 2.75 m of filtration (from 30 percent after 2.75 m to over 40 percent after 15 m) may be a combined result of biodegradation, adsorption, and groundwater dilution.

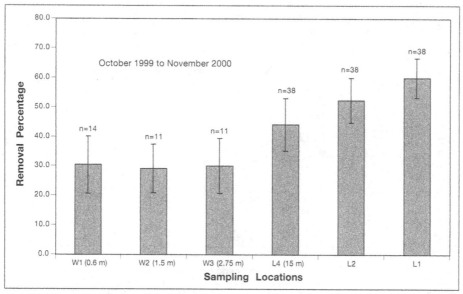

Figure 7-14. Average TOC removal for various monitoring points and collector well laterals (October 1999 to November 2000).

The seasonal variation of TOC removal through the path between the river and Lateral 4 is shown in Figure 7-15. TOC removal from river water averaged around 40 percent, fluctuating between 25 and 60 percent within the monitoring period. The change in water temperature and the percentage of groundwater dilution may be the cause of this fluctuation.

The results for biodegradable organic carbon removal are shown in Figure 7-16. About 20 percent of the DOC (or 0.56 mg/L DOC) in river water was biodegradable, which is typical for the

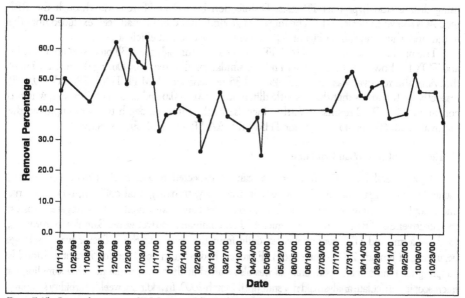

Figure 7-15. Seasonal variation of TOC removal for Collector Well Lateral 4 (October 1999 to October 2000).

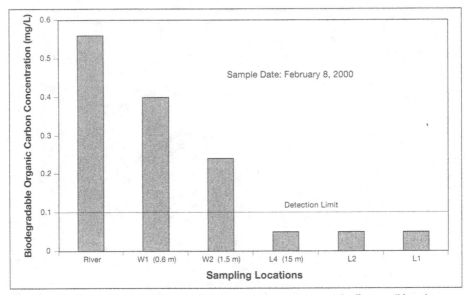

Figure 7-16. Biodegradable organic carbon results for various monitoring points and collector well laterals in February 2000.

Ohio River water in this region. The biodegradable organic carbon in the river water was rapidly removed during riverbank passage. After 0.6 m of filtration, the biodegradable organic carbon concentration decreased to 0.40 mg/L and then further decreased to 0.24 mg/L and to below detection limit (0.10 mg/L) after 1.5 m and 15 m of filtration, respectively. The results of the TOC and biodegradable organic carbon removal indicate that both biological activities and physical-chemical actions in the RBF process play an important role in biodegradable organic carbon removal.

The ability of the RBF process to remove disinfection byproduct precursors was evaluated using disinfection byproduct FP tests. Disinfection byproduct FP removal through the RBF process is shown in Figure 7-17. The impact of groundwater dilution had been excluded from the results and, thus, the data in Figure 7-17 represent true removal efficiency.

The results show that more HAA FP was removed through the RBF process than THM FP and TOX FP; however, the removal trend was similar for all three tests. During the first 2.75 m of filtration, the removal was about 25, 45, and 35 percent for THM FP, HAA FP, and TOX FP, respectively. It appears that there is little difference between 0.6 and 2.75 m of filtration, which is consistent with TOC results. A further increase in the filtration depth to 15 m improved the removal to 40, 60, and 45 percent for THM FP, HAA FP, and TOX FP, respectively.

6. Removal of Microbial Contaminants

As a natural filter, the RBF process can be expected to effectively reduce particulates, including microorganisms, in river waters. In this study, turbidity, total coliform, HPC bacteria, microscopic particulate analysis, and total aerobic spore counts were used as the surrogates to assess the effectiveness of RBF for removing microbial contaminants in river water. Samples for turbidity, total coliform, and HPC were collected daily from the river and the collector well discharge beginning in January 2000, when the RBF process had been in operation for 5 months. Monthly sampling for total aerobic spore analysis was also started at that time, while monthly sampling for microscopic particulate analysis did not start until March 2000. In addition, weekly turbidity sampling

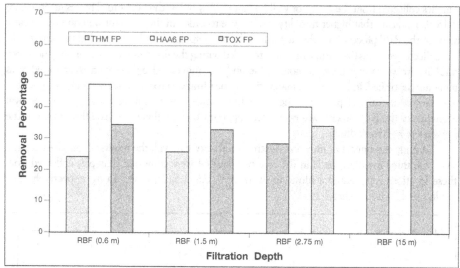

Figure 7-17. Removal of disinfection byproduct formation potential through RBF without groundwater dilution.

was conducted at the sampling wells in the riverbed (W1, W2 and W3) until the sampling facilities failed to work in February 2000 due to formation dewatering.

Turbidity Removal Through RBF

Figure 7-18 shows daily turbidity results of river water and water from the collector well. Please note that water from the collector well entered from all seven laterals in the year 2000, although mechanisms existed to shut off specific laterals. River turbidity varied during this period from 3 to 599 ntu, with a mean of 45 ntu. The median turbidity of the river water was 12.6 ntu. Although river-water turbidity fluctuated significantly during the monitoring period, the turbidity at the collector well remained stable and was consistently below 0.1 ntu. Ninety-five percent of the

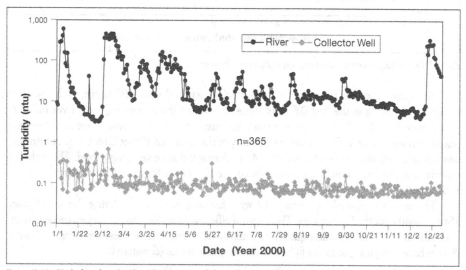

Figure 7-18. Turbidity data for the Ohio River and the collector well for the year 2000.

samples collected from the collector well had a turbidity of less than 0.2 ntu. The results in Figure 7-18 also indicate that higher turbidity results were recorded in the collector well during the early stage of the RBF process (i.e., the first 6 months of pumping). The highest turbidity measured in the collector well was 0.69 ntu, which was recorded during the first 6 months of pumping; however, turbidity in well water became more stable and less influenced by spikes in river turbidity as pumping continued, indicating the maturation of the filtration process. The results also suggest that after several months of pumping, the water travel time used to calculate the particle removal efficiency of the RBF process was no longer important because there was little fluctuation in the turbidity of riverbank filtrate.

To study the impact of filtration depth on turbidity removal, the results of samples from the river, the three sampling wells in the riverbed, and L4 were compared. Samples collected from these locations represented a filtration depth of 0, 0.6, 1.5, 2.75, and 15 m, respectively. The results are shown in Figure 7-19.

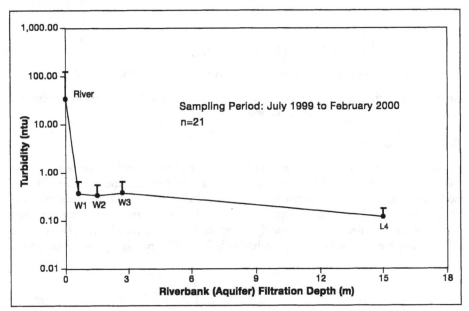

Figure 7-19. Turbidity removal as a function of filtration distance.

The data show that the reduction of turbidity increased with RBF depth, with the most reduction occurring in the first 0.6 m of filtration. During the sampling period, river turbidity varied from less than 2 ntu to greater than 400 ntu. The filtration process reduced turbidity fluctuation significantly. The maximum turbidity of the riverbank filtrate after 0.6 m of filtration was 1.5 ntu, with an average of less than 0.4 ntu during the same sampling period. The turbidity reduction in the first 0.6 m of filtration may be a result of cake filtration of large particles in the river water.

The results presented in Figure 7-19 were for samples collected during the initial stage (first 6 months) of the RBF process. The removal efficiency may increase as the process continues. Unfortunately, no further water-quality monitoring was conducted at these sampling wells after 6 months of pumping due to the failure of the sampling device (dewatered).

Removal of Indicator Microorganisms Through RBF

In drinking-water treatment, total coliform and HPC bacteria are two typical surrogates for measuring microorganisms. Both total coliform and HPC occur in surface water in relatively high concentrations. A measurement of these two parameters provides an estimation of the effectiveness of RBF for microbial pathogen removal, although the removal of total coliform and HPC may not be similar to all pathogens.

Since January 1, 2000, samples were taken from the river and the collector well daily for total coliform and HPC analysis to assess the ability of the RBF process to improve water quality with regard to microbial parameters. Composite samples of all laterals of the collector well were taken from the discharge. The results of total coliform in both the source water and riverbank-filtered water are shown in Figure 7-20, and the HPC results are shown in Figure 7-21. The results show

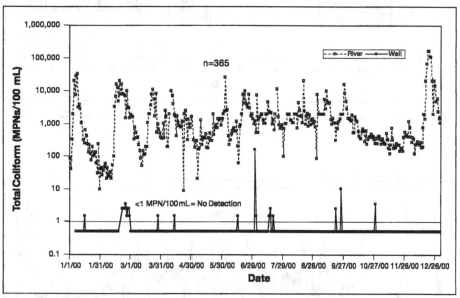

Figure 7-20. Turbidity removal as a function of filtration distance for the collector well (January 2000 to December 2000).

that total coliform and HPC concentrations in river water varied dramatically, depending on the river conditions. The total coliform concentration in river water ranged from 9 MPN/100 mL to 165,200 MPN/100 mL, while riverbank-filtered water had very few positive total coliform detections, most of which were at 1 MPN/100 mL. HPC results exhibit a trend similar to the total coliform. HPC in river water ranged from 10 to 8,820 cfu/mL. HPC counts were reduced significantly after the RBF process. HPC in the collector well ranged from 0 to 420 cfu/mL, with most of the samples under 10 cfu/mL.

The efficiency of the RBF process for these indicator microorganisms can be calculated from the results shown in Figures 7-20 and 7-21. The calculated log removal for total coliform and HPC are shown in Figures 7-22 and 7-23, respectively. The removal for total coliform ranged from 0.9 to >6.0 logs, with an average of 3.8 logs. The removal for HPC ranged from 0.4 to >4.0 logs, with an average of 2.2 logs. Strictly speaking, HPC data cannot be used to indicate the removal efficiency for microorganisms as HPC bacteria can grow during the RBF process. Some field studies (Logsdon, 2000) have also reported an increase in HPC during slow sand filtration, most likely due

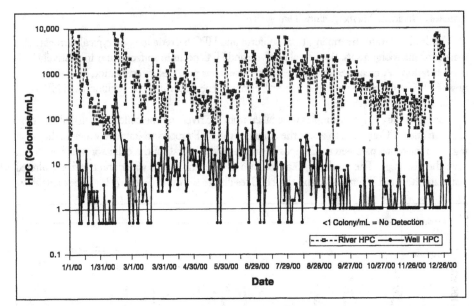

Figure 7-21. HPC concentration in both the river and collector well (January 2000 to December 2000).

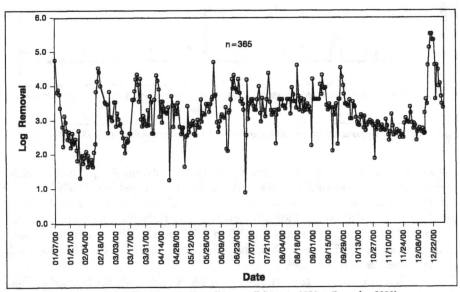

Figure 7-22. Log removal of total coliform for the collector well (January 2000 to December 2000).

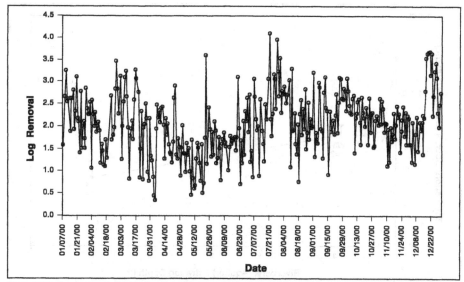

Figure 7-23. Log removal of HPC for the collector well (January 2000 to December 2000).

to bacterial growth during the filtration process. The results presented show the worst-case scenario of the RBF process.

The above results offer a direct comparison between river water and water from the collector well; thus, the groundwater dilution effect was included in the overall removal calculation. The true removal effectiveness of RBF can be estimated by excluding the impact of groundwater dilution. To accomplish the estimation, several assumptions have to be made. The assumptions are:

- Seventy percent of the well water (average for all laterals) was riverbank filtrate, based on the calculation presented in the groundwater dilution section.
- The groundwater contained no total coliform.
- The groundwater contained no HPC bacteria.

Although the third assumption did not match the field monitoring results, it is used to demonstrate the worst efficiency that can be produced from the RBF process.

Based on the above assumptions, the average removal of total coliform through RBF alone, without the dilution effect, was 3.8-log units during the monitoring period, and the average removal for HPC without the dilution effect was 2.0-log units. Both results indicate that the RBF process is an efficient method for removing microorganisms from surface source water and is similar to slow-sand filtration.

Total Aerobic-Spore Removal Through the RBF Process

The aerobic-spore analysis has been recognized as one of the best surrogates of assessing filtration performance for microbial pathogen removal (Rice et al., 1996; Nieminski et al., 2000). Unlike other microbiological parameters, spores occur at relatively high concentrations in most source waters and can be detected after the filtration process; therefore, spore concentrations can be used to assess the filtration efficiency.

The reduction of total spore counts as a function of filtration depth/distance is shown in Figure 7-24. The results show that the total aerobic-spore concentration in river water ranged from 3,000 to 15,000 cfu/100 mL, with an average of 8,031 cfu/100 mL. These data are similar to literature-reported values by Rice et al. (1996), who surveyed spore concentrations in the Ohio

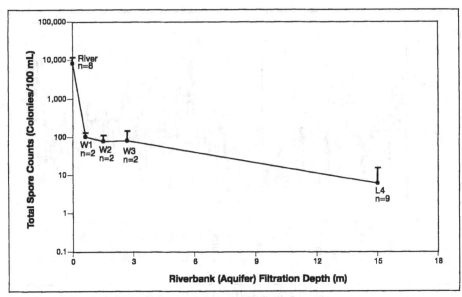

Figure 7-24. Removal of total aerobic spores as a function of filtration distance.

River and found that the yearly mean spore concentration of the river water was 12,000 cfu/100 mL. The concentration decreased to about 100 cfu/100 mL only after 0.6 m of filtration, indicating that the RBF process is effective for removing these spores. The concentration was further reduced to below 10 cfu/100 mL when the filtration depth increased to 15 m in L4 (see Figure 7-2 for location).

The log removal of total aerobic-spore counts through RBF was calculated in the same way as was total coliform and HPC. Similar assumptions were made to exclude the groundwater dilution effect, and the results presented in Figure 7-25 are log removals without dilution impact. The results show that a 1.7-log reduction in spore concentration can be achieved with a filtration depth as shallow as 0.6 m. The log removal increased with filtration depth and, after 15 m of filtration, the reduction was more than 3 logs in L4.

The spore concentrations of other lateral samples and the composite well samples are shown in Figure 7-26. The results indicate that water from all individual laterals of the collector well contained little aerobic spores (more than 3 logs removal). The typical reduction of spore counts by a conventional treatment plant is about 2 logs (Rice et al., 1996). Compared to results from conventional filtration processes and slow-sand filtration, the RBF process is more effective in removing microbial contaminants.

Microscopic Particulate Analysis of Riverbank Filtrate

Microscopic particulate analysis is used to identify surface water "bioindicators," such as plant debris, algae, diatoms, insects, rotifers, *Giardia*, and coccidian, which are characteristic of surface waters. The presence of these bioindicators in filtered water indicates the effectiveness of a filtration process or the risk of surface-water contamination. The method has been widely used as a tool to assess both groundwater under the direct influence of surface water and filter performance.

In this study, microscopic particulate analysis samples were collected monthly from the river and laterals (L1, L2 and L4) of the collector well beginning in March 2000. The results of the lateral samples were compared with river samples to calculate particle removal of the RBF process.

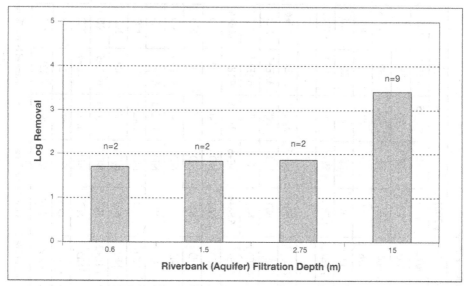

Figure 7-25. Log removal of total aerobic spore count.

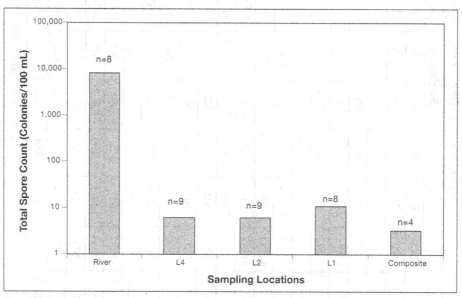

Figure 7-26. Total aerobic spore concentration in the river, laterals, and collector wells.

The detailed microscopic particulate analysis results from each location are listed in Tables 7-4 to 7-7. The results show that relatively high concentrations of algae, diatoms, and spores are frequently found in the samples collected from the Ohio River. The river water typically has about 1-million algae counts in 100 L of water; however, these particles were rarely found in well water. The results also show that no insects, microorganisms, *Giardia*, or other large-diameter pathogens were found in waters sampled from the collector well.

Table 7-4. Microscopic Particulate Analysis Results of River Water

Category	Sampling Date (February 2000 to January 2001)											
	2/23	3/28	5/16	6/14	7/20	8/15	8/29	9/19	10/16	11/15	12/19	1/17
pH	7.6	7.6	7.8	7.7	7.6	7.6	7.6	7.9	7.8	7.7	7.9	7.6
Turbidity (ntu)	180	13	15	7.3	3.0	12	4.0	11	14	8.7	249	10
Sample Volume (L)	189	280	235	348	242	208	276	252	223	299	110	307
Packed Pellet Volume (mL)	20	2	2	3	0.75	2.0	1.0	10	2.0	2.0	20	2.25
Detection Limit (No. /100 L)	2.6E+3	3.6E+2	4.2E+2	4.3E+2	8.3E+2	9.1E+2	7.1E+2	1.7E+3	4.5E+2	1.1E+3	6.1E+4	3.3E+3
Algae (No. /100 L)	1.2E+7	1.4E+6	1.1E+7	2.7E+6	5.3E+5	1.4E+7	2.3E+7	6.5E+7	2.4E+7	3.6E+6	2.7E+6	2.4E+6
Amoebae (No. /100 L)	5.3E+3	ND	ND	ND	ND	ND	ND	ND	ND	ND	3.0E+5	ND
Crustaceans/Parts (No. /100 L)	2.6E+3	ND	ND	4.3E+2	ND	9.1E+2	1.4E+3	ND	ND	ND	6.1E+4	ND
Diatoms (No. /100 L)	6.7E+6	2.2E+6	1.3E+6	9.0E+5	1.6E+5	8.1E+5	4.2E+5	6.8E+6	2.5E+6	2.0E+6	9.1E+6	4.0E+6
Insects/Parts (No. /100 L)	ND	ND	ND	ND	ND	ND	ND	ND	ND	ND	ND	ND
Invertebrates/Eggs (No. /100 L)	ND	ND	ND	ND	ND	ND	ND	ND	ND	ND	ND	ND
Nematodes/Eggs (No. /100 L)	2.6E+3	ND	ND	ND	ND	ND	ND	ND	ND	ND	1.2E+5	ND
Pollen (No. /100 L)	5.3E+3	1.1E+3	ND	ND	ND	ND	ND	ND	ND	1.1E+3	6.1E+4	ND
Protozoa (Free-Living) (No./100 L)	ND	ND	ND	ND	ND	ND	ND	ND	ND	3.4E+5	ND	ND
Rotifers/Eggs (No. /100 L)	ND	ND	ND	ND	ND	ND	ND	ND	ND	ND	ND	ND
Spores (No. /100 L)	1.1E+4	1.4E+3	4.2E+2	4.3E+2	ND	1.8E+3	7.1E+2	ND	ND	4.4E+3	6.1E+5	ND
Vegetative Debris (No. /100 L)	ND	ND	ND	ND	ND	ND	ND	ND	ND	1.1E+3	1.2E+5	ND
Giardia Cysts (No./100 L)	<30	10	<23.8	<27.0	<7.2	6.1	<2.9	15.4	6.1	<9.2	125	<13
Cryptosporidium Oocysts (No./100 L)	<30	<1.4	<23.8	<27.0	<7.2	<6.1	<2.9	<15.4	<6.1	<9.2	<125	<13

ND = No detection.   No. = Number.

Louisville Water Company   2/16/01   Samples analyzed by Clancy Environmental Consultants, Inc.   Prepared by Jack Wang.

Table 7-5. Microscopic Particulate Analysis Results of River Water Results of Water from Lateral 1

Category	Sampling Date (February 2000 to January 2001)											
	2/23	3/28	5/16	6/14	7/20	8/15	8/29	9/19	10/16	11/15	12/19	1/17
pH	7.6	7.7	7.7	7.6	7.8	7.5	7.5	7.7	7.6	7.6	7.7	7.6
Turbidity (ntu)	<0.5	<0.10	<0.10	<0.10	<0.10	0.06	0.06	0.10	<0.10	0.12	0.10	0.10
Sample Volume (L)	2,101	2,000	2,532	1,893	2,358	2,422	2,498	2,014	2,120	1,908	4,118	2,695
Packed Pellet Volume (mL)	0.05	Trace	0.04	0.05	0.10	0.10	0.05	0.05	0.05	0.10	0.10	0.1
Detection Limit (No. /100 L)	0.26	0.26	0.26	0.26	0.26	0.26	0.26	0.26	0.26	0.26	0.26	0.26
Algae (No. /100 L)	ND	ND	ND	ND	ND	ND	ND	ND	ND	0.26	ND	ND
Amoebae (No. /100 L)	ND	ND	ND	ND	ND	ND	ND	ND	ND	ND	ND	ND
Crustaceans/Parts (No. /100 L)	ND	ND	ND	ND	ND	ND	ND	ND	ND	ND	ND	ND
Diatoms (No. /100 L)	ND	ND	ND	ND	ND	ND	ND	ND	ND	ND	ND	ND
Insects/Parts (No. /100 L)	ND	ND	ND	ND	ND	ND	ND	ND	ND	ND	ND	ND
Invertebrates/Eggs (No. /100 L)	ND	ND	ND	ND	ND	ND	ND	ND	ND	ND	ND	ND
Nematodes/Eggs (No. /100 L)	ND	ND	ND	ND	ND	ND	ND	ND	ND	ND	0.52	ND
Pollen (No. /100 L)	ND	ND	ND	ND	ND	ND	ND	ND	ND	0.79	ND	0.26
Protozoa (Free-Living) (No./100 L)	ND	ND	ND	ND	ND	ND	ND	ND	ND	2,800	ND	0.78
Rotifers/Eggs (No. /100 L)	ND	ND	ND	ND	ND	ND	ND	ND	ND	ND	ND	ND
Spores (No. /100 L)	2.6	ND	0.26	ND	ND	1.1	0.53	ND	ND	2.6	ND	0.78
Vegetative Debris (No. /100 L)	ND	ND	ND	ND	ND	ND	ND	ND	ND	0.26	ND	0.52
Giardia Cysts	<0.3	<0.1	<0.7	<3.3	<0.8	<0.9	<1.0	<0.8	<0.4	<0.5	<0.3	<0.2
Cryptosporidium Oocysts	<0.3	<0.1	<0.7	<3.3	<0.8	<0.9	<1.0	<0.8	<0.4	<0.5	<0.3	<0.2

ND = No detection.   No. = Number.   Louisville Water Company   2/16/01   Samples analyzed by Clancy Environmental Consultants, Inc.   Prepared by Jack Wang.

Table 7-6. Microscopic Particulate Analysis Results of River Water Results of Water from Lateral 2

Category	Sampling Date (February 2000 to January 2001)											
	2/23	3/28	5/16	6/14	7/20	8/15	8/29	9/19	10/16	11/15	12/19	1/17
pH	7.6	7.7	7.7	7.6	7.9	7.8	7.8	7.6	7.7	7.7	7.6	7.6
Turbidity (ntu)	<0.5	<0.10	<0.10	<0.10	<0.40	0.09	0.09	0.10	<0.10	0.10	0.10	0.10
Sample Volume (L)	2,067	2,332	2,702	2,850	2,123	2,131	2,509	1,995	2,097	2,513	2,192	2,059
Packed Pellet Volume (mL)	0.05	Trace	0.05	0.10	0.40	0.10	0.10	0.05	0.05	0.02	0.05	Trace
Detection Limit (No./100 L)	0.26	0.26	0.23	0.26	0.26	0.26	0.26	0.26	0.26	0.26	0.26	0.26
Algae (No./100 L)	ND	ND	ND	0.26	ND	ND	ND	ND	ND	ND	0.26	0.26
Amoebae (No./100 L)	ND	ND	ND	ND	ND	ND	ND	ND	ND	1.0	ND	ND
Crustaceans/Parts (No./100 L)	ND	ND	ND	ND	ND	ND	ND	ND	ND	ND	ND	ND
Diatoms (No./100 L)	ND	ND	ND	ND	ND	ND	ND	ND	ND	ND	ND	ND
Insects/Parts (No./100 L)	ND	ND	ND	ND	ND	ND	ND	ND	ND	ND	ND	ND
Invertebrates/Eggs (No./100 L)	ND	ND	ND	ND	ND	ND	ND	ND	ND	ND	0.26	ND
Nematodes/Eggs (No./100 L)	ND	ND	ND	ND	ND	ND	ND	ND	ND	ND	ND	ND
Pollen (No./100 L)	0.26	ND	2.7	1.0	ND	ND	ND	ND	1.6	ND	ND	ND
Protozoa (Free-Living) (No./100 L)	ND	ND	ND	ND	ND	ND	ND	ND	ND	880	0.26	160
Rotifers/Eggs (No./100 L)	ND	ND	ND	ND	ND	ND	ND	ND	ND	ND	ND	ND
Spores (No./100 L)	1.6	ND	ND	0.52	ND	ND	ND	ND	1.3	3.3	ND	ND
Vegetative Debris (No./100 L)	ND	ND	ND	ND	ND	ND	ND	0.26	ND	ND	ND	0.26
Giardia Cysts	<0.2	<0.1	<0.9	<0.9	<2.4	<1.2	<1.0	<0.7	<0.5	<0.3	<0.2	<0.1
Cryptosporidium Oocysts	<0.2	<0.1	<0.9	<0.9	<2.4	<1.2	<1.0	<0.7	<0.5	<0.3	<0.2	<0.1

ND = No detection.   No. = Number.

Louisville Water Company   2/16/01   Samples analyzed by Clancy Environmental Consultants, Inc.   Prepared by Jack Wang.

Table 7-7. Microscopic Particulate Analysis Results of River Water Results of Water from Lateral 4

Category	Sampling Date (February 2000 to January 2001)											
	2/23	3/28	5/16	6/14	7/20	8/15	8/29	9/19	10/16	11/15	12/19	1/17
pH	7.6	7.7	7.7	7.6	7.8	7.7	7.7	7.6	7.6	7.7	7.7	7.6
Turbidity (ntu)	<0.5	<0.10	<0.10	<0.10	<0.50	0.10	0.09	0.10	<0.10	0.40	<0.50	0.10
Sample Volume (L)	1,896	2,937	2,543	1,381	2,055	871	1,662	1,972	1,628	1,151	2,244	1,007
Packed Pellet Volume (mL)	0.10	Trace	0.10	0.04	0.05	0.05	0.05	0.20	0.05	0.20	0.50	0.20
Detection Limit (No. /100 L)	0.26	0.26	0.26	0.26	0.26	0.26	0.26	0.26	0.26	0.26	0.30	0.26
Algae (No. /100 L)	ND	1.1	ND	ND	ND	ND	ND	ND	ND	ND	66	44
Amoebae (No. /100 L)	ND	ND	ND	ND	ND	ND	ND	ND	ND	ND	ND	ND
Crustaceans/Parts (No. /100 L)	ND	ND	ND	ND	ND	ND	ND	ND	0.26	0.26	ND	ND
Diatoms (No. /100 L)	ND	ND	0.26	ND	ND	ND	ND	ND	ND	ND	ND	ND
Insects/Parts (No. /100 L)	ND	ND	ND	ND	ND	ND	ND	ND	ND	ND	ND	ND
Invertebrates/Eggs (No. /100 L)	ND	ND	ND	ND	ND	ND	ND	ND	ND	ND	ND	ND
Nematodes/Eggs (No. /100 L)	ND	ND	ND	ND	ND	ND	ND	ND	ND	ND	ND	ND
Pollen (No. /100 L)	ND	ND	1.6	1.8	ND	0.26	ND	ND	ND	ND	ND	ND
Protozoa (Free-Living) (No./100 L)	ND	ND	ND	ND	ND	ND	ND	ND	ND	16,000	16,000	4,000
Rotifers/Eggs (No. /100 L)	ND	ND	ND	ND	ND	ND	ND	ND	ND	ND	ND	ND
Spores (No. /100 L)	0.26	ND	ND	0.78	ND	0.52	ND	0.26	0.52	ND	0.30	ND
Vegetative Debris (No. /100 L)	ND	ND	ND	ND	ND	ND	ND	ND	ND	0.26	ND	ND
Giardia Cysts	<0.4	<0.1	<0.9	<0.4	<0.8	<1.3	<0.7	<1.1	<0.9	<0.7	<1.5	<0.9
Cryptosporidium Oocysts	<0.4	<0.1	<0.9	<0.4	<0.8	<1.3	<0.7	<1.1	<0.9	<0.7	<1.5	<0.9

ND = No detection.   No. = Number.

Louisville Water Company 2/16/01   Samples analyzed by Clancy Environmental Consultants, Inc.   Prepared by Jack Wang.

The comparison between the river and L4 represents the least effective conditions (because of the shortest distance for flow) in the aquifer for particle removal. Table 7-8 lists the log removal results of individual particles through the path between the river and L4. The log removal for algae ranged from 4.6 logs to >8.3 logs, with most data showing removal of >7 logs. The removal for total spores ranged from 2.7 to 6.3 logs, with an average of >3.9 logs. For diatoms, the removal was >5.7 logs. All of the results indicate that the RBF process is very effective in removing microbial contaminants in surface water.

Table 7-8. Particle Removal Efficiency Through the Path Between the River and Lateral 4

	Log Removal				
	Algae	**Diatoms**	**Spores**	*Giardia*	*Cryptosporidium*
Average	>7.1	>6.7	>3.9	NC	NC
Range	4.6 to >8.3	>5.7 to >7.4	2.7 to 6.3	NC	NC

Louisville Water Company   2/16/01   Samples analyzed by Clancy Environmental Consultants, Inc.   Prepared by Jack Wang.
NC = Not calculated.

7. Summary

The results of NOM and disinfection byproduct precursor removal indicate that more than a 50-percent reduction in disinfection byproduct precursors can be achieved through the RBF process at Louisville with a filtration depth of 15 m. Biodegradation and the physical removal of particulate matter at the river/aquifer interface are the primary mechanisms for NOM and disinfection byproduct precursor removal. Adsorption may also play a role in further NOM reduction as the filtration depth increases. The results of turbidity, total coliform, HPC, total aerobic spores, and microscopic particulate analysis show that the RBF process is very effective in removing particles in surface water, suggesting that it is an effective water-treatment process for reducing the potential of microbial contamination in drinking water. The results also show that the removal efficiency increases with the filtration depth, with most of the removal occurring within the first meter of filtration. Data from total coliform, total aerobic spores, and microscopic particulate analysis suggest that a >2.5-log removal credit can be given to this RBF system.

References

Ellis, K.V. (1985). "Slow sand filtration." *CRC Critical Reviews in Environmental Control*, 15(4): 315-354.

Hubbs, S.A. (1981). "Organic reduction – Riverbank infiltration at Louisville." *Proceedings, Annual Conference*, American Water Works Association, Denver, Colorado.

Juttner, F. (1995). "Elimination of terpenoid odorous compounds by slow sand and river bank filtration of the Ruhr River, Germany." *Water Science and Technology*, 31(11): 211.

Kuhlmann, B., and Kaczmarzyk, B. (1995). "Biodegradation of the herbicides 2,4-dichlorophenoxyacetic acid, 2,4,5-trichlorophenoxyacetic acid, and 2-methyl-4-chlorophenoxyacetic acid in a sulfate-reducing aquifer." *Environmental Toxicology and Water Quality*, 10(2): 119.

Kühn, W., and Müller, U. (2000). "Riverbank filtration: An overview." *Journal American Water Works Association*, 92: 2-60

Logsdon, G. (2000). Personal Communication. Black & Veatch, Cincinnati, Ohio.

Lyverse, M.A., J.J. Stam, and M.D. Unthank (1996). "Hydrogeologicy and simulation of ground-water flow in the alluvial aquifer at Louisville, Kentucky." *Water Resource Investigation Report 91-4035*, U.S. Geological Survey, Louisville, Kentucky.

Nieminski, E.C., D.B. William, and L. Moss (2000). "Using surrogates to improve plant performance." *Journal American Water Works Association*, March 2000.

National Water Research Institute (1999). *Abstracts, International Riverbank Filtration Conference*, National Water Research Institute, Fountain Valley, California.

Rice, E.W., K.R. Fox, R.J. Miltner, D.A. Lytle, and C.H. Johnson (1996). "Evaluating plant performance with endospores." *Journal American Water Works Association*, 88(9): 122-130.

Rorabaugh, M.I. (1948). *Ground-water resources of the northeastern part of the Louisville area, Kentucky*, City of Louisville, Louisville Water Company, Louisville, Kentucky.

Rorabaugh, M.I. (1949). *Progress report on the ground-water resources of the Louisville area, Kentucky, 1945-1949*, City of Louisville and Jefferson County, Louisville, Kentucky.

Sontheimer, H. (1980). "Experience with riverbank filtration along the Rhine River." *Journal American Water Works Association*, 72: 7-386.

Wang, J., J. Smith, and L. Dooley (1995). "Evaluation of riverbank infiltration as a process for removing particles and DBP precursors." *Proceedings, Water Quality Technology Conference*, American Water Works Association, Denver, Colorado.

Wilderer, P.A., U. Foerstner, and O.R. Kuntschik (1985). "The role of riverbank filtration along the Rhine River for municipal and industrial water supply." *Artificial Recharge of Groundwater*, T. Asano, ed., Butterworth Publishers, Boston, Massachusetts.

Chapter 8. Reduction in Disinfection Byproduct Precursors and Pathogens During Riverbank Filtration at Three Midwestern United States Drinking-Water Utilities

W. Joshua Weiss
Johns Hopkins University
Baltimore, Maryland, United States

Edward J. Bouwer, Ph.D.
Johns Hopkins University
Baltimore, Maryland, United States

William P. Ball, Ph.D., P.E.
Johns Hopkins University
Baltimore, Maryland, United States

Charles R. O'Melia, Ph.D., P.E.
Johns Hopkins University
Baltimore, Maryland, United States

Harish Arora, Ph.D., P.E.
O'Brien & Gere Engineers, Inc.
Landover, Maryland, United States

Thomas F. Speth, Ph.D., P.E.
United States Environmental Protection Agency,
Cincinnati, Ohio, United States

1. Introduction

RBF is a process that subjects river water to ground passage prior to its use as a drinking-water supply. European experience with RBF (more than 20 years of literature: Doussan et al., 1997; Wilderer et al., 1985; Piet and Zoeteman, 1985; Sontheimer, 1980; Kussmaul, 1979) coupled with recent United States experience (less than 5 years of literature: Verstraeten et al., 1999; Ray, 1999; Wang et al., 1999; Bouwer et al., 1999) demonstrate that, during infiltration and underground transport, processes such as filtration, sorption, and biodegradation produce significant improvements in raw-water quality (Kühn and Müller, 2000; Kivimaki et al., 1998; Stuyfzand, 1998). Increased applications of RBF are anticipated as drinking-water utilities strive to meet increasingly stringent drinking-water regulations, especially with regard to the provision of multiple barriers for protection against microbial pathogens (e.g., *Giardia, Cryptosporidium*), and with regard to regulations for disinfection byproducts, such as THMs and HAAs.

147

C. Ray et al. (eds.), Riverbank Filtration, 147–173.
© 2002 *Kluwer Academic Publishers. Printed in the Netherlands.*

Research was undertaken to:

- Evaluate the merits of RBF for removing/controlling disinfection byproduct precursors, pathogens, and pesticides.
- Evaluate if RBF can improve finished drinking-water quality by altering NOM in a manner that is not otherwise accomplished through conventional drinking-water treatment.

The ongoing project consists of three studies that monitor the performance of three RBF systems along the Ohio, Wabash, and Missouri Rivers in the United States. The first study involved measuring a range of water-quality parameters, including TOC, DOC, ultraviolet absorbance, inorganic species, pesticides, biodegradable organic carbon, assimilable organic carbon, and disinfection byproduct FP. The second study involved the simulated treatment of river waters, including coagulation, flocculation, sedimentation, filtration, and ozonation. Parameters including TOC, DOC, and disinfection byproduct FP were compared among the treated and riverbank-filtered samples. A third study underway is employing XAD-8 resin chromatography to elucidate the changes in the character of the organic matter upon RBF, as well as the effects of these changes on the formation of disinfection byproducts.

2. Site Descriptions

Three water-treatment systems that employ RBF were investigated. These sites are owned by the American Water Works Service Company, Inc. and are located along three different major Midwestern rivers. The characteristics of the sites are given below. The sites provide three different source waters as well as a range of ground travel distances (24 to 177 m) and residence times (days to 3 months).

Indiana-American Water Company at Jeffersonville, Indiana

Jeffersonville, Indiana, lies on the Ohio River to the north of (and across the river from) Louisville, Kentucky. Drinking water for Jeffersonville is obtained from two well fields adjacent to the Ohio River. The wells investigated for this study are part of the Babbs Well Field (Figure 8-1).

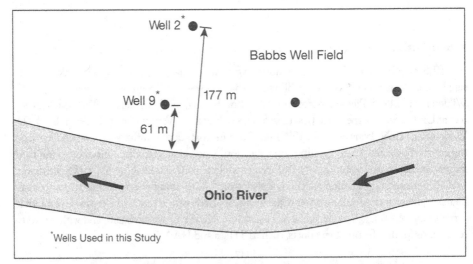

Figure 8-1. Location of the Babbs Well Field along the Ohio River. Note: One meter equals 3.28 feet.

Well 2 (177 m from the river; 20-m depth to well screen; 8-m well screen length) and Well 9 (61 m from the river; 14-m depth to well screen; 15-m well screen length) were selected for sampling. Groundwater flow analysis at this location has been performed by Stephen Champa, a consulting hydrogeologist with Eagon & Associates, Inc. (Worthington, Ohio) through the use of the groundwater flow and particle tracking program, MODPATH (Pollock, 1994). The travel time and river-water influx to the wells were analyzed, suggesting that 96 percent of the total discharge from the well field is obtained from induced infiltration from the Ohio River. The travel time to Well 9 was estimated to range between 3 and 5 days, and the travel time to Well 2 was estimated to range between 13 and 19 days.

In general, the first 3 m of subsurface material consist of brown clay, followed by an additional 3 m of clay and sand or gravel mix. A mix of coarse gravels and fine or medium sands extends to a depth of approximately 30 m. The estimated safe yields/supply capacities for the Babbs Well Field and the Ohio River are 0.23 m³/s and 0.32 m³/s, respectively (based on maximum raw-water pumping capacity). The well water is aerated, chemically treated for iron and manganese removal, and chlorinated prior to distribution.

Indiana-American Water Company at Terre Haute, Indiana

Wells at the Terre Haute, Indiana, site include a 0.53-m³/s horizontal collector well located on the bank of the Wabash River and six vertical wells, each with a capacity of 0.044 m³/s, located 122 to 762 m from the river (Figure 8-2). The collector well (horizontal arms extend out from the center; screens are 27-m deep and 24-m below the bottom of the river) and Well 3 (122-m from the river; 24-m depth to well screen; 14-m well screen length) were sampled for this study. The primary source of aquifer recharge is from the Wabash River. The temperature and hardness of the collector well discharge closely track the conditions in the Wabash River. An analysis of travel times from the river to the wells was not performed for Terre Haute.

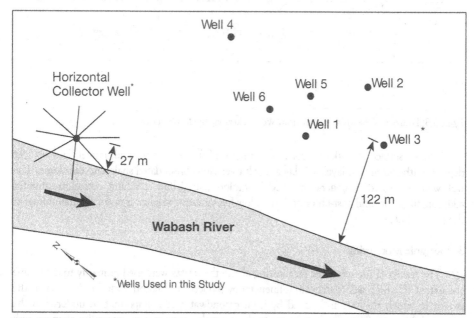

Figure 8-2. Location of the horizontal collector well and vertical wells along the Wabash River.

Aquifer material between the Wabash River and the collector well consists primarily of medium and fine sands underlain by coarser sand and gravel. The aquifer material near Well 3, located approximately 122 m from the river, consists of medium and fine sands with a sand and gravel formation located between approximately 15 and 18 m below the surface. Upon collection, well water is subjected to two parallel treatment trains. One train is a conventional treatment plant consisting of sedimentation (no flocculation) and dual media filtration. The second train is pressure filtration using dual media for iron and manganese removal. A free chlorine residual is maintained in the plant effluent.

Missouri-American Water Company at Parkville, Missouri

The Parkville, Missouri, site lies along the Missouri River, north of (and across the river from) Kansas City (Figure 8-3). Four wells are located adjacent to the Missouri River, with capacities ranging from 0.0315 m^3/s to 0.132 m^3/s and an average combined pumping rate of 0.075 m^3/s. Well 4 (37 m from the river; 17-m depth to well screen; 6-m well screen length) and Well 5 (37 m from the river; 18-m depth to well screen; 9-m well screen length) were selected for sampling during this study. No analyses of travel times or percentage of infiltration from the river are available for the Parkville site.

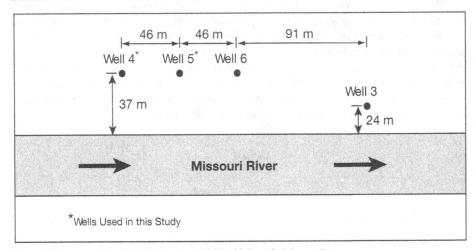

Figure 8-3. Location of the Missouri-American Well Field along the Missouri River.

The subsurface at Parkville consists primarily of fine to coarse sand, gravel, and boulder deposits with intermixed layers of clay and silt overlying consolidated shale and limestone. The well water is treated using aeration, prechlorination, partial lime softening, corrosion inhibitor addition, filtration, and post-chlorination. A low free-chlorine residual is normally maintained in the plant effluent.

3. Inorganic Monitoring

The results of the inorganic monitoring at the three sites were used primarily to determine the extent to which well waters are influenced by induced infiltration from the river versus the extent to which they are influenced by local groundwater. It is important to understand the potential dilution effects of groundwater to draw conclusions about water-quality changes with

RBF from the comparison of river and well water samples. Cation and anion analyses were performed according to U.S. Environmental Protection Agency methods 200.7, 200.8, and 300-IC (U.S. Environmental Protection Agency, 1994).

Jeffersonville, Indiana

At Jeffersonville, seasonal temperature changes in the Ohio River have a minor influence on the temperature of well waters. The aquifer dampens out the Ohio River water-temperature fluctuations during underground passage. The results for calcium, magnesium, bromide, chloride, sulfate, and total dissolved solids at Jeffersonville suggest minimal dilution effects for Well 9, in that the well waters show similar concentrations of these "conservative" parameters as Ohio River water (Figure 8-4). In most cases, the bars representing one standard deviation overlap. The water samples from the distant well (Well 2) exhibit less concentration fluctuations in comparison to changes in Ohio River chemistry, suggesting a greater contribution from regional groundwater at this well.

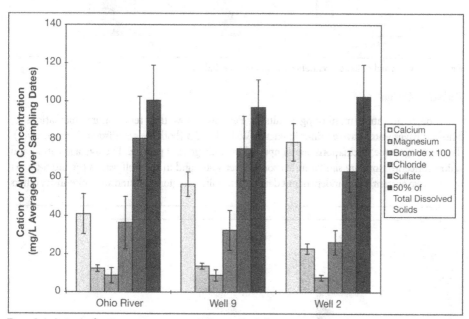

Figure 8-4. Anion and cation concentrations at Jeffersonville, Indiana.

Terre Haute, Indiana

Temperature data at Terre Haute indicate that seasonal temperature fluctuations in Wabash River water have a substantial impact at the collector well, but only a minor effect at Well 3. This reflects the short travel time for river water to reach the collector well and the long travel time to reach Well 3. The cation and anion results show that the chemistry of the collector well water closely matches the chemistry of Wabash River water (Figure 8-5). The major ion concentrations in Well 3 water differ significantly from Wabash River values, indicating a substantial influence from local groundwater.

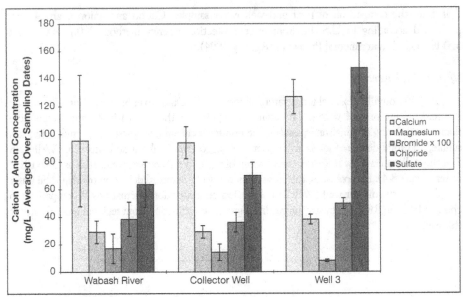

Figure 8-5. Anion and cation concentrations at Terre Haute, Indiana.

Parkville, Missouri

The temperature monitoring results at Parkville show that temperature fluctuations in Missouri River water have a minor impact at Wells 4 and 5 (both are equidistant from the river). The more extreme fluctuations are dampened by underground passage. The average cation and anion concentrations are similar in Missouri River water and in the well waters (Figure 8-6). At this site, we currently lack independent data to determine if regional groundwater chemistry differs

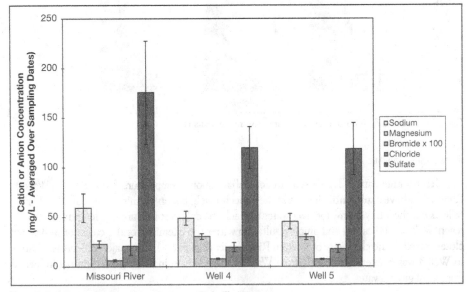

Figure 8-6. Anion and cation concentrations at Parkville, Missouri.

from Missouri River water chemistry. Nevertheless, because of the short well distance and because of evidence from other sites, we believe that infiltrated river water dominates the water being pumped from the two wells.

4. Microbial Monitoring

Several river and riverbank-filtered water samples were assayed for a number of micro-organisms. On a tri-monthly basis (seven total sampling rounds), enumerations for *Clostridium*, two bacteriophages, *Giardia*, and *Cryptosporidium* were performed. *Clostridium* and the bacteriophage were analyzed following procedures developed by the American Water Works Service Company's Quality Control and Research Laboratory in Belleville, Illinois. *Giardia* and *Cryptosporidium* were analyzed by the U.S. Environmental Protection Agency ICR method, based on the Immuno-fluorescence Assay (U.S. Environmental Protection Agency, 1996).

For analyses of *Clostridium* and bacteriophage, a constant volume of 100 mL was sampled for all locations and for all sampling rounds. *Clostridium* and the two bacteriophages (*E. coli* C and *E. coli* Famp) were detected in nearly all of the river waters, but only twice in well water samples. The frequent occurrence of non-detection in the riverbank-filtered water samples complicates the interpretation. Following the procedure of Parkhurst and Stern (1998), the average concentrations for *Clostridium* and the bacteriophages (Table 8-1) were determined by summing the number of microbial counts obtained for each sample and dividing by the sum of the volumes sampled over

Table 8-1. Clostridium and Bacteriophage Concentrations[a] and Log Removals[b]

		Bacteriophage	
	Clostridium Average Counts/100 mL	*E. Coli* C[c] pfu/100 mL	*Famp*[c] pfu/100 mL
Jeffersonville, IN			
Ohio River	122	49	12
Well 9	<0.14 [>2.9]	<0.14 [>2.5]	0.18 [1.8]
Well 2	<0.14 [>2.9]	<0.14 [>2.5]	<0.14 [>1.9]
Terre Haute, IN			
Wabash River	183	147	13
Collector Well	0.07 [3.4]	<0.14 [>3.0]	<0.14 [>2.0]
Well 3	<0.14 [>3.1]	<0.14 [>3.0]	<0.14 [>2.0]
Parkville, MO			
Missouri River	143	31	6
Well 4	<0.14 [>3.0]	<0.14 [>2.3]	<0.14 [>1.6]
Well 5	<0.14 [>3.0]	<0.14 [>2.3]	<0.14 [>1.6]

[a]Concentrations calculated as Σ (counts for all sampling rounds)/Σ (volume sampled for all sampling rounds).
[b]Log removals shown in brackets.
[c]*E. coli* C and *Famp* are the host organisms.   pfu = Plaque forming units.

all sampling rounds. Non-detects in both the river and well waters were treated as being zero. In the case that there were no detections during any sampling rounds, the average concentration was calculated as being <1 divided by the sum of the volumes sampled. Parkhurst and Stern (1998) concluded that this "sum of the counts divided by sum of the volumes" method produces the least biased average concentration for samples with frequent non-detects. Log removals were then calculated for the *Clostridium* and bacteriophage data, based on these average concentrations in river and well waters (bracketed values in Table 8-1). Based on the calculations shown in Table 8-1, the removal of *Clostridium* ranged from >2.9 to 3.4 logs, and the removal of bacteriophage ranged from >1.6 to >3.0 logs during RBF.

The monitoring results for *Giardia* and *Cryptosporidium*, including the total pathogen counts and total volume sampled, summed over seven sampling rounds are given in Table 8-2.

Table 8-2. *Giardia* and *Cryptosporidium* Monitoring Results Over Seven Sampling Rounds

	Total Volume Sampled (L)	*Giardia* Total Counts	*Cryptosporidium* Total Counts
Jeffersonville, IN			
Ohio River	6.2	0	0
Well 9	313.3	0	1
Well 2	346.3	0	0
Terre Haute, IN			
Wabash River	2.2	2	0
Collector Well	292.4	0	0
Well 3	322.5	0	0
Parkville, MO			
Missouri River	1.2	1	0
Well 4	321.1	0	0
Well 5	168.6	0	0

Giardia concentrations were above the detection limits twice in Wabash River water and once in Missouri River water, but never in any of the well waters. *Cryptosporidium* concentrations were below the detection limits in all river-water samples and in all but one well-water sample (Jeffersonville, Well 9), where it was observed at the detection limit. Since the volumes sampled for *Giardia* and *Cryptosporidium* were much lower in the river waters compared to the riverbank-filtered waters and varied among locations and from one sampling round to another, the data are not of sufficient quality for calculating the average concentrations or log removals.

At the Terre Haute facility, coliform concentrations were analyzed according to Standard Methods (American Public Health Association et al., 1998). Total coliforms were detected in 15 monthly samples of the Wabash River, but in only 2 monthly samples of the collector well, whose radial arms extend below the river at a depth approximately 21 m below the bottom (Table 8-3). Consequently, the results indicate the limited passage of coliforms during ground passage. The well waters receive disinfection prior to distribution to the consumer.

Table 8-3. Number of Monthly Sampling Rounds at Terre Haute, Indiana, with Positive Detects during Coliform Monitoring (15 Total Sampling Rounds)

	Total Coliforms	E. Coli
Wabash River	15	15
Collector Well	2	0
Well 3	0	0

5. Disinfection Byproduct Formation Potential Testing

River and riverbank-filtered water samples were collected over 2 years (19 sampling rounds each at Jeffersonville and Terre Haute, and 18 sampling rounds at Parkville) and subjected to disinfection byproduct FP testing, based on Standard Methods 5710 B) (American Public Health Association et al., 1998). Based on a 2-day chlorine demand test, each sample was dosed with three chlorine concentrations. The sample with a free chlorine residual concentration between 2 and 5 mg/L at the end of the 7-day incubation period was chosen for disinfection byproduct analysis. The test was performed at pH 7.0 using a phosphate buffer. Samples were analyzed for:

- THMs.
- Six HAAs.
- Haloacetonitriles (HAN).
- Haloketones (HK).
- Chloral hydrate (CH).
- Chloropicrin (CP).

Total THM is the sum of the four THM species ($CHCl_3$, $CHCl_2Br$, $CHClBr_2$, and $CHBr_3$) and is represented by the acronym TTHM. Total HAA is the sum of the six species measured here (CCl_3COOH, $CHCl_2COOH$, $CH_2ClCOOH$, $CHClBrCOOH$, $CH_2BrCOOH$, and $CHBr_2COOH$) and is represented by the acronym HAA6. The HAN species are represented by the following acronyms:

- Trichloroacetonitrile: TCAN.
- 1,1-dichloroacetonitrile: DCAN.
- Bromochloroacetonitrile: BCAN.
- Dibromoacetonitrile: DBAN.

TOC and DOC were measured according to Standard Methods 5310. THM and HAA analyses followed U.S. Environmental Protection Agency Method 502.2 (U.S. Environmental Protection Agency, 1992) and Standard Methods 6251B, respectively. HAN, HK, CH, and CP analyses were performed using U.S. Environmental Protection Agency Method 551 (U.S. Environmental Protection Agency, 1992).

Jeffersonville, Indiana

The Ohio River concentrations and percent reductions upon RBF of TOC, DOC, THM FP, and HAA FP averaged over the 19 sampling rounds for Jeffersonville are shown in Table 8-4. The average concentrations and reductions of HAN, HK, CH, and CP FPs are provided in Table 8-5.

Table 8-4. Average Concentrations in the Ohio River and Reductions upon RBF for Jeffersonville, Indiana, TOC, DOC, THM FP, and HAA FP

	TOC	DOC	TTHM FP	Cl-THM FP[a]	Br-THM FP[b]	HAA6 FP	Cl-HAA FP[c]	Br-HAA FP[d]
Ohio River	3.0 mg/L	2.7 mg/L	197 μg/L	155 μg/L	42 μg/L	208 μg/L	188 μg/L	20 μg/L
Well 9	60%	58%	69%	80%	30%	76%	80%	39%
Well 2	75%	74%	85%	93%	54%	84%	87%	62%

[a]Cl-THM FP = $CHCl_3$ FP.
[b]Br-THM FP = $CHCl_2Br$ FP + $CHClBr_2$ FP + $CHBr_3$ FP.
[c]Cl-HAA FP = CCl_3COOH FP + $CHCl_2COOH$ FP + $CH_2ClCOOH$ FP.
[d]Br-HAA FP = $CHClBrCOOH$ FP + $CH_2BrCOOH$ FP + $CHBr_2COOH$ FP.

TTHM FP, HAA6 FP, and HAN FP are divided into chlorinated and brominated fractions, indicated by the prefix Cl- or Br- in Tables 8-4 and 8-5. As previously discussed, inorganic data suggest that Well 9 (61 m from the river) withdraws primarily infiltrated river water, while Well 2 (177 m from the river) is likely to withdraw a substantial fraction of local groundwater.

Table 8-5. Average Concentrations in the Ohio River and Reductions upon RBF for Jeffersonville, Indiana, HAN FP, HK FP, CH FP, and CP FP

	HAN FP	Cl-HAN FP[a]	Br-HAN FP[b]	HK FP	CH FP	CP FP
Ohio River	13 µg/L	8 µg/L	5 µg/L	4 µg/L	40 µg/L	2 µg/L
Well 9	40%	64%	−1%	69%	87%	100%
Well 2	66%	81%	41%	89%	95%	97%

[a]Cl-HAN FP = TCAN FP + DCAN FP.
[b]Br-HAN FP = BCAN FP + DBAN FP.

TOC, DOC, and all of the total FP concentrations were reduced significantly upon RBF. For TTHM FP, HAA6 FP, and HAN FP, the chlorinated fractions were reduced to a significantly larger degree than the brominated fractions. With the exception of HAN FP, all of the FPs were reduced to a greater extent than TOC and DOC, suggesting a preferential removal of the halogen-reactive organic material.

Upon treatment and RBF, there is a shift in dominant species from the most chlorinated to the more brominated disinfection byproduct species (Figure 8-7). The shift is more pronounced upon RBF compared to the laboratory-simulated treatments. The reduced TOC in well water (and the resulting increased ratio of bromide to TOC) is likely to be the major reason for this shift from the chlorinated to brominated species; however, it is possible that there was preferential removal upon RBF of the precursor material for the more chlorinated species.

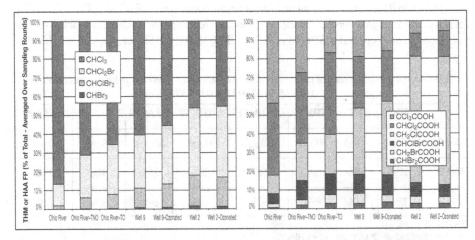

Figure 8-7. THM and HAA FP distributions at Jeffersonville, Indiana. TNO = Treated, non-ozonated. TO = Treated, ozonated.

Terre Haute, Indiana

The concentrations in the Wabash River and percent reductions upon RBF of TOC, DOC, THM FP, and HAA FP averaged over the 19 sampling rounds for Terre Haute are shown in Table 8-6. The average concentrations and reductions of HAN, HK, CH, and CP FPs are provided in Table 8-7. TTHM FP, HAA6 FP, and HAN FP are divided into chlorinated and brominated fractions, indicated by the prefix Cl- or Br- in Tables 8-6 and 8-7. It is believed that the collector well (27 m from the river) withdraws primarily riverbank-filtered water while Well 3 (122 m from the river) has a significant groundwater contribution.

Table 8-6. Average Concentrations in the Wabash River and Reductions upon RBF for Terre Haute, Indiana, TOC, DOC, THM FP, and HAA FP

	TOC	DOC	TTHM FP	Cl-THM FP	Br-THM FP	HAA6 FP	Cl-HAA FP	Br-HAA FP
Wabash River	4.7 mg/L	4.1 mg/L	326 µg/L	250 µg/l	76 µg/L	354 µg/L	321 µg/L	33 µg/L
Collector Well	67%	64%	73%	87%	24%	78%	83%	29%
Well 3	88%	88%	97%	99%	89%	93%	93%	89%

Table 8-7. Average Concentrations in the Wabash River and Reductions upon RBF for Terre Haute, Indiana, HAN FP, HK FP, CH FP, and CP FP

	HAN FP	Cl-HAN FP	Br-HAN FP	HK FP	CH FP	CP FP
Wabash River	33 µg/L	20 µg/L	13 µg/L	6.4 µg/L	70 µg/L	3 µg/L
Collector Well	69%	82%	49%	69%	91%	93%
Well 3	95%	98%	90%	98%	100%	100%

As with the Jeffersonville data, the TOC, DOC, and all of the total FP concentrations were reduced significantly upon RBF. All of the total FP concentrations were reduced to a greater extent than TOC and DOC, suggesting a preferential removal of disinfection byproduct precursor material. A shift from the more chlorinated to more brominated disinfection byproduct species upon RBF was observed. The increasing ratio of bromide to TOC from the Wabash River to the wells is likely to be the major reason for this shift from the chlorinated to brominated species; however, it is possible that there was some preferential removal upon RBF of the precursor material responsible for the more chlorinated species.

Parkville, Missouri

The Missouri River concentrations and percent reductions upon RBF of TOC, DOC, THM FP, and HAA FP averaged over the 18 sampling rounds for Parkville are shown in Table 8-8. The average concentrations and reductions of HAN, HK, CH, and CP FPs are given in Table 8-9. TTHM FP, HAA6 FP, and HAN FP are divided into chlorinated and brominated fractions, indicated by the prefix Cl- or Br- in Tables 8-8 and 8-9. Well 4 and Well 5, which are equidistant from the Missouri River (37 m from the river), are both believed to withdraw largely riverbank-filtered river water.

As with the Jeffersonville and Terre Haute data, the TOC, DOC, and various disinfection byproduct formational potential concentrations were reduced significantly upon RBF. With the

Table 8-8. Average Concentrations in the Missouri River and Reductions upon RBF for Parkville, Missouri, TOC, DOC, THM FP, and HAA FP

	TOC	DOC	TTHM FP	Cl-THM FP	Br-THM FP	HAA6 FP	Cl-HAA FP	Br-HAA FP
Missouri River	4.5 mg/L	3.6 mg/L	240 µg/L	207 µg/L	33 µg/L	228 µg/L	214 µg/L	14 µg/L
Well 4	41%	35%	57%	64%	13%	51%	54%	-5%
Well 5	40%	36%	57%	64%	16%	50%	53%	2%

Table 8-9. Average Concentrations in the Missouri River and Reductions upon RBF for Parkville, Missouri, HAN FP, HK FP, CH FP, and CP FP

	HAN FP	Cl-HAN FP	Br-HAN FP	HK FP	CH FP	CP FP
Missouri River	15 µg/L	11 µg/L	4 µg/L	4 µg/L	50 µg/L	2 µg/L
Well 4	47%	66%	−7%	27%	73%	97%
Well 5	45%	65%	−8%	49%	77%	97%

exception of HK FP in Well 4, all of the FP concentrations were reduced to a greater extent than TOC and DOC. As with the other two sites, this suggests a preferential removal of precursor material. For THM FP, HAA FP, and HAN FP, there was a shift from the chlorinated to more brominated species, as previously discussed for the other two sites.

6. Simulated Conventional Treatment

The quality of riverbank-filtered waters was compared to river waters subjected to a laboratory-simulated conventional treatment train that consisted of coagulation, flocculation, sedimentation, ozonation, and filtration. Two different trains were used, one with the ozonation step and one without the ozonation step. In addition, one set of riverbank-filtered waters was subjected to the ozonation step. The coagulation/flocculation/sedimentation step was carried out such that the alum dose represented the optimal dose for turbidity reduction during jar testing. The resulting reductions in TOC were checked for compliance with the Enhanced Coagulation Rule (White et al., 1997). The required TOC reduction for the Enhanced Coagulation Rule was achieved in 31 out of 37 total runs, where enough data were available to check for compliance. The simulated treatments were carried out over 16 sampling rounds at Jeffersonville and Terre Haute and 15 sampling rounds at Parkville.

Jeffersonville, Indiana

The average Ohio River concentrations and percent reductions upon RBF or simulated treatment for TOC, DOC, and the various FPs are provided in Tables 8-10 and 8-11. For TOC, DOC, and the FPs of THM, HAA, HK, CH, and CP, RBF provided greater reductions than simulated treatments. For HAN FP, comparable reductions were achieved by simulated treatment and RBF. In general, the reductions in FP concentrations were greater than the reductions in TOC and DOC, suggesting a preferential removal of disinfection byproduct precursors during simulated treatment as well as with RBF. Others have shown strong correlations between the humic content of water, which is preferentially removed during conventional treatment, and disinfection byproduct formation, suggesting a preferential removal of disinfection byproducts precursors upon treatment (Singer, 1999; Croué et al., 1999).

Simulated conventional treatment followed by ozonation of the Ohio River water led to an increase in the absolute concentration of chloropicrin, indicated by the negative reduction in

Table 8-10. Average Concentrations in the Ohio River and Reductions upon RBF or Simulated Conventional Treatment for Jeffersonville, Indiana, TOC, DOC, THM FP, and HAA FP

	TOC	DOC	TTHM FP	Cl-THM FP	Br-THM FP	HAA6 FP	Cl-HAA FP	Br-HAA FP
Ohio River	3.1 mg/L	2.7 mg/L	221 µg/L	181 µg/L	40 µg/L	226 µg/L	205 µg/L	21 µg/L
Ohio River - TNO[a]	36%	33%	54%	64%	5%	62%	66%	24%
Ohio River - TO[b]	46%	40%	64%	75%	13%	69%	74%	21%
Well 9	63%	58%	72%	83%	24%	77%	81%	42%
Well 9 - O[c]	63%	59%	77%	87%	30%	78%	82%	44%
Well 2	76%	74%	85%	94%	49%	83%	85%	63%
Well 2 - O[c]	77%	74%	89%	95%	60%	84%	85%	66%

[a]TNO = Treated, non-ozonated.   [b]TO = Treated, ozonated.   [c]O = Ozonated.

Table 8-11. Average Concentrations in the Ohio River and Reductions upon RBF or Simulated Conventional Treatment for Jeffersonville, Indiana, HAN FP, HK FP, CH FP, and CP FP

	HAN FP	Cl-HAN FP	Br-HAN FP	HK FP	CH FP	CP FP
Ohio River	14 µg/L	8 µg/L	6 µg/L	4 µg/L	40 µg/L	2 µg/L
Ohio River -TNO[a]	48%	54%	40%	36%	63%	76%
Ohio River - TO[b]	65%	86%	35%	28%	51%	–30%
Well 9	44%	65%	12%	57%	87%	100%
Well 9 - O[c]	68%	88%	38%	61%	84%	96%
Well 2	69%	82%	51%	80%	95%	97%
Well 2 - O[c]	90%	99%	76%	88%	94%	97%

[a]TNO = Treated, non-ozonated.   [b]TO = Treated, ozonated.   [c]O = Ozonated.

Table 8-11. These results are consistent with the results of Miltner et al. (1992), who found that while the concentrations of total THM and HAA precursors were reduced upon ozonation, the formation of chloropicrin increased upon ozonation. In addition, because the CP FP concentration in the Ohio River is so low, the increase in Ohio River treated/ozonated water is likely to fall within the range of experimental error. This error magnification is likely to be the major reason for similar increases seen subsequently in Tables 8-12, 8-13, 8-14, 8-15, 8-16, 8-17, 8-19, and 8-21.

For THM FP, HAA FP, and HAN FP, there was a shift from the more chlorinated to the more brominated species upon simulated treatment similar to that previously discussed for RBF. This is indicated by the much larger reductions in the chlorinated disinfection byproduct species than in the brominated species. Since bromide remained fairly constant upon simulated treatment or RBF while TOC was reduced, the bromide-to-TOC ratio increased, leading to the increased formation of brominated compounds upon chlorination. It is also possible that the precursor material responsible for forming the more chlorinated disinfection byproducts was preferentially removed upon treatment.

Terre Haute, Indiana

The average Wabash River concentrations and percent reductions upon RBF or simulated treatment for Terre Haute are provided in Tables 8-12 and 8-13. For TOC, DOC, and various disinfection byproduct FP concentrations, RBF provided better reductions than simulated conventional

Table 8-12. Average Concentrations in the Wabash River and Reductions upon RBF or Simulated Conventional Treatment for Terre Haute, Indiana, TOC, DOC, THM FP, and HAA FP

	TOC	DOC	TTHM FP	Cl-THM FP	Br-THM FP	HAA6 FP	Cl-HAA FP	Br-HAA FP
Wabash River	4.7 mg/L	4.0 mg/L	325 µg/L	245 µg/L	80 µg/L	320 µg/L	283 µg/L	37 µg/L
Wabash River - TNO[a]	30%	23%	44%	60%	-2%	48%	56%	–15%
Wabash River - TO[b]	45%	37%	66%	78%	30%	65%	71%	20%
Collector Well	68%	64%	72%	88%	23%	74%	79%	33%
Collector Well - O[c]	70%	66%	82%	92%	53%	80%	83%	52%
Well 3	89%	88%	97%	99%	90%	91%	91%	91%
Well 3 - O[c]	89%	87%	97%	99%	91%	91%	91%	91%

[a]TNO = Treated, non-ozonated.   [b]TO = Treated, ozonated.   [c]O = Ozonated.

Table 8-13. Average Concentrations in the Wabash River and Reductions upon RBF or Simulated Conventional Treatment for Terre Haute, Indiana, HAN FP, HK FP, CH FP, and CP FP

	HAN FP	Cl-HAN FP	Br-HAN FP	HK FP	CH FP	CP FP
Wabash River	36 µg/L	21 µg/L	15 µg/L	6 µg/L	74 µg/L	2 µg/L
Wabash River - TNO[a]	54%	68%	34%	42%	67%	74%
Wabash River - TO[b]	80%	91%	58%	52%	63%	–21%
Collector Well	70%	82%	52%	65%	91%	91%
Collector Well - O[c]	84%	95%	68%	73%	91%	89%
Well 3	95%	98%	90%	97%	100%	100%
Well 3 - O[c]	96%	98%	92%	98%	97%	97%

[a]TNO = Treated, non-ozonated.   [b]TO = Treated, ozonated.   [c]O = Ozonated.

treatments. Ozonation of the collector well water provided slightly better reductions in TOC, DOC, and various disinfection byproduct FP concentrations, with the exception of CH FP and CP FP. Reductions in most FP concentrations were higher than the observed reductions in TOC and DOC. In addition, for THM FP, HAA FP, and HAN FP, there was a significant shift from the chlorinated to more brominated species with each of the various treatments. These results are very similar to the Jeffersonville data.

Parkville, Missouri

The average Missouri River concentrations and percent reductions upon RBF or simulated treatment for Parkville are provided in Tables 8-14 and 8-15. For TOC, DOC, and various disinfection byproduct FP concentrations, RBF provided comparable or better reductions than simulated conventional treatments. Upon ozonation, higher reductions were observed for TOC, DOC, TTHM FP, HAA6 FP, and HAN FP. With the exception of HK FP, the reductions in total FP concentrations were greater than the reductions in TOC and DOC, suggesting a preferential removal of FP precursors with the various treatments. In addition, for THM FP, HAA FP, and HAN FP, there was a shift from the more chlorinated to more brominated species upon treatment due primarily to the increasing bromide-to-TOC ratio upon treatment. These results are very similar to the Jeffersonville and Terre Haute data.

Table 8-14. Average Concentrations in the Missouri River and Reductions upon RBF or Simulated Conventional Treatment for Parkville, Missouri, TOC, DOC, THM FP, and HAA FP

	TOC	DOC	TTHM FP	Cl-THM FP	Br-THM FP	HAA6 FP	Cl-HAA FP	Br-HAA FP
Missouri River	4.1 mg/L	3.4 mg/L	225 µg/L	193 µg/L	32 µg/L	220 µg/L	206 µg/L	14 µg/L
Missouri River - TNO[a]	36%	27%	49%	55%	13%	45%	49%	-4%
Missouri River - TO[b]	48%	38%	66%	71%	35%	61%	64%	17%
Well 4	36%	31%	53%	60%	12%	49%	52%	3%
Well 4 - O[c]	48%	45%	65%	73%	21%	56%	60%	-4%
Well 5	35%	32%	53%	60%	15%	47%	50%	1%
Well 5 - O[c]	49%	45%	66%	74%	21%	56%	59%	1%

[a]TNO = Treated, non-ozonated.   [b]TO = Treated, ozonated.   [c]O = Ozonated.

Table 8-15. Average Concentrations in the Missouri River and Reductions upon RBF or Simulated Conventional Treatment for Parkville, Missouri, HAN FP, HK FP, CH FP, and CP FP

	HAN FP	Cl-HAN FP	Br-HAN FP	HK FP	CH FP	CP FP
Missouri River	16 µg/L	12 µg/L	4 µg/L	5 µg/L	52 µg/L	2 µg/L
Missouri River - TNO[a]	49%	57%	27%	24%	58%	60%
Missouri River - TO[b]	65%	85%	8%	32%	55%	-112%
Well 4	51%	69%	0%	37%	74%	94%
Well 4 - O[c]	65%	82%	15%	27%	71%	58%
Well 5	47%	65%	-5%	41%	77%	97%
Well 5 - O[c]	55%	75%	-1%	24%	74%	68%

[a]TNO = Treated, non-ozonated.   [b]TO = Treated, ozonated.   [c]O = Ozonated.

7. Uniform Formation Conditions Testing

During four sampling rounds, uniform formation conditions (UFC) testing was performed on raw and treated river and well-water samples. While the objective of the FP test is to provide an indication of the total concentrations of disinfection byproduct precursor material in source waters, the objective of the UFC test is to more realistically evaluate concentrations of disinfection byproducts that could be expected to form when the residence time in the treatment and distribution system is on the order of one decay. The UFC test was developed based on surveys of treatment systems to simulate average conditions (including pH, residence time, free chlorine residual concentration, and temperature) in actual utilities (Summers et al., 1996). The UFC testing was performed with the chlorine dose based on the sample TOC concentration to provide a 24-hour free chlorine residual concentration of 1 mg/L. The UFC test was performed at a pH of 8.0, in contrast to the FP test pH of 7.0.

Jeffersonville, Indiana

Linear regressions between Jeffersonville FP and UFC concentrations of total THM and HAA6 were obtained by forcing the "best fit" line through the point (0,0). The slope was found to be greater than 1.0 for both the TTHM and HAA6 regressions (THM FP/THM UFC = 1.7;

HAA6 FP/HAA6 UFC = 3.1). This was expected because of the higher chlorine dose and longer contact time during the FP test. The much higher slope for the HAA6 data indicates that the formation of these acidic compounds was even more strongly affected by FP conditions, which include a lower pH (7.0) and a much longer contact time (7 days) than the UFC test (pH 8.0 and 1 day, respectively).

The average Ohio River TOC, DOC, and disinfection byproduct UFC concentrations and percent reductions upon RBF or simulated treatment for Jeffersonville are provided in Tables 8-16 and 8-17. The TOC and DOC data presented here include only those months during which UFC testing took place. The data show a significant reduction in total disinfection byproduct UFC concentrations along with the individual species upon RBF. There was an apparent increase in the absolute brominated HAN UFC concentrations upon both simulated treatment and RBF as well as the HK UFC concentrations upon simulated treatment, as indicated by the negative reductions in Table 8-17. In general, reductions of UFC concentrations upon RBF were larger than reductions in Ohio River water subjected to simulated conventional treatment. TTHM and HAA6 UFC reductions were higher than the corresponding TOC and DOC reductions, indicating a preferential removal of precursor material.

Table 8-16. Average Concentrations in the Ohio River and Reductions upon RBF or Simulated Conventional Treatment for Jeffersonville, Indiana, TOC, DOC, THM UFC, and HAA UFC

	TOC	DOC	TTHM UFC	Cl-THM UFC	Br-THM UFC	HAA6 UFC	Cl-HAA UFC	Br-HAA UFC
Ohio River	3.2 mg/L	2.6 mg/L	143 µg/L	117 µg/L	26 µg/L	86 µg/L	76 µg/L	10 µg/L
Ohio River - TNO[a]	38%	32%	59%	68%	16%	71%	77%	24%
Ohio River - TO[b]	55%	47%	67%	76%	22%	89%	93%	61%
Well 9	67%	60%	78%	92%	14%	88%	95%	31%
Well 9 - O[c]	71%	65%	77%	95%	–4%	93%	98%	37%
Well 2	77%	72%	83%	99%	11%	90%	97%	41%
Well 2 - O[c]	82%	78%	86%	99%	25%	97%	100%	70%

[a]TNO = Treated, non-ozonated.   [b]TO = Treated, ozonated.   [c]O = Ozonated.

Table 8-17. Average Concentrations in the Ohio River and Reductions upon RBF or Simulated Conventional Treatment for Jeffersonville, Indiana, HAN UFC, HK UFC, CH UFC, and CP UFC

	HAN UFC	Cl-HAN UFC	Br-HAN UFC	HK UFC	CH UFC	CP UFC
Ohio River	9 µg/L	6 µg/L	3 µg/L	1 µg/L	19 µg/L	2 µg/L
Ohio River - TNO[a]	27%	61%	–50%	–5%	59%	81%
Ohio River - TO[b]	44%	82%	–41%	–21%	55%	–33%
Well 9	44%	93%	–66%	100%	93%	100%
Well 9 - O[c]	57%	97%	–33%	100%	91%	100%
Well 2	59%	100%	–32%	100%	100%	100%
Well 2 - O[c]	67%	100%	–4%	100%	100%	100%

[a]TNO = Treated, non-ozonated.   [b]TO = Treated, ozonated.   [c]O = Ozonated.

In general, the percent reductions in TTHM and Cl-THM concentrations were very similar between the FP and UFC tests, both on a mass and molar basis. Generally, the HAA6 and Cl-HAA reductions were slightly higher in the UFC testing results than in the FP results. There was no clear relationship for the Br-THM and Br-HAA concentrations between FP and UFC testing at Jeffersonville.

The data in Figure 8-8 show the distribution of the individual THM UFC and FP species as a percentage of the total THM UFC or FP at Jeffersonville. The calculations were based on molar concentrations of the various compounds averaged over 4 months of UFC testing. As observed with the FP data, the UFC data showed a shift from the chlorinated to the more brominated THM species during RBF; however, the shift was much more pronounced in UFC data. The distribution of the HAA UFC and FP species at Jeffersonville is shown in Figure 8-9. Again, the shift from the chlorinated to the more brominated compounds was more pronounced with UFC data than with FP data. This effect was much more significant with HAA data than with THM data.

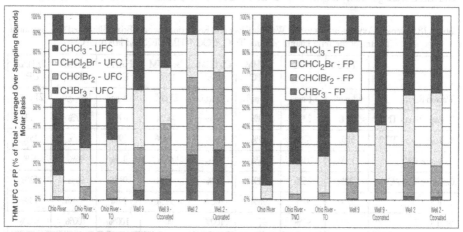

Figure 8-8. THM UFC and FP distributions at Jeffersonville, Indiana. TNO = Treated, non-ozonated. TO = Treated, ozonated.

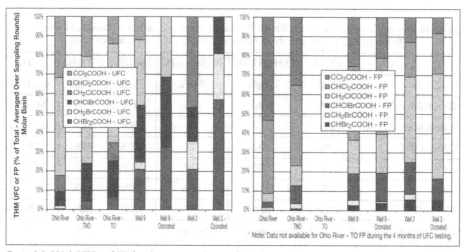

Figure 8-9. HAA UFC and FP distributions at Jeffersonville, Indiana. TNO = Treated, non-ozonated. TO = Treated, ozonated.

Terre Haute, Indiana

Linear regressions between Terre Haute FP and UFC concentrations of total THM and HAA6 were obtained by forcing the "best fit" line through the point (0,0). As was the case with Jeffersonville, the slope was greater than 1.0 for both TTHM and HAA6 regressions (THM FP/THM UFC = 1.1; HAA6 FP/HAA6 UFC = 2.7) because of the higher chlorine dose and longer contact time during the FP test.

The average Wabash River TOC, DOC, and disinfection byproduct UFC concentrations and percent reductions upon RBF or simulated treatment for Terre Haute are provided in Tables 8-18 and 8-19. The data show a significant reduction in both the total amounts and the individual species, with RBF providing larger reductions in disinfection byproduct UFC concentrations than simulated conventional treatment. In general, the percent reductions in TTHM and Cl-THM concentrations were very similar between the two tests. The HAA6 and Cl-HAA reductions were generally slightly higher in the UFC testing results than in the FP results.

Table 8-18. Average Concentrations in the Wabash River and Reductions upon RBF or Simulated Conventional Treatment for Terre Haute, Indiana, TOC, DOC, THM UFC, and HAA UFC

	TOC	DOC	TTHM UFC	Cl-THM UFC	Br-THM UFC	HAA6 UFC	Cl-HAA UFC	Br-HAA UFC
Wabash River	6.1 mg/L	5.1 mg/L	270 µg/L	197 µg/L	73 µg/L	205 µg/L	184 µg/L	21 µg/L
Wabash River - TNO[a]	34%	23%	39%	69%	−41%	69%	77%	−1%
Wabash River - TO[b]	51%	42%	53%	81%	−24%	79%	86%	15%
Collector Well	75%	71%	81%	95%	43%	91%	96%	43%
Collector Well - O[c]	79%	75%	83%	97%	45%	94%	99%	52%
Well 3	92%	91%	98%	100%	92%	99%	100%	93%
Well 3 - O[c]	92%	91%	97%	100%	90%	100%	100%	100%

[a]TNO = Treated, non-ozonated.   [b]TO = Treated, ozonated.   [c]O = Ozonated.

Table 8-19. Average Concentrations in the Wabash River and Reductions upon RBF or Simulated Conventional Treatment for Terre Haute, Indiana, HAN UFC, HK UFC, CH UFC, and CP UFC

	HAN UFC	Cl-HAN UFC	Br-HAN UFC	HK UFC	CH UFC	CP UFC
Wabash River	20 µg/L	11 µg/L	9 µg/L	2 µg/L	43 µg/L	6 µg/L
Wabash River - TNO[a]	-28%	47%	−52%	7%	67%	64%
Wabash River - TO[b]	29%	75%	2%	38%	55%	68%
Collector Well	59%	91%	27%	100%	95%	100%
Collector Well - O[c]	72%	100%	42%	100%	98%	100%
Well 3	90%	100%	80%	100%	100%	100%
Well 3 - O[c]	89%	100%	78%	100%	100%	100%

[a]TNO = Treated, non-ozonated.   [b]TO = Treated, ozonated.   [c]O = Ozonated.

The distributions of TTHM UFC and HAA6 UFC concentrations among the individual species at Terre Haute (Figures 8-10 and 8-11) show the same trend that was observed at Jeffersonville. When compared to FP distributions, UFC distributions show a larger shift to the brominated THM and HAA species.

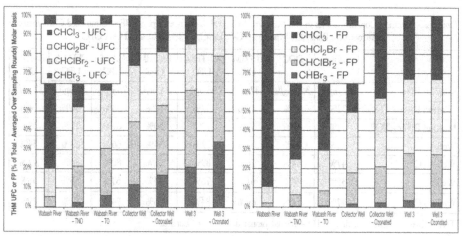

Figure 8-10. THM UFC and FP distributions at Terre Haute, Indiana.

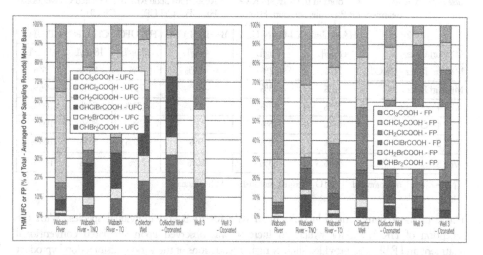

Figure 8-11. HAA UFC and FP distributions at Terre Haute, Indiana. Note: HAA6 UFC for Well 3-Ozonated was 0 µg/L.

Parkville, Missouri

Linear regressions for Parkville TTHM and HAA6 data were obtained by forcing the "best fit" lines through the point (0,0). The correlation of FP with UFC concentrations for Parkville HAA6 data was much less strong than for the other sites (a low correlation coefficient was obtained) and was also weak at low UFC concentrations for Parkville TTHM data. Although the linear fit to the data was poor, the slopes were calculated to gain a general idea of the relationship between FP and UFC data. The slope was greater than 1.0 for both TTHM and HAA6 regressions (THM FP/THM UFC = 1.8; HAA6 FP/HAA6 UFC = 3.3) because of the higher chlorine dose and longer contact time during the FP test.

The average Missouri River TOC, DOC, and disinfection byproduct UFC concentrations and percent reductions upon RBF or simulated treatment for Parkville are provided in Tables 8-20 and 8-21.

Table 8-20. Average Concentrations in the Missouri River and Reductions upon RBF or Simulated Conventional Treatment for Parkville, Missouri, TOC, DOC, THM UFC, and HAA UFC

	TOC	DOC	TTHM UFC	Cl-THM UFC	Br-THM UFC	HAA6 UFC	Cl-HAA UFC	Br-HAA UFC
Missouri River	3.7 mg/L	3.3 mg/L	132 µg/L	98 µg/L	34 µg/L	78 µg/L	66 µg/L	12 µg/L
Missouri River - TNO[a]	32%	30%	45%	57%	10%	53%	56%	34%
Missouri River - TO[b]	44%	45%	57%	70%	20%	67%	72%	38%
Well 4	35%	43%	62%	72%	34%	66%	68%	54%
Well 4 - O[c]	47%	54%	78%	89%	48%	80%	86%	42%
Well 5	35%	44%	61%	70%	34%	63%	68%	38%
Well 5 - O[c]	47%	54%	69%	84%	26%	74%	82%	29%

[a]TNO = Treated, non-ozonated.　[b]TO = Treated, ozonated.　[c]O = Ozonated.

Table 8-21. Average Concentrations in the Missouri River and Reductions upon RBF or Simulated Conventional Treatment for Parkville, Missouri, HAN UFC, HK UFC, CH UFC, and CP UFC

	HAN UFC	Cl-HAN UFC	Br-HAN UFC	HK UFC	CH UFC	CP UFC
Missouri River	11 µg/L	7 µg/L	4 µg/L	0.8 µg/L	18 µg/L	0.8 µg/L
Missouri River - TNO[a]	21%	52%	−34%	−7%	55%	75%
Missouri River - TO[b]	30%	73%	−44%	−73%	30%	−96%
Well 4	37%	57%	4%	−131%	86%	100%
Well 4 - O[c]	67%	86%	33%	2%	80%	100%
Well 5	43%	63%	9%	−60%	85%	100%
Well 5 - O[c]	56%	83%	10%	11%	75%	100%

[a]TNO = Treated, non-ozonated.　[b]TO = Treated, ozonated.　[c]O = Ozonated.

The data show significant reductions upon RBF for total and individual species. With the exception of HK UFC, for which an increase was observed upon simulated conventional treatment and RBF, RBF provided slightly higher reductions of the various disinfection byproduct UFC concentrations than simulated treatment. In contrast, the average FP concentrations showed similar reductions between the simulated treatments and RBF. There was an increase in the actual concentrations of HK for all but Well 4 (ozonated). There was also an increase in the actual concentrations of brominated HAN UFC upon simulated treatment. The Br-HAN increase was not observed in riverbank-filtered waters. Because the observed concentrations in river water are very small, the increases observed upon treatment represent only very small changes in the concentrations and are likely to be in the range of experimental error.

The percent reductions in chlorinated TTHM and Cl-THM concentrations were very similar between the two tests, although Cl-THM UFC reductions upon RBF were slightly higher than Cl-THM FP reductions. Generally, HAA6 and Cl-HAA reductions were slightly higher in UFC testing results than in FP results. There was no clear relationship for the Br-THM and Br-HAA concentrations between FP and UFC testing at Parkville.

The distributions of TTHM UFC and HAA6 UFC concentrations among the individual species at Parkville (Figures 8-12 and 8-13) show the same trend that was observed at the other sites. When compared to FP distributions, UFC distributions show a larger shift to the brominated THM and HAA species.

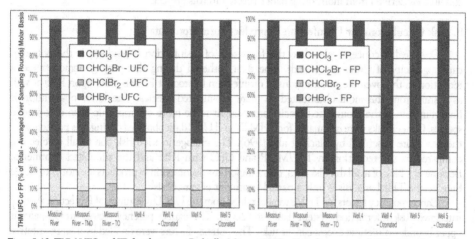

Figure 8-12. THM UFC and FP distributions at Parkville, Missouri. TNO = Treated, non-ozonated. TO = Treated, ozonated.

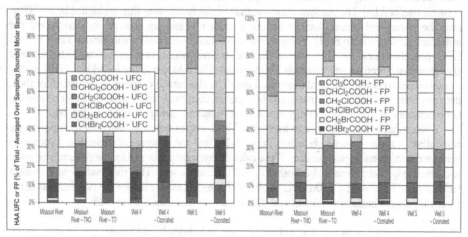

Figure 8-13. HAA UFC and FP distributions at Parkville, Missouri. TNO = Treated, non-ozonated. TO = Treated, ozonated.

Conclusions from UFC Testing

The comparison between the FP and UFC results revealed important differences in terms of TTHM and HAA6 concentrations that are formed from precursor materials and the relative distributions of chlorinated and brominated disinfection byproducts. First, the UFC test resulted in lower TTHM and HAA6 concentrations. The average HAA6 FP-to-HAA6 UFC ratios for the three sites were 3.1, 2.7, and 3.3. Consequently, the HAA6 FP concentrations were approximately three times larger than the HAA6 UFC concentrations. The average TTHM FP-to-TTHM UFC ratios for the three sites were 1.7, 1.1, and 1.8, indicating approximately 75 percent higher concentrations of TTHMs with the FP test at two of the sites. Second, the UFC test resulted in greater relative amounts of the brominated disinfection byproducts in comparison to FP results. The differences between FP and UFC results stem mainly from the differences in chlorine dose, contact time, and pH.

Both the FP and UFC data demonstrate that RBF can provide reductions in TOC, DOC, and disinfection byproduct precursors greater than or comparable to those from simulated conventional treatment. Thus, the use of either the FP or UFC test does not alter the major conclusions derived from monitoring data collected in this project.

8. Risk Calculations for Disinfection Byproduct Formation Data

One health concern related to the formation of disinfection byproducts is the potential carcinogenicity of these compounds. The data gathered during this project demonstrated a reduction in the overall concentrations of disinfection byproduct FP and UFC upon RBF as well as a shift from the more chlorinated to more brominated species. The latter observation is important, since the brominated species are thought to be more carcinogenic than chlorinated species. Cancer risks were calculated for THM FP and UFC data obtained during this project using the procedure described by Black et al. (1996). The change in total risk due to all species was compared to the reduction in TTHM FP and UFC with RBF and the simulated treatments. Because of a lack of available risk data, other important disinfection byproducts are not considered here, so the THM analysis is only a partial treatment of cancer risk.

Cancer potency factors for THMs were obtained from the U.S. Environmental Protection Agency Integrated Risk Information System database (2000). The cancer potency factors are 95-percent upper bounds on lifetime excess cancer risks. The potency factors were divided by a representative body weight of 70 kg and a lifespan of 70 years, and multiplied by an average daily water consumption of 2 L/d. The resulting annual individual risk factors were then multiplied by 1×10^6 to give the normalized cancer potency factors in units of cases per year per million people per milligram per liter (Table 8-22). These factors were then multiplied by the various THM FP and UFC concentrations for the river, well, and treated water samples to calculate the risk corresponding to consumption of each THM species. The risks due to the individual THM species were summed to obtain the total theoretical risk. The theoretical cancer risk calculated for these data represents an "excess risk" that would be expected due to the consumption of water containing concentrations of THMs in the range of what was measured during FP and UFC testing.

Theoretical cancer risks due to individual THM species were calculated by multiplying the normalized cancer potency factors (Table 8-22) by the average disinfection byproducts concentrations in milligrams per liter. These calculations were performed for the river, treated river, and riverbank-filtered waters for both FP and UFC data sets. The river values and percent reductions in TTHM FP, TTHM UFC, and total theoretical cancer risk for the two data sets at the three sites upon simulated treatment and RBF are shown in Tables 8-23 to 8-25.

Table 8-22. Cancer Potency Factors for Individual THM Species

THM	Cancer Potency Factor[a] (cases/person/lifetime/[mg/kg body weight/d])	Normalized Cancer Potency Factor[b] (cases/year/million people/[mg/L])
$CHCl_3$	0.0061	2.5
$CHCl_2Br$	0.062	25.3
$CHClBr_2$	0.084	34.3
$CHBr_3$	0.0079	3.2

[a]Potency factors obtained from U.S. Environmental Protection Agency Integrated Risk Information System database.
[b]Normalized based on 70 kg body weight, 70-year lifespan, 2 L/d water consumption, and 10^6 people.

Table 8-23. Theoretical Upper Bound THM Cancer Risks (for Average Concentrations) of Ohio River Water and Reductions in Calculated Risk at Jeffersonville, Indiana

	TTHM FP	Total Risk (FP)	TTHM UFC	Total Risk (UFC)
Ohio River	221 μg/L	1.5 cases/yr/10^6 people	143 μg/L	1.0 cases/yr/10^6 people
Ohio River - TNO[a]	54%	21%	59%	29%
Ohio River - TO[b]	64%	30%	67%	36%
Well 9	72%	40%	78%	37%
Well 9 - O[c]	77%	45%	77%	31%
Well 2	85%	61%	83%	48%
Well 2 - O[c]	89%	70%	86%	57%

[a]TNO = Treated, non-ozonated.   [b]TO = Treated, ozonated.   [c]O = Ozonated.

Table 8-24. Theoretical Upper Bound THM Cancer Risks (for Average Concentrations) of Wabash River Water and Reductions in Calculated Risk at Terre Haute, Indiana

	TTHM FP	Total Risk (FP)	TTHM UFC	Total Risk (UFC)
Wabash River	325 μg/L	2.8 cases/yr/10^6 people	270 μg/L	3.1 cases/yr/10^6 people
Wabash River - TNO[a]	44%	11%	39%	-18%
Wabash River - TO[b]	66%	40%	53%	2%
Collector Well	72%	36%	81%	58%
Collector Well - O[c]	82%	61%	83%	62%
Well 3	97%	92%	98%	95%
Well 3 - O[c]	97%	92%	97%	94%

[a]TNO = Treated, non-ozonated.   [b]TO = Treated, ozonated.   [c]O = Ozonated.

Table 8-25. Theoretical Upper Bound THM Cancer Risks (for Average Concentrations) of Missouri River Water and Reductions in Calculated Risk at Parkville, Missouri

	TTHM FP	Total Risk (FP)	TTHM UFC	Total Risk (UFC)
Missouri River	225 μg/L	1.3 cases/yr/10^6 people	132 μg/L	1.2 cases/yr/10^6 people
Missouri River - TNO[a]	49%	27%	45%	19%
Missouri River - TO[b]	66%	47%	57%	29%
Well 4	53%	28%	62%	40%
Well 4 - O[c]	65%	38%	78%	56%
Well 5	53%	29%	61%	40%
Well 5 - O[c]	66%	38%	69%	38%

[a]TNO = Treated, non-ozonated.   [b]TO = Treated, ozonated.   [c]O = Ozonated.

The theoretical upper bound THM cancer risks (for average concentrations) for Jeffersonville simulated treatment THM FP data are given in Figure 8-14. The percent reductions in average TTHM FP, TTHM UFC, and the average theoretical cancer risk corresponding to the two data sets are given in Figure 8-15.

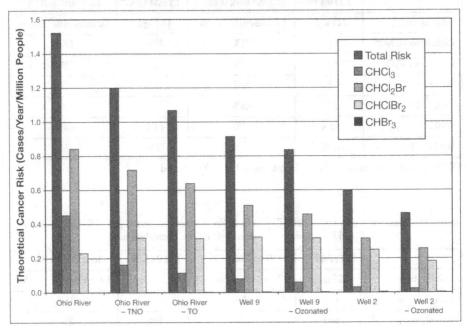

Figure 8-14. Theoretical upper bound cancer risk induced by THM species for Jeffersonville, Indiana, simulated treatment THM FP data. TNO = Treated, non-ozonated. TO = Treated, ozonated.

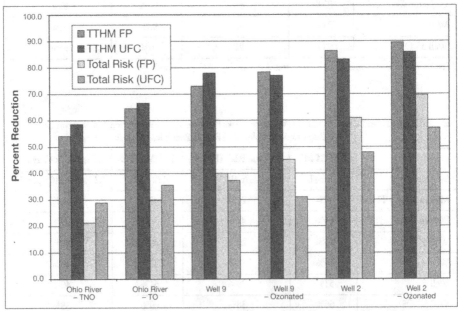

Figure 8-15. Average percent reduction in TTHM FP, TTHM UFC, and total theoretical cancer risk calculated from FP and UFC data sets. TNO = Treated, non-ozonated. TO = Treated, ozonated.

At Jeffersonville, the theoretical upper bound cancer risk (for average concentrations) decreased with simulated treatment and RBF relative to the raw Ohio River water (see Figure 8-14). Given the relative distribution of brominated and chlorinated components, the percent reduction in total risk upon treatment or RBF was lower than the corresponding reduction in TTHM FP or UFC (see Figure 8-15). RBF was more effective at reducing the total risk than the simulated treatment of the Ohio River water. Similarly, at Terre Haute, RBF performed significantly better at reducing the theoretical cancer risk than the simulated treatment (see Table 8-24). In the case of the treated, non-ozonated Wabash River water, the total risk calculated from the UFC data set increased relative to the raw Wabash River water, as indicated by the negative reduction in Table 8-24. This corresponds to the increase that was observed in the actual concentration of the brominated THM UFC concentrations for the Wabash River (treated/non-ozonated) (see Table 8-18). At Parkville, total risk calculated from the FP data set gave comparable reductions between the simulated treatments and RBF (see Table 8-25). Using the UFC data set, RBF was significantly more effective at reducing the total theoretical cancer risk than the simulated treatments. The data also show that ozonation reduces the theoretical cancer risk.

Because of the changes in the speciation of disinfection byproduct formation upon RBF and the difference in toxicity among the various species, it is important to look at the changes in toxicity of the disinfection byproducts upon RBF. The data show that while the reductions in total theoretical risk were not necessarily as high as the corresponding reductions in TTHM FP or UFC because of the shift to the brominated species, there was still a significant reduction in total risk associated with the reduction in THM FP or UFC upon RBF.

9. Conclusions

The three sites investigated during this project demonstrated the ability of RBF to effectively reduce a variety of microbial contaminants, TOC, DOC, disinfection byproduct precursor material, and total theoretical THM cancer risk. TOC and DOC reductions ranged from 35 to 67 percent for the closest wells at the three sites. Total THM FP and HAA FP reductions ranged from 57 to 73 percent and 50 to 78 percent, respectively. The higher reduction in disinfection byproduct precursor concentrations compared to TOC or DOC suggests a preferential removal of precursor material. In addition, a shift was observed upon RBF from the chlorinated to the more brominated disinfection byproduct species.

Further, it was demonstrated that RBF can provide reductions in TOC, DOC, disinfection byproduct precursor material, and theoretical cancer risk that match or exceed those provided by conventional treatment processes, even though the distribution of disinfection byproducts favors the brominated species in riverbank-filtered water. Total THM FP and HAA FP reductions upon simulated treatment ranged from 44 to 66 percent and 45 to 69 percent, respectively. Total THM FP and HAA FP reductions upon RBF ranged from 53 to 82 percent and 47 to 80 percent, respectively. Reductions in the theoretical cancer risk due to the THM FP concentrations ranged from 11 to 47 percent and 28 to 45 percent for the treated and riverbank-filtered waters, respectively. For two of the sites, ozonation of the riverbank-filtered water provided the largest reduction in theoretical cancer risk. As utilities respond to increasingly stringent regulations regarding disinfection byproducts and microbial contaminants, the water-quality improvements commensurate with RBF can help meet those regulations.

Acknowledgements

We gratefully acknowledge the support of the U.S. Environmental Protection Agency Office of Research and Development (Project CR-826337). We also thank Esther Van Zundert for her work in The Johns Hopkins laboratory, performing the UFC testing during the "practical period" of her graduate studies. The views expressed herein have not been subjected to Agency review and, therefore, do not necessarily reflect the views of the Agency, and no official endorsement should be inferred.

While this manuscript was in press, the cancer potency factor for chloroform used here in calculating theoretical upper bound cancer risks was removed from the U.S. Environmental Protection Agency's IRIS database. A dose of average daily water consumption of 0.01 mg/kg/day (or 350 μg/L for a 70-kg person with an average daily water consumption of 2 L/d) is now considered protective against cancer risk. Consequently, our calculations with the earlier potency factor are conservative with respect to theoretical upper bound excess cancer risk.

References

American Public Health Association, American Water Works Association, and Water Environment Federation (1998). *Standard methods for the examination of water and wastewater*, 20th Edition. American Public Health Association, Washington, D.C.

Black, B.D., G.W. Harrington, and P.C. Singer (1996). "Reducing cancer risks by improving organic carbon removal." *Journal American Water Works Association*, 88(6): 40-52.

Bouwer, E., J. Weiss, W. Ball, C. O'Melia, and H. Arora (1999). "Water quality improvements during riverbank filtration at three midwest utilities." *Abstracts, International Riverbank Filtration Conference*, National Water Research Institute, Fountain Valley, California, 49-52.

Croué, J.P., J.F. Debroux, G.L. Amy, G.R. Aiken, and J.A. Leenheer (1999). "Natural organic matter: Structural characteristics and reactive properties." *Formation and control of disinfection by-products in drinking water*, P.C. Singer, ed., American Water Works Association, Denver, Colorado, 65-93.

Doussan, C., G. Poitevin, E. Ledoux, and M. Detay (1997). "River bank filtration: Modeling of the changes in water chemistry with emphasis on nitrogen species." *Journal of Contaminant Hydrology*, 25: 129-156.

Kivimaki, A.L., K. Lahti, T. Hatva, S.M. Tuominen, and I.T. Miettinen (1998). "Removal of organic matter during bank filtration." *Artificial Recharge of Groundwater*, J.H. Peters, ed., A.A. Balkema, Rotterdam, The Netherlands, 107-112.

Kühn, W., and U. Müller (2000). "Riverbank filtration: An overview." *Journal of American Water Works Association*, 92(12): 60-69.

Kussmaul, H. (1979). "Purifying action of the ground in the treatment of drinking water." *Oxidation techniques in drinking water treatment*, W. Kuhn and H. Sontheimer, eds., U.S. Environmental Protection Agency, Washington, D.C., 597-607.

Miltner, R.J., H.B. Shukairy, and R.S. Summers (1992). "Disinfection by-product formation and control by ozonation and biotreatment." *Journal of American Water Works Association*, 84(11): 53-62.

Parkhurst, D.F., and D.A. Stern (1998). "Determining average concentrations of *Cryptosporidium* and other pathogens in water." *Environmental Science and Technology*, 32(21): 3424-3429.

Piet, G.J., and B.C.J. Zoeteman (1985). "Bank and dune infiltration of surface water in The Netherlands." *Artificial recharge of groundwater*, T. Asano, ed., Butterworth Publishers, Boston, Massachusetts, 529-540.

Pollock, D.W. (1994). "User's Guide for MODPATH/MODPATH-PLOT, Version 3: A particle tracking post-processing package for MODFLOW, the U.S. Geological Survey finite-difference ground-water flow model." *U.S. Geological Survey Open-File Report 94-464*, U.S. Geological Survey, Reston, Virginia.

Ray, C. (1999). "Pesticide and nitrate issues at selected bank filtration sites in Illinois." *Abstracts, International Riverbank Filtration Conference*, National Water Research Institute, Fountain Valley, California, 35-38.

Singer, P.C. (1999). "Humic substances as precursors for potentially harmful disinfection by-products." *Water Science and Technology*, 40(9): 25-30.

Sontheimer, H. (1980). "Experience with riverbank filtration along the Rhine River." *Journal of American Water Works Association*, 72(7): 386-390.

Stuyfzand, P.J. (1998). "Fate of pollutants during artificial recharge and bank filtration in The Netherlands." *Artificial recharge of groundwater*, J.H. Peters, ed., A.A. Balkema, Rotterdam, The Netherlands, 119-125.

Summers, R.S., S.M. Hooper, H.M. Shukairy, G. Solarik, and D. Owen (1996). "Assessing DBP yield: Uniform formation conditions." *Journal of American Water Works Association*, 88(6): 80-93.

Verstraeten, I.M., J.G. Miriovsky, and E.C. Lee (1999). "Herbicide removal through bank filtration using horizontal collector wells at Ashland, Nebraska." *Abstracts, International Riverbank Filtration Conference*, National Water Research Institute, Fountain Valley, California, 17-20.

Wang, J., R. Song, T. Sweazy, and S. Hubbs (1999). "Determination of water travel time of riverbank filtration process." *Abstracts, International Riverbank Filtration Conference*, National Water Research Institute, Fountain Valley, CA, 55.

White, M.C., J.D. Thompson, G.W. Harrington, and P.C. Singer (1997). "Evaluating criteria for enhanced coagulation compliance." *Journal American Water Works Association*, 89(5): 64-77.

Wilderer, P.A., U. Förstner, and O.R. Kuntschik (1985). "The role of riverbank filtration along the Rhine River for municipal and industrial water supply." *Artificial recharge of groundwater*, T. Asano, ed., Butterworth Publishers, Boston, Massachusetts, 509-528.

U.S. Environmental Protection Agency (1992). "Methods for the determination of organic compounds in drinking water, Supplement II." *EPA/600/R-92/129*, U.S. Environmental Protection Agency, Washington, D.C.

U.S. Environmental Protection Agency (1993). "Methods for the determination of inorganic substances in environmental samples." *EPA/600/R-93/100*, U.S. Environmental Protection Agency, Washington, D.C.

U.S. Environmental Protection Agency (1994). "Methods for the determination of metals in environmental samples, Supplement I." *EPA/600/R-94/111*, U.S. Environmental Protection Agency, Washington, D.C.

U.S. Environmental Protection Agency (1996). "ICR microbial laboratory manual." *EPA/600/R-95/178*, U.S. Environmental Protection Agency, Washington, D.C.

U.S. Environmental Protection Agency Integrated Risk Information System (IRIS) database (2000). U.S. Environmental Protection Agency website, <http://www.epa.gov/iriswebp/iris/index.html>.

Chapter 9. Occurrence, Characteristics, Transport, and Fate of Pesticides, Pharmaceuticals, Industrial Products, and Personal Care Products at Riverbank Filtration Sites

Ingrid M. Verstraeten, Ph.D.
United States Geological Survey
Baltimore, Maryland, United States

Thomas Heberer, Ph.D.
Institute of Food Chemistry
Technical University of Berlin
Berlin, Germany

Traugott Scheytt, Ph.D.
Technical University of Berlin
Berlin, Germany

1. Introduction

There is a high level of concern among scientists and policy makers throughout the world because of the presence of anthropogenic substances that can potentially affect the endocrine systems of organisms living in aquatic environments as well as affect human health from aquatic recreational activities and drinking water (Ghijsen and Hoogenboezem, 2000). It was found in the early 1990s that the presence of these substances was linked to a decrease in the sperm count of men for over the last 50 years (Ghijsen and Hoogenboezem, 2000) and to the premature physical development of women; however, these symptoms may most likely be caused by other environmental factors.

An article by Beer (1997) clarifies the importance of the quality of our drinking water. Along the Thames River in England, water passes at least six times through people before reaching the estuary, demonstrating the existing enormous pressure on water resources and the importance of keeping water, including wastewater, as uncontaminated as possible. Given that the population is expected to double worldwide in about 20 years, Beer's finding (1997) will become even more significant. Not only do problems related to microorganisms and nitrates need attention, but also a variety of organic chemicals should be evaluated. These organic chemicals include chemicals used on crops, in animal feed, by industry, and by humans. Numerous studies are being conducted to develop lists of priority pollutants for:

- Toxicological characteristics studies.
- Occurrence in the environment.
- Fate and transport.

At the end of the nineteenth century, water-pollution issues became more and more evident in the growing metropolitan centers of Europe. New concepts for wastewater treatment and disposal, such as sewage-irrigation farms, were developed and, later, sewage-treatment plants were

C. Ray et al. (eds.), Riverbank Filtration, 175–227.
© 2002 *Kluwer Academic Publishers. Printed in the Netherlands.*

designed. New technologies also were necessary to provide sufficient amounts of high-quality drinking water for public-water supply. For almost 100 years, bank filtration has been used as a method of drinking-water treatment at Lake Müggelsee in Berlin, Germany. RBF has been used along the Rhine River for more than 75 years (Sontheimer, 1980). Originally, riverbank-filtered water along the Rhine River could be used as drinking water without further treatment but, as pollution increased (resulting from an increase in population and industrial activities), water treatment increased extensively. Bank filtration reduces organics in water, including:

- Pesticides (Verstraeten et al., 1999b).
- Personal care products, like fragrance compounds (Jüttner, 1999).
- Pharmaceuticals (Heberer, in press).
- Hydrocarbons (Jüttner, 1999).

In recent years, most large utilities in Europe and the United States have begun using additional treatment with filtration, ozonation, chlorination, and/or granular activated carbon to reduce and, ultimately, remove undesirable organic chemicals in drinking water. This has led to an increased interest in the potential reuse of treated wastewater; however, treatments such as ozonation and chlorination may lead to the formation of chlorinated degradates, shifting the health risk associated with consuming drinking water from the consumption of the parent organic compounds in part to the consumption of their degradates (Verstraeten et al., in press). The risks to human health from the consumption of water contaminated with pesticides are related not only to the concentration of pesticides in the water, but also to the number of people consuming the water. For example, it is expected that 75 percent of the population of the State of Nebraska in the United States will live in the Omaha-Lincoln metropolitan area by the year 2010. Drinking water for both these cities is impacted by herbicide concentrations through RBF (Verstraeten and Ellis, 1994; Verstraeten et al., 1999a).

The possibility of eliminating or reducing pollutants during RBF has been studied (Stuyfzand, 1989; Kruhm-Pimpl, 1993; Verstraeten et al., 1999a and in press). The quality of riverbank-filtered water depends not only upon the contaminant itself, but also on the hydraulic and chemical characteristics of the bottom sediment and aquifer, the local recharge/discharge conditions, and biochemical processes. The fate and transport of organic chemicals are complex and can include microbial degradation, adsorption/desorption, photolysis, oxidation, and transport from one medium in the hydrologic system (e.g., surface-water bodies and riverbanks) to another medium. Organic contaminants that are not readily biodegradable may occur in riverbank-filtered water, often in lower concentrations than those in the river. Simple models can be used to evaluate the reactive transport of contaminants, such as THMs, during RBF and artificial recharge (Stuyfzand, 1998b).

Generally, organic chemicals do not behave as ideal tracers because they tend to interact with the solid media. Tracer studies with these reacting chemicals can establish the adsorption/desorption characteristics locally and can show the heterogenous nature of the hydraulic properties of the aquifer, sometimes indicating the presence of preferential flow paths (Verstraeten et al., 1999b).

The occurrences of pharmaceuticals, personal care products (like synthetic musk fragrances), and endocrine-disrupting chemicals, including certain pesticides, pesticide degradates, biocides (nonagricultural pesticides, including wood preservatives, cooling water biocides, and anti-coating biocides), phenols, and polar aromatic sulfonates, have been identified in:

- Wastewater from many countries (Eschke et al., 1995; Heberer, 1995; Stan and Heberer, 1997; Halling-Sørensen et al., 1998; Heberer and Stan, 1998; Heberer et al., 1998; Heberer et al., 1999; Ternes, 1998; Daughton and Ternes, 1999; Möhle et al., 1999; Barber et al., 2000; Wilken et al., 2000a; Heberer, in press; Heberer et al., 2001).

- Runoff from confined animal feeding operations, which are common in the United States (Lindsey and Thurman, 2000; Meyer et al., 2000).
- Surface water (Frank et al., 1990; Jani et al., 1991; Nondek and Frolikova, 1991; Thurman et al., 1991; Djuangsih, 1993; Eschke et al., 1995; Heberer, 1995; Heberer et al., 1997; Stan and Heberer, 1997; Buser et al., 1998; Halling-Sørensen et al., 1998; Heberer and Stan, 1998; Heberer et al., 1998; Kalajzic et al., 1998; Kalkhoff et al., 1998; International River Waterworks, 1998; Daughton and Ternes, 1999; Heberer and Dünnbier, 1999; Barber et al., 2000b; Groshart and Balk, 2000; Lange et al., 2000; Heberer, in press; Heberer et al., 2001).
- Marine environments (Buser et al., 1998; Siegener and Chen, 2000).
- Groundwater (Stan and Linkerhägner, 1992; Burkart and Kolpin, 1993; Heberer, 1995; Barbash and Resek, 1996; Dünnbier et al., 1997; Heberer and Stan, 1997; Heberer et al., 1997; Heberer et al., 1998; Kolpin et al., 1998; Burkart et al., 1999a, 1999b; Heberer and Dünnbier, 1999; Hodiaumont et al., 1999; Verstraeten et al., 1999b; Lange et al., 2000; Heberer, in press).
- Drinking water (Ang et al., 1989; Jani et al., 1991; Stan et al., 1994; Heberer and Stan, 1996; Tsipi and Hiskia, 1996; Kalajzic et al., 1998; Verstraeten et al., 1999b; Lange et al., 2000; Heberer, in press.; Verstraeten et al., in press).

For example, a study of the occurrence of pesticides in shallow groundwater (1.5- to 5-m below land surface) near a river in Denmark has shown that 23 of 46 compounds were present in groundwater, including atrazine and its degradates: bentazone, (4-chloro-2-methylphenoxy) acetic acid (MCPA), metamitron, isoproturon, and simazine (Spliid and Køppen, 1998). According to Van Genderen et al. (1999), 1,328 different organic compounds have been identified in the Rhine and Meuse Rivers, the Ijselmeer and Haringvleet Lakes, or associated drinking water. Fifty-eight compounds in the environment are suspected to have some carcinogenic or mutagenic capacities (Van Genderen et al., 1999). For about one-third of these compounds, the carcinogenic and mutagenic capabilities are unknown. Recent attention also has been placed on the presence of oestrogenic chemicals (endocrine disrupting chemicals) in the environment, including xeno-estrogens, which are hormone-like substances that can possibly influence hormonal processes in animals and humans.

Pesticides in the Aquatic Environment

The use of pesticides for pest control has increased over the last five decades, replacing manual or mechanical treatment methods with chemical treatment. The use and number of pesticides have grown steadily worldwide since the 1960s, but declined slightly in Germany by the late 1990s. In the United States, the national use of herbicides and insecticides grew from 86-million kilograms (kg) of active ingredient in 1964 (Barbash and Resek, 1996) to an estimated 300-million kg in 1993, nearly a 350-percent increase in only 29 years, which has raised concerns about adverse effects on the environment and human health.

In the United States, agriculture is the most intense in the Midwest, where corn and soybeans are major row crops. In 1991, approximately two-thirds of all the herbicides were used for agriculture; one-half of the herbicides were represented by triazine and chloroacetanilide compounds, which are pre-emergent herbicides (Boyd, 1999). In 1991, about 490-million kg of pesticides (active and inactive ingredients) were used, including 285-million kg of herbicides (58 percent), 113-million kg of insecticides (23 percent), and 54-million kg of fungicides (11 percent), according to the U.S. Environmental Protection Agency (Aspelin et al., 1992). In 1994, the use of less than 200 (Denmark) to 400 (France) pesticides was permitted in the

European Union member states (Zullei-Seibert, 1996b). Herbicides were used most often in
Denmark, Germany, Great Britain, and Austria, whereas fungicides were more important in
France, Greece, Italy, The Netherlands, and Portugal, and insecticides dominated in Spain
(Zullei-Seibert, 1996b). Statistically, between 2 and 8 kg of pesticides were applied per hectare
(ha) per year in the European Union member states for agricultural purposes, with the exception
of The Netherlands (18 kg/ha). In Germany alone, about 30-million kg of pesticides are used
annually in agriculture. More than 50 percent of these pesticides are herbicides (Figure 9-1).

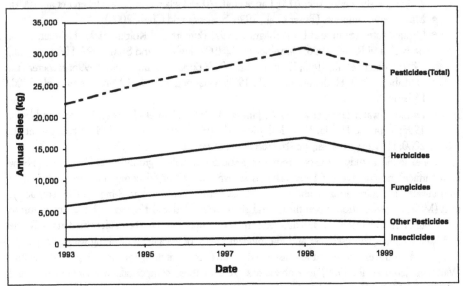

Figure 9-1. Annual sales of various pesticide classes in Germany between 1993 and 1999 (Industrieverband Agrar, 2000).

The Industrieverband Agrar compiles the most recent estimates on pesticide sales in
Germany in annual reports based on data from its member states (Table 9-1). These estimated
quantities of pesticide sales include most of the pesticides used in Germany (as much as
95 percent) and indicate the importance of a pesticide in agriculture. The actual amounts sold for
the individual pesticides are reported cumulatively by the Industrieverband Agrar (Table 9-1).
Based on the estimates, the top eight herbicides make up more than one-half of the total amount
of herbicides used in Germany. For example, in Germany, more than 4-million kg of isoproturon
and glyphosate were sold in 1999 (29 percent of total herbicide sales). In Germany, a federal
commission determined pesticide target values for drinking water and aquatic habitats. These
values are compiled in Table 9-2, together with some data concerning their occurrence in water
samples. The occurrence data have been compared with target values to indicate compounds of
special environmental interest. The determination of target values may also be seen as a guideline
to identify problematic pesticides and to assess potential environmental risks.

The use of several pesticides has been banned in Germany and other countries because of
their environmental persistence, combined with a tendency for bioaccumulation (e.g., organo-
chlorinated pesticides, such as dichlorodiphenyl trichloroethane [DDT] or aldrin) or of their great
potential to leach into groundwater (e.g., the use of atrazine has been banned since 1991).

Typically, herbicides are applied to fields to control a variety of grasses and broadleaf weeds.
There also is a substantial use of herbicides outside of agriculture (e.g., urban uses and other

Table 9-1. Ranking of Herbicides in Germany According to Annual Sales in 1999

Ranking	Annual Amount Sold (t)	Herbicide	Cumulative Weight (t)	Cumulative Weight Compared to the Total Weight (%)
1	>1,000	Isoproturon	4,073	29
2		Glyphosate		
3	>500	Metamitrone		
4		Dichlorprop		
5		Metazachlor	7,778	56
6		Pendimethaline		
7		Bentazone		
8		Glyphosate-trimesium		
9	>200	Mecoprop		
10		Metolachlor		
11		MCPA		
12		Terbuthylazine		
13		Prosulfocarb	10,923	78
14		Ioxynil		
15		Pyridat		
16		Bromoxynil		
17		Aclonifen		
18	>100	Chloridazon		
19		Fluoroxypyr		
20		Ethofumesat		
21		Bifenox		
22		Dimethachlor		
23		Diflufenican	12,704	91
24		Quinmerac		
25		Flurtamone		
26		Trifluraline		
27		Sulcotrione		
28		Flufenacet		
29		Cyanamide		
Total Number of Relevant Herbicides: 76			13,994	100

Adapted with permission from Industrieverband Agrar (2000).
t = Ton = 1,000 kg.

Table 9-2. Percentage of Surface-Water Samples in which Pesticide Residue Concentrations Exceeded the Target Value Set by the German "Länderarbeitsgemeinschaft Wasser und Pflanzenschutzmittel-Wirkstoffe," Germany, 1996 to 1998

Compound Name	Target Value for Drinking-Water Supply (µg/L)	Target Value for Aquatic Habitats (µg/L)	Samples Exceeding the Target Values				Percentage Values Not Available*
			0%	0 to ≤10%	>10 to ≤25%	>25%	
2,4-D	0.1	2	A		D		
α-Endosulfane	0.1	0.005	D			A	
β-Endosulfane	0.1	0.005	D		A		
Ametryne	0.1	0.5	A		D		
Atrazine	0.1	—	D				
Azinphos-Methyl	0.1	0.01	D		A		
Azinphos-Ethyl	0.1	—	D				
Bentazone	0.1	70	A		D		
Bromacile	0.1	0.6	A,D				
Chloridazone	0.1	10	A		D		
Chlortoluron	0.1	0.4			A,D		
Dichlorprop	0.1	10	A		D		
Dichlorvos	0.1	0.0006	D		A		
Dimethoate	0.1	0.2		A	D		
Diurone	0.1	0.05				A,D	
Etrimphos	0.1	0.004	D			A**	
Fenitrothion	0.1	0.009	D				A
Fenthion	0.1	0.004			A,D		
Hexazinone	0.1	0.07		A,D			
Isoproturon	0.1	0.3				A,D	
Lindane	0.1	0.3	A,D				
Linuron	0.1	0.3	A,D				
Malathion	0.1	0.02	D		A		
MCPA	0.1	2	A		D		
Mecoprop	0.1	50	A			D	
Metazachlor	0.1	0.4	A	D			
Methabenzthiazuron	0.1	2	A	D			
Metolachlor	0.1	0.2	A	D			
Parathion-Methyl	0.1	0.02	D				A
Parathion-Ethyl	0.1	0.005	D				A
Prometryn	0.1	0.5	A,D				
Propazine	0.1	—		D			
Simazine	0.1	0.1		A,D			
Terbuthylazine	0.1	0.5	A	D			
Triazophos	0.1	0.03				D**	A
Tributyltin	0.1	0.0001	D		A**		
Trifluralin	0.1	0.03	A,D				
Triphenyltin	0.1	0.0005	No Data				

— = No target value established.   A = Protection of "aquatic habitats/organisms."   D = Protection of "drinking-water supply."
* = Target values exceeded. Percentage data not available because at 50 percent of all sampling locations the determination limits were larger than the target values.
** = Statistically not validated because only investigated at less than 15 sampling locations.
Data from Umweltbundesamt (2000).

right-of-ways for highways and utility lines). The use of pesticides has led to their presence in many hydrologic systems in the world and numerous compounds are found in a wide variety of environments, including surface water, groundwater, wastewater, and drinking water. This causes some concern, even though manufacturers must document the human and ecotoxicological safety of manufactured chemicals. Pesticides are transported into rivers by runoff from urban and rural areas, by groundwater discharge, along drainage tiles, and by atmospheric deposition. Pesticide concentrations in rivers are governed by local usage, farming practices, rainfall, and physico-chemical properties of the pesticides. Pronounced seasonal variations exist in the occurrence of pesticides in rivers. Typically, the largest concentrations are present in rivers shortly after pesticides are applied to the land and following a rainstorm that produces runoff.

In some cases, the presence of numerous herbicides and other organic compounds in groundwater has been attributed to surface water that was contaminated with herbicides by periodic flooding, bank storage of river water, artificial recharge by impoundments, and induced infiltration (Exner, 1990; Duncan et al., 1991; Thurman et al., 1991; Taylor, 1994; Wang and Squillace, 1994; Frycklund, 1998; Burkart et al., 1999a; Verstraeten et al., 1999b; Lange et al., 2000; Mazounie et al., 2000; Verstraeten et al., in press). In other cases, pesticides have been identified in surface water during base-flow conditions and have been attributed to inflow from contaminated groundwater (Barbash and Resek, 1996). Concentrations of herbicides in surface water and groundwater have been related to land use, soils, climate, and hydrogeological characteristics of the watersheds or recharge zones (Kolpin, 1997; Burkhart et al., 1999b).

The promulgation of regulations has lagged behind the formulation of new pesticides. The World Health Organization issued drinking-water quality guidelines for 33 pesticides (World Health Organization, 1993). The U.S. Environmental Protection Agency has defined health advisories for 71 pesticides (U.S. Environmental Protection Agency, 1994). In the United States, pesticide regulations have been established for some parent compounds, but not degradates — the U.S. Environmental Protection Agency maximum contaminant level (MCL) for atrazine is 3.00 µg/L and for alachlor is 2.00 µg/L; an MCL of 20 µg/L for total triazines is currently under discussion. In Europe, the regulations for pesticides and several other parameters are not based on toxicological aspects, but on the "precautionary principle." Thus, the maximum tolerance levels for pesticide residues in drinking water have been set at 0.10 µg/L for the sum of concentrations of the parent compounds and their degradates in drinking water (e.g., atrazine and degradates) and at 0.50 µg/L for the sum of all pesticides and their degradates present in drinking water.

The toxicity of the pesticide degradates and additives or of the synergistic effects of combinations of compounds and their degradates remain almost totally unknown. A study done on parent and degradate compounds in drinking water by Upham et al. (1997) suggested that the pesticides alachlor, atrazine, carbofuran, and 2,4-dichlorophenoxyacetic acid (2,4-D) reacted to both ozone and chlorine and that the degradates did not exhibit more or less toxicity than their parent compounds. A study done by Taets et al. (1998) did not identify a synergistic effect of the presence of atrazine, simazine, and cyanazine in water when exposed to Chinese hamster ovary cells; however, a study done on the synergistic activation of estrogen receptors with combinations of environmental chemicals, such as dieldrin, endosulfan, or toxaphene, suggested that they were 1,000 times more potent in human estrogen receptor-mediated transactivation than any single chemical (Arnold et al., 1996). This synergistic interaction of chemical mixtures with human estrogen receptors may have profound environmental implications, suggesting an interplay between environmental signals and biological systems, and may represent an uncharacterized level of receptor-mediated gene regulation (Arnold et al., 1996). Regulations for other organic compounds exist both in the United States and Europe; however, the presence of other endocrine disruptors in drinking water remains unregulated, but Europe drinking-water regulations state that, when technically feasible, anthropogenic compounds in drinking water should be reduced to a minimum.

The detection of pesticides in the environment and drinking water depends not only on the occurrence of pesticides in the environment (local agricultural practices, climate, geohydrologic setting, well selection criteria, sampling protocols, and timing of sampling), but also on the analytical method used and its sensitivity. Studies conducted by the U.S. Geological Survey in the Midwestern United States on water from near-surface aquifers showed an increase in detection from 28.4 percent during the 1991 study to 59.0 percent during 1992 (Kolpin et al., 1995). This increase in the frequency of detection was attributed to a decrease in reporting limits of as much as a factor of 20 (from 0.20 to 0.01 μg/L) and an increase in the number of degradates of pesticides for which samples were analyzed. Atrazine detections more than doubled when the analytical reporting limit was decreased. The study also showed that pesticide degradates were more common than parent compounds. In the United States, drinking-water regulations do not exist for these degradates. Nevertheless, studies completed during the last decade demonstrate that analyses of both parent and degradate compounds are important to more fully understand the environmental fate and transport of pesticides in a hydrologic system (Kalkhoff et al., 1998; Kolpin et al., 1996). In the Midwestern United States, when atrazine is detected, other parent herbicides also tend to be present (Kolpin et al., 1995), but generally at lower concentrations (Verstraeten et al., 1999b).

Zullei-Seibert (1996b) reported results from a study requested by the European Union in 1994. One objective of this study was to estimate the loads of pesticides in surface water and groundwater in Europe. The number of pesticides analyzed in surface and groundwater monitoring studies varied considerably between the European member states (Zullei-Seibert, 1996b). While more than 200 compounds were monitored in Germany, between 100 and 200 were monitored in Great Britain and The Netherlands, between 50 and 100 were monitored in France, Italy, Austria, and Spain, and about 50 were monitored in Denmark and Greece. It is difficult to directly compare the monitoring results. Nevertheless, Zullei-Seibert (1996b) found that pesticide residues significantly impacted water quality in the European member states. Surface water was more prone to contain these residues than groundwater. More than 100 compounds, mostly herbicides, were detected in European surface water used for drinking-water production (Zullei-Seibert, 1996a). Atrazine, simazine, MCPA, and 2,4-D were detected at concentrations greater than 0.1 μg/L. Zullei-Seibert (1996b) estimated that about 11-billion m^3/yr of raw water used for drinking-water production in Europe are contaminated with pesticide residues at concentrations greater than 0.1 μg/L. This amount of raw water is equal to approximately 30 percent of the drinking-water production of the European member states. In an investigation of 14 water catchment areas in Germany, Skark and Zullei-Seibert (1995) evaluated different input sources for pesticides in groundwater, such as:

- Agricultural use.
- Pesticide loads in surface water.
- Application on railroad tracks.
- Use in forestry.
- Use in private gardens.
- Application on sporting grounds and sealed surfaces.

Agricultural pesticide application was the main contamination source for 47 percent of all subsurface sampling points, but 14 percent of the sampling points were also influenced, mainly by the pesticide loads of surface water. A combination of both sources was often determined.

In Germany, most of the pesticide detections in groundwater have been reported for atrazine, deethylatrazine, and simazine (Skark and Zullei-Seibert, 1995; UBA-Umweltbundesamt, 2000) (Table 9-3). The numerous detections prompted the ban on the agricultural use of atrazine and simazine in Germany in 1991 and 1999, respectively.

Table 9-3. Pesticide Detections in Groundwater in Germany, 1996 to 1998

Pesticide/Degradate	Number of Pesticide Detections								
	1996			1997			1998		
	Non-Detections	<0.1 µg/L	>0.1 µg/L	Non-Detections	<0.1 µg/L	>0.1 µg/L	Non-Detections	<0.1 µg/L	>0.1 µg/L
Deethylatrazine	6,494	1,442	536	4,080	792	32	2,782	736	332
Atrazine	7,041	1,322	273	4,482	725	143	3,176	661	143
Simazine	7,981	374	38	4,909	200	29	3,732	158	14
Bromacil	5,355	67	137	2,740	32	80	2,283	36	59
Hexazinone	5,738	71	76	2,962	24	9	2,288	17	1
Deisopropylatrazine	4,871	124	25	2,725	76	5	3,201	77	16
Propazine	7,572	93	14	4,284	76	8	3,327	52	6
Diuron	3,981	7	25	3,316	24	27	1,624	18	16
Terbuthylazine	8,041	74	7	4,795	49	14	3,551	25	9
Bentazone	3,537	49	19	2,382	28	13	985	10	19
Isoproturon	4,461	43	12	5,409	29	12	1,918	18	8
Deethylterbuthylazine	5,322	43	9	2,765	30	2	2,623	19	3
Lindane + Isomers	2,524	97	11	2,456	109	17	2,487	150	34
2,6-Dichlorobenzamide	85	22	13	2,113	85	51	258	48	53
Metolachlor	5,805	22	6	2,975	14	6	2,893	5	6
Sebutylazine	3,470	22	4	4,823	24	4	1,824	9	9
Metazachlor	7,788	14	2	4,488	10	10	3,498	4	3
Metalaxyl	2,756	8	8	545	5	3	1,259	4	2
Dikegulac	303	6	10	115	2	1	103	13	4
1,2-Dichloropropane	426	4	11	303	4	5	186	5	4
Mecoprop	1,818	9	5	2,638	19	14	1,197	10	5
Hexachlorobenzene	666	14	0	610	8	0	621	16	0
Monuron	2,381	11	2	174	2	2	471	0	1
Chlortoluron	3,470	5	6	2,606	4	3	1,285	4	5

Data from Umweltbundesamt (2000).

One potential reason for the common detection of these compounds in the aquatic environment, other than their frequent use and long half-life in water, could be attributed to their "good analytical properties." The analysis of atrazine and simazine residues in water samples is relatively easy, inexpensive, and highly sensitive compared to other organic compounds. Atrazine, deethylatrazine, and simazine are extracted by applying liquid/liquid or solid-phase extraction. The concentrated sample extract then is analyzed by high-performance liquid chromatography (HPLC) or gas chromatography (GC) using conventional detectors (ultraviolet detection or diode array detection for HPLC or a nitrogen-phosphorous detector for GC analysis).

Complicated sample preparation steps, such as derivatization procedures, or hyphenated instrumentation, such as GC or HPLC applying mass spectrometry (MS), or tandem mass spectrometry (GC-MS/MS, HPLC-MS/MS, or HPLC-MS), are not necessary to analyze for atrazine or simazine residues and some of their degradates. Nevertheless, the use of GC-MS for the confirmation of atrazine and simazine results obtained by conventional methods is highly recommended, especially if no additional method is used for confirmation; therefore, some detections of atrazine, especially those reported in earlier years, may be false positives. Today, GC-MS, GC-MS/MS, or HPLC-MS/MS are readily available for trace analysis at the parts per trillion level in many routine laboratories in Europe and the United States. Consequently, the number of compounds that can be analyzed in complicated sample matrices, such as sewage or matrix-prone surface water, is growing steadily. Nevertheless, several compounds are still difficult to analyze at levels less than 0.1 μg/L.

Because of high analytical costs, only a limited spectrum of compounds are routinely measured in monitoring programs. To reduce the analytical costs of pesticides and other trace organics, they are analyzed using so-called "multi-methods" to detect as many compounds as possible in a single analytical procedure. Compounds such as glyphosate or pesticide degradates such as dischlorophenyl acetic acid (DAA) cannot be analyzed by those multi-methods; therefore, they are often not measured. Skark and Zullei-Seibert (1995) compared the pesticide application and pesticide analyses of water samples from 14 water catchment areas in Germany. Between 1986 and 1991, 5,772 samples were analyzed from these areas and provided more than 200,000 data for pesticide residues. Between 20 and 50 percent of all pesticides applied in Germany were monitored in these areas. Considering compounds with an annual application amount of >0.5 kg/ha, the ratio between analyzed and applied pesticides increased from 40 to 60 percent. The previously mentioned aspects of analytical chemistry should be taken into account when interpreting the results from monitoring studies. The most important compound present in an aquatic environment is not always the compound most frequently detected.

Since 1996, glyphosate is the only pesticide used by the Deutsche Bahn AG (a federal railway company) for weed control on railway tracks in Germany. The specific amounts of pesticides applied on railway tracks can be higher by a factor of six than the application rate used in agriculture, but the retaining capacity of railway tracks for pesticides is low (Schweinsberg et al., 1999). The herbicides applied formerly in Germany included 2,4-D; 2,4,5-trichlorophenoxyacetic acid (2,4,5-T); and triazine derivatives (e.g., atrazine and urea derivatives, such as diuron). Traces of almost all of the herbicides applied could be detected in samples of groundwater and drinking water in the vicinity of railway tracks. Along railway tracks, in the vicinity of the application site, glyphosate was found at concentrations of 0.6 and 0.2 μg/L in surface water and groundwater, respectively (Schweinsberg et al., 1999). Glyphosate is a nonselective, systemic leaf herbicide and its main degradate is aminomethylphosphonic acid (AMPA). AMPA also is formed as a degradation product of amino-(trimethylenephosphonic) acid (ATMP), which is widely used for industrial purposes. Thus, the detection of AMPA can indicate the potential presence of glyphosate (Gledhill and Feijtel, 1992). Glyphosate can be degraded by immobilized bacteria in wastewater (Heitkamp et al., 1992). In the United States, a watershed study of glyphosate transport in runoff was done in the late 1970s (Edwards et al.,1980). The occurrence of glyphosate was documented in well water when applied near well-water substations to control weeds (Smith et al., 1996). Occurrences of glyphosate in riverbank-filtered water in the United States have not been documented.

Pharmaceuticals and Personal Care and Industrial Products in the Aquatic Environment

Pharmaceuticals are used in large quantities in human medical care, as veterinary drugs, and as feed additives in animal production. They also are sprayed on orchards. Drugs used in human medical care may enter the environment via discharges of municipal sewage or hospital effluents, sewage sludges, landfill leachates (waste disposal), domestic septic tanks, or production residues. Animal drugs or drugs used as feed additives may be discharged into the environment by agricultural use (such as liquid manure), municipal sewage, sewage sludges, landfill leachates (solid-waste disposal), production residues, or direct introduction via aquacultures.

In human medical care, the sales of drugs reached $210.7 billion worldwide from March 1999 through February 2000 (www.epa.gov/nerlesd1/chemistry/pharma/faq.htm, accessed March 12, 2001). The top therapeutic markets were cardiovascular, alimentary, antidepressants, anti-infectives, and respiratory. In a 12-month period from February 1999 to February 2000, sales in the United States were about $92 billion; in Europe, $54 billion; and in Japan, $49 billion. In 1997, the United States per capita expenditure on drugs was the third highest in the world ($319), followed by Belgium, Japan, and France. About 1 to 2 percent of the sales were used for veterinary purposes. In 1997, United States pharmaceutical sales were 1.4 percent of the gross national product. While pharmaceutical sales amount to hundreds of kilograms per year, the production of personal care products are an order of magnitude larger (www.epa.gov/nerlesd1/chemistry/pharma/faq.htm, accessed March 12, 2001).

In Germany, the exact figures for the consumption of pharmaceuticals are not accessible. It is possible, however, to calculate the approximate figures of pharmaceutical prescriptions in Germany's human medical care. Several compounds, such as carbamazepine, diclofenac, and ibuprofen, are prescribed at amounts of up to 100 tons per year (t/yr) in Germany (Stan and Heberer, 1997). An annual total prescription of 50 kilograms per year (kg/yr) was estimated by Ternes et al. (1999) for the oral contraceptive, 17α-ethinyloestradiol. Taking into account over-the-counter sales and their applications in hospitals, the amounts of several pharmaceuticals, such as ibuprofen and aspirin, used in medical care are much higher (up to 1,000 t/yr). This implies that the annual amounts of some pharmaceuticals used are similar to those of the most important herbicides. Thus, several of these compounds may be present in receiving surface water at concentrations similar to, or even higher than, those of the pesticide residues. Pharmaceuticals enter surface water as point sources at high concentrations, whereas pesticides normally are sprayed extensively onto agricultural areas and tend to enter the streams as nonpoint sources; however, pharmaceuticals also enter the environment in a dispersed manner as nonpoint sources when manure that contains traces of pharmaceuticals is applied to the field.

Data on the occurrence of pharmaceutical residues in the environment were often only random byproducts from other investigations, such as studies on the occurrence of pesticide residues. In 1990, after the reunification of East and West Germany, the Senate of Berlin commissioned the monitoring of water collected from wells near the former Berlin Wall. The purpose of this monitoring effort was to identify and quantify pesticide residues caused by the intensive use of these compounds by border troops of the former German Democratic Republic to see clearly and have a free field of fire. In 1991, as a result of these investigations, 2-(4)-chlorophenoxy-2-methyl propionic acid (clofibric acid) was first detected in Berlin groundwater samples (Stan and Linkerhägner, 1992). Clofibric acid is the pharmacologically active metabolite of the drugs clofibrate, etofyllin clofibrate, and etofibrate, used as blood-lipid regulators in human medical care. Its detection in these groundwater samples was only a random finding, because it also could be detected in an analytical method for the analysis of phenoxyacid herbicides. In fact, clofibric acid is a structural isomer of the herbicide mecoprop, frequently used as a pre-emergent herbicide in agriculture. Initially, its occurrence in groundwater samples was not linked with its medical application; however, between 1992 and 1995, clofibric acid also was detected at concentrations up to the microgram-per-liter level in groundwater samples collected from former sewage irrigation fields near Berlin and in Berlin tap water samples (see the case study in Section 4) (Heberer, 1995; Heberer and Stan, 1997). It became evident that these residues were caused by the infiltration of sewage effluents into the soil and that clofibric acid is a very mobile compound that does not substantially adsorb in the subsoil, but does easily leach into the aquifer. It was also found in samples collected from the fourth aquifer at depths greater than 70 m. In Germany, the first findings of clofibric acid focused on drug residues as new emerging residues in the aquatic system (Umweltbundesamt, 1996) and led to several new studies investigating the occurrence and fate of pharmaceutical residues in the aquatic environment.

To date, more data exist on the presence of antibiotics in the environment than any other therapeutic drug (Daughton and Ternes, 1999). These drugs are present in the environment as a result of the releases of treated and untreated waste into surface water worldwide from wastewater-treatment plants, from individual land owners with or without private septic systems, and from animal feeding operations. The first detection of antibiotics in wastewater was reported in the 1970s in the United States (Daughton and Ternes, 1999), although this finding was largely ignored. In Europe, information has been gathered since the 1980s whereas, in the United States, studies have been developed since the early to mid-1990s (www.epa.gov/nerlesd1/chemistry/pharma/faq.htm, accessed April 18, 2001). More data on pharmaceuticals have been collected because of their potential effect on native biota and the development of resistance in potential human pathogens. An overview of the current knowledge of the occurrence of pharmaceutical and personal care products in the environment and the potential implications of this emerging problem is given by Daugthon and Ternes (1999). References relevant to pharmaceuticals and personal care products in the environment can be found at www.epa.gov/nerlesd1/chemistry/ppcp/reference.htm.

In urban areas, drug residues originate mainly from the use of pharmaceuticals in human medical care. The major sources for these residues are discharges of municipal sewage and hospital effluents. According to current knowledge, these residues originate mainly from the therapeutic use of pharmaceuticals and, only to a lesser extent, from the improper disposal of expired drugs via the toilet (Umweltbundesamt, 1996; Heberer and Stan, 1998). Important factors for the occurrence of pharmaceutical residues in the environment include the individual amounts prescribed and the fate of the individual compounds, both in the sewage-treatment works and in the aquatic environment. The following questions need to be answered:

- Is the compound, ultimately, biodegradable?
- Can it be metabolized in several different ways?
- Can it be degraded or (de-)conjugated?
- Is it persistent?

Some pharmaceuticals are not eliminated in the human body. They are excreted by the human body (only slightly transformed or unchanged) and mostly conjugated to polar molecules, such as the glucoronides. These conjugates are easily cleaved in the raw sewage or during sewage treatment. Thus, some pharmaceutically active compounds are discharged almost unchanged from municipal sewage treatment plants into the receiving water (Stan and Heberer, 1997; Halling-Sørensen et al., 1998; Heberer and Stan, 1998; Ternes, 1998; Daughton and Ternes, 1999; Möhle et al., 1999; Wilken et al., 2000; Heberer, in press; Heberer et al., 2001). The term "pharmaceutically active compounds" was introduced by Daugthon and Ternes (1999) and comprises pharmaceuticals and their pharmaceutically active degradates. Pharmaceuticals have been detected in wastewater, surface water, groundwater, and drinking water in Switzerland, Germany, The Netherlands, and Belgium. According to Mons et al. (2000), concentrations of pharmaceuticals varied from 0.31-μg/L carbamazepine in surface water to 0.90-μg/L erythromicin in effluent of a wastewater-treatment plant. Pharmaceuticals were not detected in drinking water in that study; however, actual concentrations may be 2 to 10 times larger because of the performance characteristics of the analytical methods, such as low recoveries.

In the United States, little is known about the presence of pharmaceuticals in the aquatic environment. Baseline data have been collected from streams in Iowa, but results are not yet available (Kolpin et al., 2000). A national study has been conducted to evaluate the presence of antibiotics and other organic compounds in surface water in the United States (Kolpin et al., in press). These investigators report that pharmaceuticals, hormones, or other emerging contaminants were found in 86.3 percent of the water samples collected from 139 suspicious stream sites in 32 states in the United States. Detergent degradates, plasticizers, and fecal steroids were detected in larger concentrations than other compounds analyzed. Nonprescription drugs were detected in 80 percent of the samples, prescription drugs in 36 percent of the samples, and reproductive hormones in 20 percent of the samples. Tetracyclines was the most frequently detected class of antibiotics, followed by sulfonamides, macrolides, and beta-lactams in liquid waste from hog lagoons in the United States (Meyer et al., 2000). Concentrations varied from less than 1 μg/L to more than 700 μg/L; therefore, a concern exists in the United States that these contaminants can be transported from lagoons or after the application of manure on the land into a riverine environment and, ultimately, into drinking water at trace levels.

In the United States, the use of antibiotics for livestock production is a concern. About 90 percent of approximately 2.5-million kg of antibiotics sold are given as growth-promoting and prophylactic agents (Kolpin et al., 2000). They are given in sub-therapeutic doses instead of treating active infections, thereby lowering the cost of animal care (Kolpin et al., 2000). Researchers report that antibiotics have been detected in the United States at trace levels (0.05 to 2.1 μg/L) in surface and groundwater (Meyer et al., 2000) and that tetracycline and sulfonamide antibiotics have been detected in samples collected from groundwater and wastewater of confined animal feeding operations (Thurman and Hostetler, 2000; Lindsey and Thurman, 2000). Tetracycline and sulfonamide antibiotics are believed to occur at levels on the order of 100 μg/L in wastewater from confined animal feeding operations (Meyer et al., 2000; Thurman and Lindsey, 2000). Wastes generated at confined animal feeding operations, in turn, are generally applied directly onto fields to recycle nutrients. Thurman and Lindsey (2000) hypothesized that the transport of sulfonamide antibiotics (acidic compounds) will be more rapid through soil to groundwater than that of tetracyclines. The presence of caffeine and pharmaceuticals, such as chlorpropamide, phensuximide, and carbamazepine, has been used to indicate that domestic waste was a source of nitrate in groundwater in wells from three communities near Reno, Nevada (Seiler et al., 1999).

At this time, no information has been published about the presence of pharmaceuticals in riverbank-filtered and drinking water created from RBF in the United States. A study has been completed that analyzed surface water used for drinking water (Kolpin et al., in press). Recently, the U.S. Geological Survey initiated a study in cooperation with the U.S. Environmental Protection Agency to evaluate the transport of pharmaceuticals from a river through an alluvial aquifer into a municipal water supply. Evaluating the transport of antibiotics at the river/aquifer interface will be emphasized, and temporal variations in the presence of antibiotics in the surface water, groundwater, and drinking water will be determined.

On September 9, 1999, during a reconnaissance, samples were collected from surface water, well water, and treated drinking water at the RBF site in Lincoln, Nebraska. Twenty-four antibiotics were quantified. Sulfamethoxazole and trimethoprim were detected at concentrations of <1 µg/L in riverbank-filtered and raw water. Both compounds are components of the Bactrim antibiotic. No traces of pharmaceuticals were detected in the drinking-water sample.

Another small reconnaissance was conducted by the U.S. Geological Survey in 2001 to evaluate the presence of pharmaceuticals in surface and riverbank-filtered water at three well fields along the Platte River in Nebraska. The U.S. Geological Survey also collected raw-water samples of groundwater and surface water from a network of 76 drinking-water sources across 25 states and Puerto Rico in 2001 at locations such as river intakes or raw-water sampling ports used by individual water purveyors. All samples are currently being analyzed in U.S. Geological Survey laboratories.

The occurrence of hormone compounds, such as 17-β-oestradiol, in the environment also has been a concern (Barber et al., 2000). Analyses of samples for steroid hormones indicated the presence of androgens and estrogens at parts per trillion concentrations (Barber et al., 2000b). Analyses of water from 24 streams in 19 states in the United States indicated that wastewater contaminants, such as nonylphenol and triclosan, were detected in 50 percent of the samples at parts per billion concentrations (Barber et al., 2000a).

2. Physical and Chemical Characteristics of Selected Classes of Organics

The behavior of organics, including pesticides and pharmaceuticals, in the subsurface depends on hydrologic conditions and on physical, chemical, and biological processes in the soil and in the vadose and saturated zones. Advective transport and hydrodynamic dispersion, in combination with natural processes, such as precipitation, coprecipitation, sorption/desorption, filtration, biotic and abiotic degradation, volatilization, and metabolism, lead to the retardation or even complete removal of anthropogenic compounds. The occurrence of pesticides in the aqueous environment is also influenced by local agricultural practices, pedology, and climate, leading to a large variety in pesticide occurrences with large spatial and temporal variabilities in the occurrence of a specific herbicide. The removal efficiencies of organic contaminants during RBF can vary from 6 to 100 percent (Crites, 1985; Wilderer et al., 1985; Verstraeten et al., 1999b).

Sorption and Desorption

Sorption is one of the most important processes in the removal of organics and depends upon:
- The structure and position of functional groups of the sorbate.
- The presence and degree of molecular unsaturation of the sorbent.
- The chemical characteristics of the sorbate, such as acidity, water solubility, charge distribution, polarity or lipophility, and the ability to polarize.
- The mineralogical composition, organic matter content, cation exchange capacity, and microbial activity of the sorbent.
- The duration of substance influx, infiltration velocity, particle transport, residence time, and groundwater hydraulics (mixing with other water) (Crites, 1985; Verstraeten et al., 1996; Burkart et al., 1999a, 1999b).

Many organic pollutants are hydrophobic, which indicates that these substances have a low affinity for solutions in water and prefer solutions in apolar liquids. These pollutants are readily adsorbed by organic matter or sediments. Except for herbicides, most pesticides tend to be hydrophobic, rather than lipophobic. The tendency to become sorbed can be related to the distribution coefficient of the chemical and an apolar liquid, like octanol. In general, the higher the octanol/water partition constant (K_{ow}), the higher the affinity of a pollutant to be sorbed on soil material.

$$K_{ow} = \frac{C_{n-octanol}}{C_{water}} \qquad (1)$$

where C = Concentration of substance (mg/L).

Values of K_{ow} for selected pesticides and pharmaceuticals are shown in Tables 9-4 and 9-5, respectively. Because the partitioning of an organic solute between water and octanol is, conceptually, not much different between itself and water, a correlation exists between K_{ow} and solubility (Chiou et al., 1982). Pesticides and pharmaceuticals differ widely in their overall solubilities. Some solutes, such as the pesticide glyphosate (1.06 to 106 mg/L solubility) or the pharmaceutical phenazone (51 to 900 mg/L solubility), are very soluble, whereas others, such as pendimethaline or diclofenac (Tables 9-4 and 9-5), are sparingly soluble.

Table 9-4. Physical and Chemical Characteristics of Selected Pesticides

Compound	Vapor Pressure (mPa) at 20°C	Solubility (mg/L) at 20°C	Log K_{ow}	K_{oc}	K_d (mL/g)
Alachlor	2.9 (25°C)	212	3.09	No Data	No Data
Aldicarb	13	6,000	1.14	No Data	3.2
Atrazine	4×10^{-7}	30	2.49	No Data	No Data
Bitertanol	2.2×10^{-6}	2.9	4.1	20 to 40	No Data
Chlortoluron	17	70	2.29	No Data	0.2 to 1.7
Dichlofluanid	0.014	130	3.58	No Data	No Data
Dichlorprop	3.7×10^6	0.2 (In Percent)	1.9	No Data	No Data
Glufosinate-Ammonium	<0.1	1.37×10^6	<1	No Data	No Data
Glyphosate	<0.002	1.06×10^6	−3.2	<900	No Data
Isoproturon	3.3×10^{-3}	55	2.5	460	No Data
Mecoprop	0.31	620	0.09	No Data	No Data
Methabenzthiazuron	<1	59	2.64	No Data	$K_F = 0.9$
Methidathion	0.19	250	2.52	127 to 860	No Data
Parathion-Ethyl	7.4×10^5	25	3.83	No Data	0 to 72
Parathion-Methyl	2.0	55	3.00	No Data	No Data
Penconazol	0.21	70	3.70	No Data	No Data
Pendimethaline	1.25	0.3	5.18	No Data	5.5 to 97
Simazine	8.1×10^{-2}	5	2.11	No Data	No Data
Terbuthylazine	0.15	8.5	3.05	No Data	0.8 to 2.0
Triadimefon	<1	70	3.18	No Data	No Data
Trifluralin	0.26	<1	5.07	No Data	No Data
Vinclozolin	0.13	3.4	3.00	No Data	No Data

Data from Perkow (1992); Industrieverband Agrar e.V. (2000); Biologische Bundesanstalt für Land-und Forstwirtschaft (1995); Matthess et al. (1997); Zullei-Seibert (1996a).

mPa = 10^{-3} Pascal.

K_{ow} = Octanol/water partition constant.

K_{oc} = Organic carbon and water partition coefficient.

K_d = Distribution coefficient.

K_F = Partition coefficient from Freundlich isotherm.

Table 9-5. Physical and Chemical Characteristics of Selected Pharmaceuticals

Compound	Vapor Pressure (mm Hg)	Solubility (mg/L)	Log K_{ow}	K_d (mL/g)
Acetyl salicylic acid	No Data	3,330[e]	1.19[h]	2.2[d]
Carbamazepine	1.84×10^{-7} [a]	17.66[j]	2.25[l]	No Data
Clofibrate	No Data	No Data	3.62[l]	No Data
Clofibric acid	1.13×10^{-4} [a]	582.5[j]	2.84[f] 2.57[c]	0.3
Diazepam	2.78×10^{-8} [a]	50.0[b]	2.82[i]	102[d]
Diclofenac	6.14×10^{-8} [a]	2.37[m]	4.51[n] 4.02[l]	No Data
Fenofibrate	No Data	No Data	5.19[l]	No Data
Fenoprofen	No Data	2,500[g] (37°C)	3.90[l]	No Data
Ibuprofen	1.86×10^{-4} [a]	21[b]	3.50[h] 3.97[n]	251[k]
Paracetamol (i.e., Acetaminophenol)	7.0×10^{-6} [a]	14,000[b]	0.49[f] 0.46[y]	0.4[d]
Phenazone	3.06×10^{-5} [a]	51,900[b]	0.38[y]	No Data
Primidone	3.64×10^{-9} [a]	500[b]	0.91[c] 0.73[l]	No Data
Propyphenazone	5.2×10^{-6} [a]	2,400[e]	2.32[g]	No Data

Temperature 25°C, unless stated otherwise.

mm Hg = Millimeters of mercury.

K_{ow} = Octanol water partition constant.

K_d = Distribution coefficient.

1 Pa = 7.500617×10^{-3} mm Hg.

[a]Calculated value from Neely and Blau (1985). [b]Yalkowsky and Dannenfelser (1992). [c]Hansch et al. (1995). [d]Calculated K_d Value (mL/g) from KOW (Stuer-Lauridsen et al., 2000). [e]Arzneibüro der Bundesvereinigung Deutscher Apothekerverbände (2000). [f]Henschel et al. (1997). [g]Holm et al. (1995). [h]Stuer-Lauridsen et al. (2000). [i]Syracuse Science Center (2000). [j]Calculated value from Meylan et al. (1996). [k]Experimental K_d-values (mL/g) for sewage sludge (Stuer-Lauridsen et al., 2000). [l]Syracuse Science Center (2000), Calculated value according to Meylan and Howard (1995). [m]Fini et al. (1993). [n]Avdeef et al. (1998). [r]Rafols et al. (1997). [s]Budavari (1996). [y]Moeder et al. (2000).

Sorption is modeled by fitting experimentally derived isotherms to theoretical equations. One of the most common models is the Freundlich isotherm:

$$S = K_F \; C^N \tag{2}$$

where S is the concentration of substance sorbed (mg/kg); K_F is the partition coefficient of Freundlich isotherm; C is the concentration of substance in water (mg/L); and N is the constant.

If the Freundlich isotherm has N = 1, the isotherm is linear and S is related to C by the distribution coefficient K_d (Henry isotherm). When sorption is absent, K_d is zero; however, in practice, almost any chemical can be sorbed on the solid particles of soils or sediments. The stronger sorption is, the higher the K_d value. Several studies (e.g., Karickhoff et al., 1979; Schwarzenbach and Giger, 1985) have shown how the distribution coefficient can be expressed as the product of K_{oc} and the weight fraction of organic carbon. K_{oc} and K_d values for pesticides are presented in Table 9-4, and K_d values for pharmaceuticals are tabulated in Table 9-5. A good correlation exists between log K_{oc} and log K_{ow}, with regression equations presented by Karickhoff et al. (1979), Schwarzenbach and Westall (1981), and others.

The tracer velocity of organic compounds is lower than groundwater flow velocity. This behavior is described by the retardation factor R_d, whereby the transport velocity of a substance can be calculated with respect to groundwater flow velocity.

$$R_d = 1 + \frac{(1-n)}{n} \rho_b \ K_d \qquad (3)$$

where n is the porosity; ρ_b is the bulk density of the solid (g/cm^3); and K_d is the distribution coefficient (mL/g) (Henry isotherm).

K_d and R_d values have been determined both in the laboratory and field for several pesticides and for a variety of porous rock aquifers. According to Matthess and Pekdeger (1985), the retardation factors are between 0.4 and 10.

Filtration and Colloids

Colloids are particles varying in size from 1 to 1,000 nanometers (nm) and include microorganisms, large macromolecules, and inorganic fragments. The migration of particles provides a way for mass transport in the subsurface, either as contaminants themselves or by contaminants sorbed onto these particles (McCarthy and Zachara, 1989). Electrostatic forces between negatively charged particles and negatively charged solids may repulse colloids and can result in a colloid transport velocity that exceeds groundwater flow velocity. On the other hand, physical and chemical processes may lead to a filtration of colloids (Matthess et al., 1997).

Biotic and Abiotic Degradation

Organic substances can be transformed into simpler inorganic forms by biotic and abiotic degradation. The most important abiotic transformation reaction for many organic substances (e.g., carbamates, triazines) is hydrolysis, which involves a reaction of an organic molecule with water or a component ion of water. The degradation of organic compounds is often related to redox processes, especially oxidation reactions. These reactions are referred to as biodegradation because they are microbiologically catalyzed. Degradation in the subsurface often follows the first-order rate law, yielding Equation 4 after integration:

$$C = C_0 \ e^{-\lambda \ t} \qquad (4)$$

where C is the concentration (mg/L); λ is the rate constant (1/s); C_0 is the initial concentration (mg/L); and t is time (s).

In only a few cases, biodegradation leads to a complete mineralization of organic substances to compounds such as carbon dioxide and water. The pharmaceutical, acetyl salicylic acid, is an example of the complete degradation of such a compound in groundwater. In most cases, metabolites are formed that can increase, decrease, or exert a negligible effect on the toxicity of compounds (Barbash and Resek, 1996). The formation of metabolites is influenced by pH, ionic strength, redox conditions (Stuyfzand, 1998a), the presence of catalysts or other reactive species, the presence of surfactants, the potential for biochemical transformations mediated by microorganisms, the mineralogy and chemistry of subsurface sediments (Barbash and Resek, 1996), and the chemical characteristics of the compound, such as adsorption behavior (Roberts, 1985; Van Hoorick et al., 1998).

Volatilization

Volatilization is controlled by the physical and chemical characteristics of the compound, the sorptive characteristics of the river bed and aquifer media, the concentration of the compound, competition with other compounds in relevant concentrations, soil-water content, air movement, air temperature, and diffusion processes. This process of liquid- or solid-phase evaporation amounts to a loss of pesticides on the order of 5 to 10 percent of the total mass in surface water and is less important in the unsaturated zone. Volatilization is most important after the application of pesticides, depending on the weather and agricultural methods (Taylor and Spencer, 1990). Vapor pressures of pesticides vary over seven orders of magnitude (see Table 9-4) and over 20 orders of magnitude for pharmaceuticals. In general, vapor pressures of pharmaceuticals are low, and volatilization is almost nonexistent when compared to that of pesticides.

Application of Theory

Artificial groundwater recharge can eliminate up to 100 percent of the concentration for hydrophobic substances with high sorption tendencies, such as DDT and heptachlor (Zullei-Seibert, 1996a). Retention percentages can vary from 10 percent (such as atrazine, simazine) to 100 percent (such as MCPA), depending upon the conditions listed in Table 9-6. Infiltration was found to be accompanied by a smoothing or reduction of pesticide peak concentrations found in the infiltrated water (Zullei-Seibert, 1996a).

Table 9-6. Efficacy of Artificial Groundwater Recharge for the Removal of Selected Pesticide Residues from Surface Water and Physico-Chemical Properties of the Individual Compounds

| Compound | Reduction During Artificial Groundwater Recharge | | | |
	Input Concentration (µg/L)	Mean Reduction (%)	Water Solubility (mg/L)	Log K_{ow}
Atrazine	0.01 to 0.30	12	30	2.49
Simazine	0.01 to 0.20	13	5	2.11
Diuron	0.03 to 1.02	90	42	2.82
Chlortoluron	0.03 to 0.77	100	70	No Data
Isoproturon	0.03 to 1.30	20	55	2.50
MCPA	0.03 to 0.52	100	825	No Data
Deethylatrazine	0.01 to 0.10	0	600	No Data
Dichlorprop	0.03 to 0.05	100	No Data	No Data
Lindane	0.008 to 0.04	100	10	No Data
Methabenzthiazuron	0.03 to 0.04	100	59	No Data
Terbuthylazine	0.01 to 0.21	90	8.5	3.05

Data from Zullei-Seibert (1996a).

K_{ow} = Octanol/water partition coefficient.

Kuhlmann et al. (1995) studied the behavior of three phenoxyacetic acid herbicides (2,4-D, 2,4,5-T, and MCPA) during subsurface transport and RBF in a model system consisting of laboratory filter columns filled with natural underground materials. These test filters were operated at different redox environments applying natural aerobic and anaerobic groundwater. In the presence of oxygen, the biodegradation of the three herbicides was observed to start after an initial lag time; however, no degradation was observed under sulfate-reducing conditions. In other experiments, Kuhlmann et al. (1995) varied the concentrations of herbicides, time, and the nutrient content to assess the factors that may influence microbial degradation in the anaerobic environment, but these variations did not affect degradation rates. The maximum retention of 2,4-D, 2,4,5-T, and MCPA in the filters was calculated at 30 percent, mainly attributed to adsorption to the filter material.

Preuss et al. (1998) investigated the behavior of the hydroxybenzonitrile herbicides, bromoxynil and ioxynil, and the phenoxy acid herbicide, mecoprop, during artificial groundwater recharge and RBF. The adsorption and biodegradation of these compounds was investigated in batch cultures and model ecosystems to allow a risk assessment. In their experiments, Preuss et al. simulated worst-case situations during slow sand filtration, aerobic and anaerobic subsurface transport, and residence in the unsaturated and saturated sediments. The experiments showed that bromoxynil, ioxynil, and mecoprop can pass through the slow sand filters, especially under continuous input situations. The tested compounds could be partially degraded during slow sand filtration and underground passage, but only under specific environmental conditions. Under natural groundwater conditions, herbicides were persistent up to more than 4 weeks. The adaptation times and degradation rates of bromoxynil, ioxynil, and mecoprop indicated a contamination risk during groundwater recharge and RBF. Preuss et al. assumed that this risk potential was caused by the low adsorption of these compounds, their ability to inhibit microbial activities, and their biological stability under natural groundwater conditions for as much as 28 days.

The occurrence of pharmaceuticals in sewage effluent and surface water (Stan and Heberer, 1997; Halling-Sørensen, et al., 1998; Heberer and Stan, 1998; Ternes, 1998; Daughton and Ternes, 1999; Möhle et al., 1999; Mons et al., 2000; Wilken et al., 2000; Heberer, in press; Heberer et al., 2001) as well as in groundwater (Heberer, 1995 and in press; Heberer et al., 1997; Heberer and Stan, 1997) shed light on the transport behavior of these compounds. Many pharmaceutically active compounds are water-soluble with low to moderately high log K_{ow} values, indicating that they are mobile and migrate with groundwater in the aquifer; however, pharmaceuticals with the highest solubility and with low log K_{ow} values, such as acetyl salicylic acid or paracetamol, are easily degraded (Scheytt, 2002). Thus, the occurrence of these compounds in groundwater has not been, and is unlikely to be, reported. Moreover, the degradation of pharmaceutical residues may not be restricted to microbial processes. Buser et al. (1998) observed the photolytic degradation of the drug, diclofenac, in natural water in laboratory experiments; therefore, a significant concentration decrease of this drug in surface water can be expected depending on sunlight intensities, exposure time, and turbidity of the water. In the aquifer, photolytic reactions are of minor importance; therefore, this type of reaction is restricted to surface water and the uppermost parts of the unsaturated zone.

Pharmaceuticals with the highest known mobility are clofibric acid and carbamazepine, in spite of their moderately high log K_{ow} values. Laboratory column experiments revealed that clofibric acid exhibited no retardation (Scheytt et al., 1998) and carbamazepine exhibited low retardation (Scheytt, 2002), whereas diclofenac, ibuprofen, and propyphenazone were significantly retarded (Scheytt et al., in preparation). In the case of propyphenazone, retardation was most likely caused by a weak sorption of this compound onto colloids. After flow conditions in the column were changed, a marked increase of propyphenazone was detected at the outflow of the column, indicating a possible mobilization of colloids and of propyphenazone sorbed on these particles (Scheytt, 2002).

3. The Presence of Pesticides, Pharmaceuticals, Industrial Products, and Personal Care Products in Riverbank-Filtered Water

In the United States, the Surface Water Treatment Rule under the 1986 Amendment to the Safe Drinking Water Act required that public-water supplies be evaluated for susceptibility to surface-water effects (U.S. Environmental Protection Agency, 1989). Alluvial aquifers adjacent to large streams are important sources of water for many municipalities. Alluvial aquifers can have large transmissivities and hydraulic conductivities, which make them very desirable for water supply because large amounts of water can be withdrawn. The U.S. Environmental Protection Agency outlined a method to determine if a groundwater source is groundwater under the direct influence of surface water (Vasconcelos and Harris, 1992).

In general, the number of detected compounds decreases as the compounds move from surface water to groundwater and, ultimately, into finished drinking water. Concentrations of organics in riverbank-filtered water, generally, are less than those observed in rivers (Schaffner et al., 1987; Ray et al., 1998; Verstraeten et al., 1999b; Kühn and Müller, 2000; Ray et al., in press), but the length of exposure to contaminants increases (Verstraeten et al., 1999b, in press; Kühn and Müller, 2000) with distance from the river (Roger and Fontenelle, 1999); however, certain degradates (e.g., deethylatrazine) can occur in higher concentrations in riverbank-filtered water than in river water (Verstraeten et al., in press). Chauveheid et al. (1999) found that water from alluvial aquifers was more vulnerable to pesticide contamination than water from limestone aquifers; however, in areas where macroporosity, such as fractured rock, was present, pesticides could be detected even at great depths in limestone aquifers. In general, karst limestone aquifers tend to have fractures that increase transport to larger depths.

Herbicides have been identified in riverbank-filtered water in the United States, Europe, and Newly Independent States among other countries worldwide (Szabo et al., 1994; Verstraeten et al., 1999b, in press; Boyd, 2000); however, pesticide detections in the aquatic environment have not been reported or studied in many countries, possibly because of a lack of funding, a lack of analytical techniques at lower parts per billion levels, or a lack of political interest in this type of environmental pollution.

In countries where environmental studies are conducted to assess the presence of pesticides in aquatic environments, herbicides have been detected more commonly than other types of pesticides, with atrazine, degradates of atrazine, and degradates of acetanilides reported more frequently than any other herbicides in river and riverbank-filtered water (Gojmerac et al., 1994; Notenboom et al., 1999; Verstraeten et al., 1999b; Kolpin et al., 2000; Mazounie et al., 2000). Atrazine, deethylatrazine, simazine, diuron, isoproturon, glyphosate, chlidazon, metolachlor, linuron, bromacil, pentachlorophenol, and others have been found in concentrations generally less than 0.5 µg/L in river water, groundwater, or riverbank-filtered water contained in sandy alluvial and limestone aquifers of the Meuse, Ijzer, and Schelde Rivers, and near the City of Liège in Belgium (Belgaqua and Phytofar, 1999; Hodiaumont et al., 1999; Roger and Fontanelle, 1999). In drinking water obtained from alluvial sediments partly influenced by the infiltration of surface water, atrazine detections of more than 0.1 µg/L decreased from 20 percent (1991 to 1996) to 5 percent (1998) because the use of atrazine was banned in Belgium. Deethylatrazine detections decreased from 3 to 2 percent in the same time frame. Similarly, atrazine detections of 0.05 to 0.1 µg/L decreased from 29 percent from 1991 to 1996 to 21 percent in 1998. Deethylatrazine detections decreased from 31 to 19 percent in the same time frame (Belgaqua and Phytofar, 1999).

Roger and Fontanelle (1999) suggested that exposure to deethylatrazine (a degradate of atrazine that may be just as toxic) may be more important than exposure to atrazine itself because deethylatrazine is much more soluble (about 100 times) and its adsorption potential by soil is less than atrazine (about 10 times). In addition, the breakthrough of deethylatrazine will be much faster than the breakthrough of atrazine when water is treated with organic carbon to reduce organics in drinking water (Roger and Fontenelle, 1999). Zullei-Seibert (1996a) stresses that strong hydrophobic compounds (e.g., DDT, heptachlor, and diuron) can be eliminated during RBF, but that herbicides with higher solubilities, including atrazine and simazine, show a retention of only 10 percent.

Even though studies done in the United States and Europe have shown significant transport of a variety of compounds into riverbank-filtered water, Dreher and Gunatilaka (1998) found no detections of pesticides (alachlor, atrazine, chlordane, and heptachlor) and other micropollutants (total polycyclic aromatic hydrocarbons and polychlorinated biphenyls) at levels greater than 0.1 µg/L in the Danube River and no presence of these contaminants in riverbank-filtered water. The study does not indicate, however, the detection limit or reporting limit of the analyses; therefore, these nondetections may be related to the study design, including such factors as sample timing, sampling methodology, analytical methods and, most importantly, method reporting levels.

A research study carried out in Germany between 1987 and 1992 investigated the pesticide pollution of groundwater and public-drinking water caused by artificial groundwater recharge or RBF (Mathys, 1994). Samples from public-drinking water, raw water, groundwater, and surface water in an area with intensive agriculture were analyzed for pesticide and nitrate residues. The objective was to monitor the degree of pesticide pollution in public drinking-water supplies and to characterize the pathways by which these substances reach potable water. According to Mathys, monitoring revealed that only potable water from waterworks using artificial groundwater recharge were polluted by pesticide residues. For more than 12 months, almost all surface water tested, including canals, contained pesticides at highly fluctuating concentrations; therefore, they were always a potential source for groundwater recharge. Apart from agricultural runoff, there were notable detections of river water contaminated with the herbicide diurone (caused by effluent from municipal sewage) during the summer.

Artificial groundwater recharge was identified as the main factor for the input of pesticides into the aquifer and drinking water. It was possible to track the influence of surface-water quality on the degree of pesticide contamination in water from wells; thus, well water influenced by RBF or infiltration contained a larger number and significantly higher amounts of compounds than water in "groundwater" wells not influenced by surface water (Mathys, 1994). Among the pesticides, triazines and phenylurea herbicides were frequently detected. Several other polar pesticides, such as glyphosate, bentazone, bromoxynil, ioxynil, phenoxy acids (2,4-D, MCPA, mecoprop, dichlorprop, etc.), were not analyzed in this study; thus, no data were provided to assess the presence of these other environmentally important compounds.

Additional infiltration experiments and the seasonal changes of pesticides in raw and infiltrated water showed the great mobility of these compounds during their movement through the subsoil as well as the vulnerability of the aquifers; however, no correlation was found between the occurrence of pesticides and nitrate in raw and infiltrated water. Thus, Mathys (1994) concluded that nitrate is not suitable as an indicator of pesticide pollution. Regarding the insufficient removal of herbicides, such as triazines and phenylureas, during RBF or infiltration and, because of the large pesticide loads of surface water, Mathys recommended minimizing pesticide losses in the whole catchment area, especially as runoff into surface water, and abstaining from using slowly degradable herbicides in cities, on railways, or on private yards.

Until recently, very few studies on the behavior of pharmaceutically active compound residues during bank filtration have been carried out. Generally, polar pharmaceutically active compound residues have been recognized as problematic compounds for RBF (Heberer and Stan, 1996, 1997; Heberer et al., 1997; Brauch et al., 2000; Kühn and Müller, 2000), but more comprehensive studies are currently underway to examine the behavior of polar pharmaceutically active compound residues in Germany and the United States.

The occurrence of pharmaceuticals in rivers shows some evidence of temporal and spatial variations and depends upon the time of year, the proximity to a wastewater-treatment plant, the proportion of discharges from the wastewater-treatment plant to the river, and other factors. Of concern in riverbank-filtered water and drinking water are water-soluble pharmaceuticals, such as diclofenac, ibuprofen, bleomycin, clofibric acid, fenofibrate, diazepam, and carbamazepine (Zullei-Seibert, 1998; Kühn and Müller, 2000). Generally, they are lipid-reducing agents and antirheumatics.

Jüttner (1999) assessed the efficacy of RBF for removing fragrance compounds and aromatic hydrocarbons. Fragrance compounds included menthol, limonene, a-terpineol, 4-tertiary butylcyclohexanol, and 4-tertiary butylcyclohexanone. The concentrations of these compounds were reduced from 0.1 to about 0.01 µg/L. This was mainly because of RBF and, in part, because of microbial degradation potentially facilitated by the presence of facultative denitrifying bacteria. Jüttner reported that most organics were removed in the first few meters of the alluvium away from the river. Concentrations of hydrocarbons were similarly reduced.

The International Association of River Waterworks did a study to assess the presence of endocrine disruptors in river water, processed water, and drinking water (Ghijsen and Hoogenboezem, 2000). Oestron (1 to 4 nanograms per liter [ng/L]) was detected in the Rhine River; bisphenol A (as much as 150 ng/L) in the Meuse River; phthalates (0.1 to 0.35 µg/L) in Dutch surface water; and alkylphenol polyethoxylates (0.9 to 15 µg/L) in the Rhine and Meuse Rivers. They evaluated processed water, including riverbank-filtered water at Lekkerkerk or Nieuw Lekkerland in The Netherlands, that contained bisphenol A (30 and 130 ng/L) and nonylphenol ethoxylates (1.6 µg/L). In drinking water, phthalates were detected along the Rhine and Meuse Rivers. Oestrogenic activity was detected in drinking water in one sample, with detections of the more soluble compounds in processed and raw surface water a bit more frequent and containing higher concentrations. The oestrogenic activity of water from the Meuse River was higher than that from the Rhine River, and large fluctuations in concentrations were found. The study also showed that a more effective removal of oestrogenic activity can be achieved by infiltration basins rather than by riverbank infiltration. Compounds studied during the project included 17α-oestradiol, 17β-oestradiol, oestron, and 17β-ethinyloestradiol, as well as industrial compounds, such as bisphenol-A, phthalates, and some ethoxylates. These detections suggest that the presence of these potential endocrine disrupting chemicals should be evaluated at RBF sites, especially near rivers where most of the river water is derived from wastewater sources.

The behavior of organic micropollutants was studied at two sites during the infiltration of water from the Glatt River to groundwater in Switzerland (Schwarzenbach et al., 1983; Schaffner et al., 1987). The micropollutants studied included nonpolar volatile compounds and pentachlorophenol, nonylphenol, nonpethoxylate, nonphenyl diethoxylate, NTA, and EDTA. Many compounds were removed up to 98 percent after only 7 m, mainly in the first 2.5 m, from the riverbank (Schaffner et al., 1987). Grischek et al. (1994) confirmed the findings of Schaffner et al. (1987) with respect to EDTA along the Elbe River near Dresden, Germany. They suggested that EDTA, combined with isotopes such as tritium, can be used to determine flowpaths and to evaluate mixing and redox reactions. Grischek et al. (1994) also showed that the reduction of organic compounds occurred within the first few decimeters of the infiltration path; however, caution should be used when using EDTA as a conservative tracer; Kühn and Müller (2000) reported some reduction in EDTA concentrations in riverbank-filtered water as compared to river water along the Rhine River.

Several studies have been conducted that identified the presence of polar aromatic sulfonates, including naphthalenesulfonates and stilbenesulfonates, in riverbank-filtered water along the Rhine, Elbe, and Danube Rivers. The presence of polar aromatic sulfonates reflected the great production volumes at the well fields, plus the mobility and recalcitrant nature of the compounds (Lange et al., 2000). Naphthalenesulfonates have been identified in drinking water from waterworks along the Elbe River.

The behavior of trace organics, including tetrachloroethylene, lindane (γ-BHC), and benzo(a)pyrene, was studied by Herrman et al. (1986) at a site in northern Bavaria. These researchers determined that polycyclic aromatic compounds were removed within the first few decimeters of infiltration, but it was demonstrated that a strong hydrophobic compound, such as γ-BHC, moved rapidly with the infiltrating water to the groundwater. The significance of the organic matter for the retention of hydrophobic compounds depends on the sorptive characteristics of the solutes. At low organic matter content, the importance of inorganic surfaces and their cation exchange capacity increases. A decrease in redox potential will induce the reduction of ferric oxides to soluble ferrous irons, which have a great exchange capacity, potentially leading to increased sorption sites.

4. Case Studies

The 10 case studies that follow provide a broad overview of different types of contaminants present in rivers that can be transported into riverbank-filtered water and, ultimately, into drinking water. Several other case studies have been discussed in a recent paper by Ray et al. (2002) and have not been duplicated in this chapter. These authors report on four case studies:

- Torgau, Germany.
- Düsseldorf, Germany.
- Louisville, Kentucky, United States.
- Jacksonville, Illinois, United States.

Contaminants discussed include organics (adsorbable organic halogen, adsorbable organic sulfur, selected pesticides, halogenated hydrocarbons, aromatic amines, and sulfonic acids) at the Torgau site, disinfection byproduct precursors at the Louisville site, and atrazine and nitrate at the Jacksonville site. The removal of organics was not discussed at the Düsseldorf site. Findings of those case studies are in agreement with findings presented herein.

Generally, the removal of organic substances occurs in the first few meters at the river/aquifer interface (Grischek et al., 1997). The removal of organics tends to be on the order of tens of percent caused by a combination of several mechanisms, including dilution with other water; microbiological, chemical, and physical degradation; volatilization; adsorption/desorption processes or dispersion; and diffusion.

The 10 case studies presented herein focus on:

- The removal of pesticides and pharmaceuticals.
- The occurrence and fate of these contaminants during RBF.
- Their use as potential tracers.

Transport of Triazine and Acetamide Herbicides and Their Degradates from the Platte River into Drinking Water at the Lincoln RBF Site in Nebraska, United States

The water supply in the City of Lincoln, Nebraska, is affected by herbicides from the Platte River in riverbank-filtered water in late spring and early summer (Verstraeten et al., 1999b). The City's well field consists of 38 active production wells. Two wells are horizontal collector wells (Sites 2 and 3) on an island in the Platte River, and 36 wells are vertical wells on the west bank of the Platte River (Figure 9-2). These wells provide water for about 220,000 residents of Lincoln, Nebraska. The water demand varies from a winter rate of about 121-million L/d to a summer rate of about 295-million L/d (Verstraeten and Miriovsky, 1999). Generally, about 30 to 50 percent of the municipal water for the City of Lincoln is contributed by one or two collector wells and the remaining 50 to 70 percent of the water by a variable selection of vertical wells. The City's collector wells consist of a main vertical caisson and seven horizontal laterals, with total lengths of 393 m (Site 2) and 432 m (Site 3) and at depths of 24 m (Site 2) and 26 m (Site 3). Each collector well is equipped with three large pumps with a total capacity of 98.4-million L/d. The City's production wells are developed in Quaternary-age alluvial sediments consisting of sand, gravel, silt, and clay.

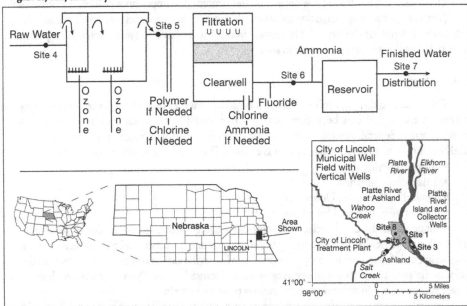

Figure 9-2. Location of sampling sites and the municipal drinking-water treatment plant near Lincoln, Nebraska. Modified from Verstraeten et al. (in press).

During a study in 1996, the relationship between river/aquifer interaction and the management of collector well laterals was evaluated by performing injections with dye (Verstraeten et al., 1999b). The results indicated that the quality of river water affects the quality of riverbank-filtered water and that several management schemes can be used to minimize the effect of the quality of river water on drinking water. Additional research has shown that 50 to almost 100 percent of water in the collector wells consists of induced surface water, depending on the management of the wells in the well field and the laterals of the collector wells (Steele and Verstraeten, 1999) in addition to the pumping rate.

The specific objectives of the studies conducted at the well field were to:
- Determine the extent to which selected herbicides and degradates in the Platte River are removed through RBF while operating two collector wells during two spring runoff events.
- Evaluate the water-quality impact on the operation of the well field with special emphasis on the operation of the collector wells during spring runoff events.
- Evaluate the effect of RBF, ozonation, and chlorination on the presence of triazines and chloroacetanilides in drinking water.

The studies were conducted to ensure that the U.S. Environmental Protection Agency disinfection and inactivation (concentration multiplied by time) criteria were met (U.S. Environmental Protection Agency, 1991, 1996). During spring runoff events, the impact of river-water quality on raw and finished water in the treatment plant is great and imposes a burden on ozone demand.

Selected herbicide and degradate concentrations were studied in water samples from the Platte River and both collector wells during 1995 and 1996 spring runoff events. A variety of parent herbicides were detected in water from the collector wells. Concentrations of herbicides and degradates increased or decreased, depending upon the chemistry and characteristics of the compound (Verstraeten et al., 1999b). During spring runoff events, with the collector wells pumping at their maximum pumping rate, herbicides in the Platte River at Site 2 were generally transported through the alluvial aquifer into collector wells in 6 to 7 days (Figure 9-3, only atrazine data are shown).

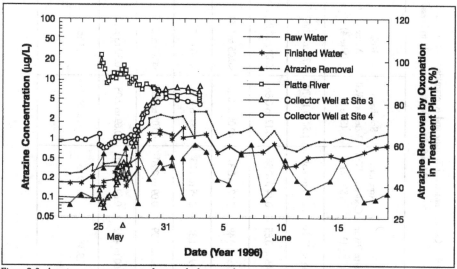

Figure 9-3. Atrazine concentrations and removal of atrazine by ozonation in a treatment plant in Lincoln, Nebraska, in Spring 1996 (Verstraeten and Miriovsky, 1999).

During two spring runoff events in 1995 and 1996, atrazine concentrations in water from the collector wells reached approximately 7 µg/L (70 times more than the background concentration of 0.10 µg/L in groundwater). Concentrations of herbicides and degradates in the collector wells during these pumping scenarios were generally one-half to one-fifth the concentrations of herbicides in the river for atrazine, alachlor, alachlor ESA, metolachlor, cyanazine, and acetochlor. For example, alachlor ESA was detected in water from the Platte River at a maximum concentration of 2.1 µg/L and in water from the collector wells at a maximum concentration of 1.2 µg/L. Concentrations of herbicides and degradates were attenuated during RBF, either by dilution mixing, adsorption/desorption processes, or microbial and chemical degradation, physical, chemical, and biological processes (Table 9-7).

Table 9-7. Concentrations of Triazine and Acetamide Herbicides and Degradates, 1997 to 1999
(Modified from Verstraeten et al., In Press)

Analyte	Year	Platte River (Site 1)	Northern Collector Well (Site 2)	Southern Collector Well (Site 3)	RBF Water (Site 4)	Post-Ozonation (Site 5)	Post-Chlorination and Filtration (Site 6)	Treated Water (Site 7)	Ground-water (Site 8)
Atrazine	1997 to 1999	9.64 to 12.50	1.50 to 9.39	1.50 to 10.70	0.77 to 11.60	0.50 to 4.40	0.50 to 3.50	0.50 to 4.10	<0.05 to 0.11
Deethylatrazine	1997 to 1999	0.46 to 0.85	0.14 to 0.86	0.12 to 0.86	0.08 to 0.85	0.16 to 1.60	0.17 to 1.41	0.18 to 1.55	<0.05
Deisopropylatrazine	1997 to 1999	0.23 to 0.48	0.07 to 0.37	<0.05 to 0.42	<0.05 to 0.41	0.09 to 0.69	0.07 to 0.61	0.09 to 0.64	<0.05
Cyanazine	1997 to 1999	3.49 to 1.36	0.35 to 1.03	0.33 to 0.96	0.16 to 1.15	0.19 to 0.52	0.18 to 0.39	0.19 to 0.50	<0.01 to 0.09
Alachlor	1997 to 1999	0.60 to 0.99	0.09 to 0.20	<0.05 to 0.24	<0.05 to 0.19	<0.05 to 0.08	<0.05 to 0.06	<0.05 to 0.07	<0.05
Metolachlor	1997 to 1999	3.60 to 4.72	0.69 to 1.80	0.69 to 2.08	0.30 to 1.62	0.13 to 0.64	0.12 to 0.48	0.12 to 0.56	<0.05
All Triazine and Acetamide Herbicides	1999	29.35	No Data	22.29	22.86	11.79	9.24	10.15	No Data

Concentrations in µg/L.

In 1999, the concentrations of herbicides in finished water were greater than in previous years (Verstraeten et al., in press). At that time, changes in triazine and chloroacetanilide concentrations during natural and artificial treatment by RBF, ozonation, filtration, and chlorination were measured in water for the Lincoln public-water supply during abnormal, worst-case conditions. During this test, only laterals that extended under the river and the shortest possible subsurface travel distances were used, thereby:

- Maximizing the presence of surface water in the collector wells.
- Maximizing concentrations of herbicides in riverbank-filtered water.
- Minimizing the formation of degradates through chemical, biological, and photolytic degradation during RBF.
- Maximizing the effect of additional water treatment in the treatment plant on concentrations of herbicides and degradates.

The 1999 results indicated that parent compounds in treated water decreased by 76 percent of the concentration present in river water (33 percent by RBF) (Verstraeten et al., in press). Atrazine concentrations decreased 14 percent from atrazine concentrations observed in the Platte River during RBF. In 1997, when less surface water was present in the water from collector wells, atrazine concentrations decreased 84 percent during RBF and additional treatment processes.

Overall, degradates of herbicides for which analytical techniques existed decreased by 21 percent during treatment (an increase of 26 percent during RBF, a decrease of 23 percent during ozonation, and another decrease of 24 percent during chlorination, including minor filtration). After RBF, increases in numerous degradates were measured (e.g., deethylatrazine, deisopropylatrazine, cyanazine amide, cyanazine acid, and deethylcyanazine). After ozonation, concentrations of degradates either increased (e.g., deethylatrazine) or decreased (e.g., alachlor ESA).

The results obtained at this RBF study illustrate that both RBF and the ozonation of water, during which degradation products are created, can shift the risk to human health associated with the consumption of water containing herbicides in part from the parent compounds to their degradation products (Verstraeten et al., in press). The selective operation of a well field at RBF sites during and shortly after spring runoff events can minimize the presence of herbicides in surface water on the quality of finished water. The quality of drinking water can be improved by encouraging longer subsurface travel times, by enhancing filtration and degradation of the herbicides, and by establishing early-warning systems based on real-time water-quality monitoring of the river. These measures will allow water-supply managers to manage the well field most affected by the quality of river water. The City of Düsseldorf in Germany has such a complex early warning system in place, and the City of Lincoln in Nebraska is considering such a system.

Despite these management actions, it remains to be seen whether long-term exposure to smaller levels of herbicides is safer than short-term exposure to greater concentrations of herbicides. In 1999, whereas large concentrations of herbicides in river water passed by the well field in several (number unknown) days, smaller concentrations of herbicides appeared to be present for a period of 1 month at levels above background concentrations in riverbank-filtered water and finished drinking water. No known study has been conducted at a RBF site using a mass-balance analysis to evaluate the loss of actual parent and degradate herbicide mass during RBF. In fact, "herbicide removal" is a term used by engineers to describe the attenuation of herbicides during transport. The use of the term "removal" may lead to erroneous interpretations and risk assessments of the presence of contaminants in drinking water.

Occurrence and Fate of Polar DDT Degradates in Surface Water and Bank-Filtered Water
in Berlin, Germany

Nonpolar contaminants are easily adsorbed to sediments or in the fatty tissues of birds, fishes, or mammals; therefore, the major problem associated with these compounds is their potential for bioaccumulation in aquatic biota. They are not recognized as problematic contaminants during bank filtration because bank filtration has been proven to be an effective technique, especially for the removal of nonpolar contaminants. In the environment, nonpolar contaminants may be converted into more polar degradates that have a lower K_{ow} and a different, much more problematic, leaching behavior. Over time, surface-water contamination by nonpolar organics originating from toxic waste sites may also become relevant for neighboring bank filtration sites. This is demonstrated by the example of some very persistent DDT residues found in surface-water and groundwater samples in Berlin, Germany.

More than 60 years after the introduction of the insecticide DDT and more than 25 years after its ban in most developed countries, 2,2-bis(chlorophenyl)acetic acid (DDA) formed by the degradation of DDT (2,2-bis[chlorophenyl]-1,1,1-trichloroethane) residues was identified for the first time as an important environmental contaminant (Dünnbier et al., 1997; Heberer and Dünnbier, 1999). DDA was detected as the main contaminant up to the microgram-per-liter level in surface water of the Teltowkanal, a canal south of Berlin (Heberer and Dünnbier, 1999). Several other intermediates from the degradation of DDT were also detected at low concentrations in surface water of the Teltowkanal (Heberer and Dünnbier, 1999) (Figure 9-4). These include:

- DDD (2,2-bis[chlorophenyl]-1,1-dichloroethane).
- DDE (2,2-bis[chlorophenyl]-1,1-dichloroethylene).
- DDMU (2,2-bis[chlorophenyl]-1-chloroethylene).
- DDOH (2,2-bis[p-chlorophenyl]-ethanol).
- DDMS (2,2-bis[chlorophenyl]-1-chloroethane).
- DDCN (2,2-bis[chlorophenyl]-acetonitrile).
- DBP (dichlorobenzophenone).

These results (Table 9-8) added some new facets to the ongoing controversy over the fate of DDT in the natural environment (Pereira et al., 1996; Quensen et al., 1998; Renner, 1998). They also confirmed several laboratory experiments about the natural remediation of DDT residues (Marei et al., 1978; Ware et al., 1980).

As far as bank filtration and its use for drinking-water production is concerned, the polar degradation product DDA has been shown to be a potential problem for public drinking-water suppliers. DDA was the only DDT derivative that leached into groundwater at a drinking-water treatment plant downstream from a toxic waste site (Dünnbier et al., 1997; Heberer and Dünnbier, 1999). This particular drinking-water plant uses more than 60 percent of bank-filtered water in drinking-water production.

DDA was found at maximum concentrations of up to 1,000 ng/L in surface water collected from the Teltowkanal (Figure 9-5). DDA was the dominant DDT derivative in surface-water samples (Table 9-8). In water from wells at a bank filtration area near the Teltowkanal and downstream from the DDT Superfund site, o,p'-DDA and p,p'-DDA were detected at concentrations as much as 0.28 and 1.7 μg/L, respectively. The chromatogram of a derivatized extract of a groundwater sample, containing 65 ng/L of o,p'-DDA and 220 ng/L of p,p'-DDA (Figure 9-6), was recorded using capillary GC/MS with selected ion monitoring. In some of the well water, the concentrations of DDA exceeded the European maximum tolerance levels for pesticides and their degradates in drinking water (Dünnbier et al., 1997; Heberer and Dünnbier,

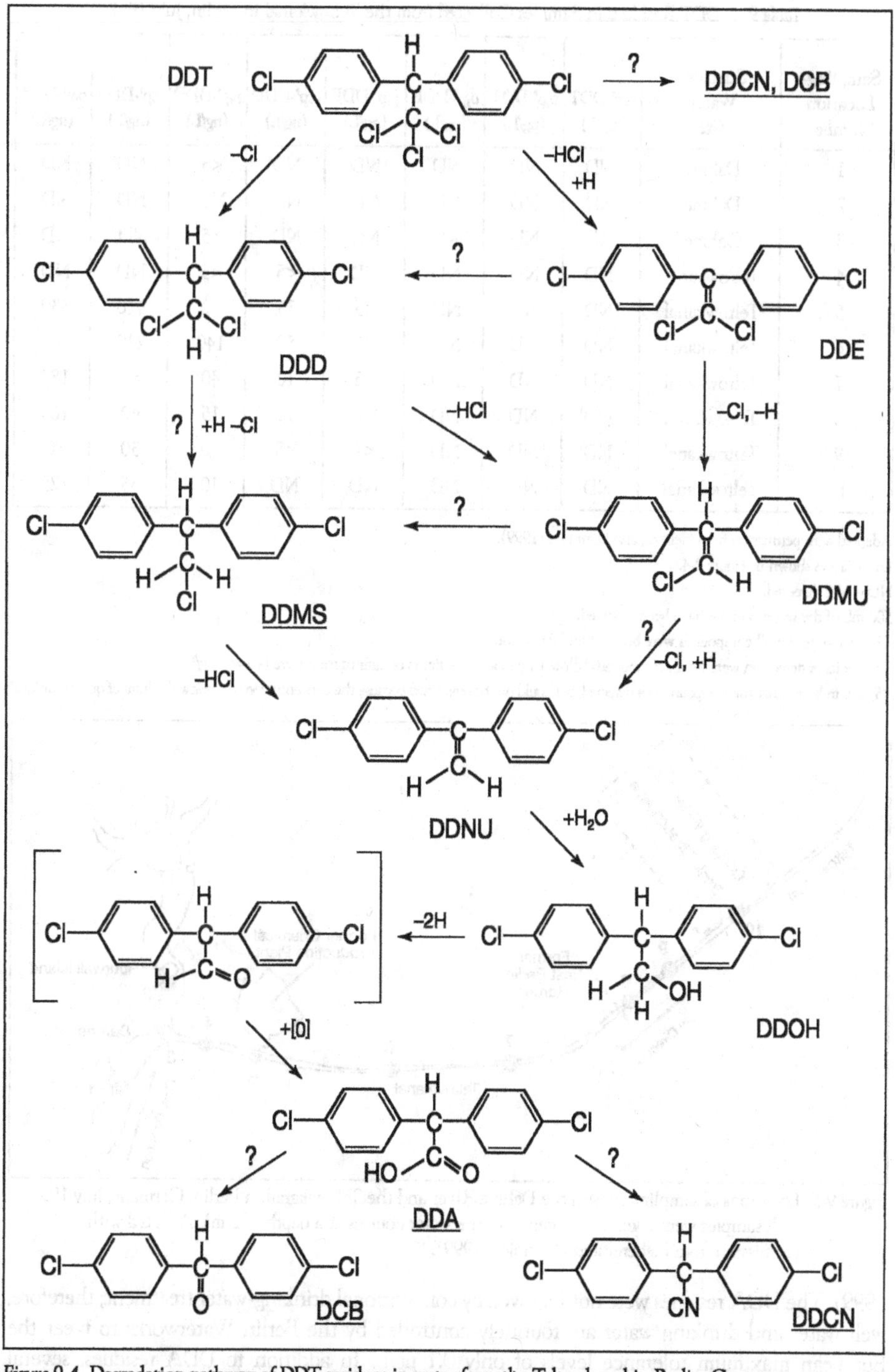

Figure 9-4. Degradation pathway of DDT residues in water from the Teltowkanal in Berlin, Germany
(Names of DDT derivatives identified as key degradates in the aqueous phase are underlined).
Adapted with permission from Heberer and Dünnbier (1999).

Table 9-8. DDT Residues in Samples Collected from the Teltowkanal in Berlin, July 1997

Sampling-Location Number	Surface-Water Site	o,p'-DDT (ng/L)	p,p'-DDT (ng/L)	o,p'-DDE (ng/L)	p,p'-DDE (ng/L)	o,p'-DDD (ng/L)	p,p'-DDD (ng/L)	o,p'-DDA (ng/L)	p,p'-DDA (ng/L)
1	Dahme	ND	ND	ND	ND	ND	<5	ND	ND
2	Dahme	ND	ND	ND	ND	ND	ND	ND	ND
3	Dahme	ND	ND	<5	ND	ND	<5	ND.	ND
4	Teltowkanal	ND	ND	ND	ND	<5	<5	ND	ND
5	Teltowkanal	ND	ND	ND	<5	35	80	110	330
6	Teltowkanal	ND	ND	ND	5	50	140	210	760
7	Teltowkanal	ND	ND	ND	<5	10	40	55	180
8	Teltowkanal	ND	ND	ND	ND	10	15	40	165
9	Teltowkanal	ND	ND	ND	<1	<5	10	30	80
10	Teltowkanal	ND	ND	ND	ND	ND	10	35	125

Adapted with permission from Heberer and Dünnbier (1999).

Locations as shown in Figure 9-4.

ND = Not detected.

500 mL of the water samples have been analyzed.

The recoveries for all compounds were better than 70 percent.

The limits of detection were for all compounds below 1 ng/L, and the limits of quantitation were below 5 ng/L.

<5 ng/L indicates that the compound was detected, but could not be quantified because the concentration was below the limit of quantification.

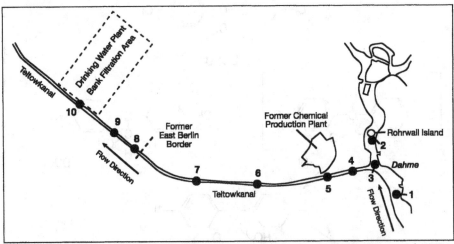

Figure 9-5. Locations of sampling sites in the Dahme River and the Teltowkanal in Berlin, Germany, July 1997 (All samples were taken in the center of the water courses at a depth of 2 m). Adapted with permission from Heberer and Dünnbier (1999).

1999). The DDA residues were not removed by conventional drinking-water treatment; therefore, well water and drinking water are routinely controlled by the Berlin Waterworks to meet the European maximum tolerance levels of only 0.1 µg/L. In addition to DDA residues, several pharmaceuticals and other polar contaminants also were found in water from these wells (Heberer et al., 1997).

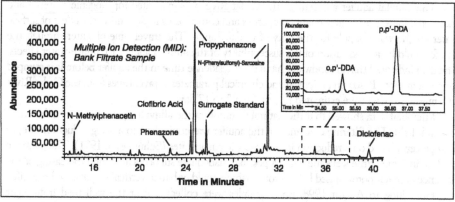

Figure 9-6. Chromatogram of the derivatized extract of a groundwater sample collected from a bank filtration area at the Teltowkanal in Berlin, Germany (recorded with GC/MS in selected ion monitoring mode).

If the situation of the Teltowkanal is similar or can be compared to other DDT-contaminated Superfund sites, DDA also should be found at other locations. In the Berlin study, DDA accounted for more than 60 percent of the total DDT residues identified in surface water of the Teltowkanal. The concentrations of DDA, on average, were five times greater than those detected for DDD. It is probable that DDA may have been present at considerable concentrations in those groundwater and drinking-water samples reported to contain residues of DDD; thus, DDA should be recognized as an important parameter for which surface-water, groundwater, and drinking-water samples should be analyzed when examining DDT residues.

Pesticides in Riverbank-Filtered Water at Cedar Rapids, Iowa, in the United States

The City of Cedar Rapids is in east-central Iowa and has a population of about 110,000. Cedar Rapids is located along a flood plain of the Cedar River and uses an alluvial aquifer adjacent to the Cedar River for its drinking-water supply. The alluvial aquifer is hydraulically connected to the Cedar River, bedrock, and upland areas in the study area. The City has three well fields in use along the Cedar River, with a total of 55 municipal wells (53 vertical wells and two horizontal collector municipal wells) (Schulmeyer, 1991) with an average discharge of 108-million L/d in 1992.

The Cedar River drainage basin has a well-developed stream pattern that drains an area approximately 16,850 km[2] above the well field. Overlying the Silurian- and Devonian-age limestone/dolomite bedrock is a 2- to 30-m layer of unconsolidated glacial till, loess, and alluvium (Boyd, 1999; Hansen, 1970; Prior, 1991). Recharge to the alluvial aquifer typically occurs as infiltration from the Cedar River is induced by the pumping of municipal wells, the infiltration of precipitation, and flow from both adjacent hydrogeologic units and the river when the stage is higher than the groundwater level (Wahl and Bunker, 1986). In areas influenced by the pumping of municipal wells, the gradient extends from the river to the well field; in areas outside this cone of depression, the water-table gradient, generally, is towards the river (Boyd, 1999). Results from a regional groundwater model (Schulmeyer and Schnoebelen, 1998) suggest that about 74 percent of the water pumped from the alluvial aquifer is from the Cedar River, about 21 percent from adjacent and underlying hydrogeologic units, and about 5 percent from precipitation. Upstream land use in the Cedar River Basin is over 80 percent agricultural. In this area, corn and soybeans are the major crops (U.S. Department of Agriculture, 1976).

The alluvial aquifer adjacent to the Cedar River was evaluated for biogenic material and monitored for selected water-quality properties and constituents to determine the effect of surface water on the water supply for the City of Cedar Rapids. The travel time of water through the aquifer could be an indication of the susceptibility of the alluvial aquifer to surface-water effects. The data indicated that groundwater has a short residence time in the aquifer before it is pumped for consumption. Based on biological and chemical parameters, travel times from the Cedar River to a municipal well were estimated to vary from 7 to 17 days.

Microbial data showed that the natural filtration of the alluvial aquifer was very effective at this well field. The filtering efficiency of the aquifer was equivalent to a 3-log reduction rate, or 99.9 percent, based on a reduction in microscopic particulates (Schulmeyer, 1991). The presence of algae in some of the wells indicated a possible inadequate surface seal around the casing or the presence of macropores caused by tree roots, which could enhance vertical seepage to the aquifer.

From June to August 1998, water samples were collected near the well field from eight monitoring wells completed at depths from 3 to 13 m, installed at about 1-m intervals from the river in alluvial sediments. The samples were analyzed for selected triazine and acetanilide herbicides and degradates (Boyd, 1999). Atrazine was the most frequently detected and at greatest concentrations (a maximum of 2.71 μg/L); also detected were acetochlor (a maximum of 0.17 μg/L), cyanazine (a maximum of 0.16 μg/L), and metolachlor (a maximum of 0.33 μg/L) (Table 9-9). Deethylatrazine, deisopropylatrazine, and hydroxyatrazine concentrations were less than 0.50 μg/L and decreased with time during the growing season.

Table 9-9. Ranges of Maximum Concentrations of Selected Herbicides and Organic Carbon Detected in Water from Municipal and Observation Wells and Surface-Water Sites Near Cedar Rapids, Iowa, United States

Sampling Site	Cedar River	Municipal and Observation Wells	Drinking Water
Deethylatrazine	0.66	0.03 to 0.50	0.30
Atrazine	8.16	<0.05 to 1.64	1.15
Alachlor	1.42	<0.05 to 0.18	0.16
Cyanazine	2.07	<0.05 to 0.68	0.64
Metribuzin	0.13	<0.05 to 0.07	0.07
Acetochlor	1.17	<0.05 to 0.27	0.15
Organic Carbon (mg/L)	6.90	0.90 to 6.40	2.30

In μg/L, unless otherwise specified.

At the site, the data showed that acetanilide degradates were transported into the alluvial aquifer in a manner similar to that indicated for atrazine and deethylatrazine. Atrazine and deethylatrazine in the Cedar River probably were transported into the alluvial aquifer with infiltration induced by the pumping of municipal wells, based on the deethylatrazine-to-atrazine ratio. Deethylatrazine-to-atrazine ratio values in samples from deeper observation wells were larger than in shallower wells, indicating that more degradation occurred during transport to deeper wells than to the shallower wells. Furthermore, data of the degradates of acetanilides suggest that relatively little degradation of parent acetanilide compounds occurred during transport from the river into the alluvial aquifer because concentrations of chloroacetanilide herbicides were similar in river and groundwater. Generally, herbicides enter the alluvial aquifer with riverbank-storage water during periods of high river stage and, afterwards, water contaminated with herbicides is released back to the river from riverbank storage during low river stage; however, when induced recharge occurs by pumping municipal wells near a river, the release of riverbank-storage water contaminated with herbicides during low flow can be prevented (Boyd, 1999).

Bentazone as a Possible Tracer of Riverbank-Filtered Water from the Rhine River in Germany

In the water catchment area, Eich, owned by the public utility of Mainz, Germany, groundwater is reclaimed from the Rhine River with an unknown amount of riverbank filtrate. At the RBF site, fluvial Quaternary sediments overlie marine tertiary silts and clays with very low permeabilities. The Quaternary aquifer has a thickness of 70 to 80 m and is divided by an aquitard consisting of fine-grain sand, silt, and clay. The upper aquifer has a thickness of 10 to 15 m, whereas the locally defined middle aquifer has a thickness of about 60 m. The average linear velocity of groundwater is about 10 m/d and can be as low as a few centimeters per day. Figure 9-7 presents the distribution of sediments within the Quaternary sediments and the location of the wells in addition to the likely direction of groundwater flow.

Pesticide residues from the Rhine River are transported via RBF into the aquifer. Pesticides analyzed were bentazone, mecoprop, dichlorprop, and 2,4-D. Only mecoprop in groundwater exceeded the maximum contaminant level of German drinking-water regulations, set at 100 ng/L for a single compound. Bentazone, detected at high concentrations in all samples between 1989 and 1996, was identified as the most important pesticide detected at concentrations of 0.5 µg/L in blended raw water and as much as 2 µg/L in the raw water of deep wells (Meitzler et al., 1996). The results of a geologic evaluation of borehole sediment samples, combined with the findings from a study of herbicides in groundwater, have been used to create a model for riverbank-filtrate flow regimes.

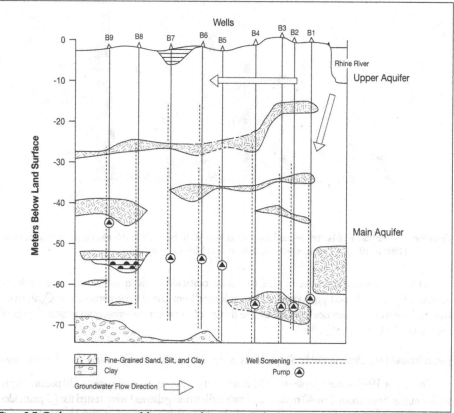

Figure 9-7. Geologic cross-section of the water catchment area, Eich, showing groundwater flow directions. Reproduced with permission from Meitzler et al. (1996).

Concentrations of bentazone differ significantly in Wells B1 to B9 (Figure 9-8). Meitzler et al. (1996) described how different portions and different ages of riverbank-filtered water lead to these varying concentrations. Well B1 receives the highest portion of riverbank filtrate. Because there is no aquitard near the Rhine River, groundwater flow is directed to the deeper parts of the aquifer, leading to a pesticide pool beneath the aquitard at B1 (about 20 to 35 m below land surface) where another layer with low permeability is located. Because these layers have low permeability, Wells B2 and B3 are somewhat protected against contamination with bentazone. Groundwater from Well B4 exhibited high concentrations, most likely because the well is screened from 26 m below the land surface and, thus, falls within the bentazone pool. The portion of riverbank-filtered water decreases along the flow path and, by this, so does the amount of bentazone in groundwater; therefore, bentazone concentrations decrease in Wells B5 and B6 (Meitzler et al., 1996). The other wells (Wells B7 to B9) have high portions of naturally recharged groundwater free of bentazone, and the concentrations are low because of the dilution of naturally and artificially recharged groundwater.

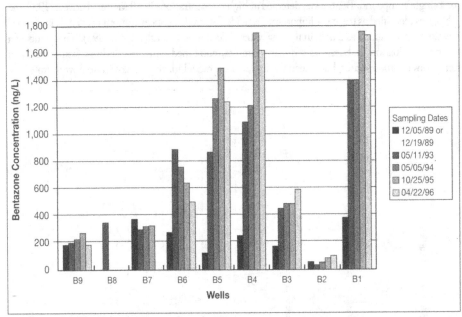

Figure 9-8. Concentration of bentazone residues in water from Wells B1 to B9 in the water catchment area, Eich (1989 to 1996). Reproduced with permission from Meitzler et al. (1996).

The management practices needed to reduce pesticide loads in individual wells could be achieved by an increased production of groundwater from the deeper parts of the Quaternary aquifer; however, this practice would raise the danger of saltwater intrusion from greater depths of the aquifer (Meitzler et al., 1996).

Pesticides and Degradates in Public-Supply Wells in the City of Schenectady in New York, United States

In August 1999, water samples from 32 community water-supply well systems (30 public-supply wells ranging from about 7 to 40 m deep and two infiltration galleries) were tested for 60 pesticides and degradates in central and western New York (Table 9-10). Sixteen of the 32 sites have the potential to be affected by induced infiltration based on several variables and their proximity (less

Table 9-10. Maximum Concentrations of Selected Compounds in Water from the Mohawk River and 32 Wells of the City of Schenectady in New York, United States

Compound	Mohawk River (µg/L)	Wells (µg/L)
Alachlor	0.46	Not Detected
Alachlor ESA	0.46	0.21
Metolachlor	0.679	E 0.003
Metolachlor ESA	0.57	0.31
Metolachlor Oxanilic Acid	0.35	0.05
Atrazine	0.581	0.019
Deethylatrazine	E 0.0321	E 0.012
Deethyldeisopropylatrazine	E 0.01	E 0.01
Hydroxyatrazine	E 0.023	E 0.005
Prometon	E 0.004	E 0.002
Simazine	0.061	Not Detected
Cyanazine	0.038	Not Detected
2,4-D	E 0.03	Not Detected
Diazinon	0.014	Not Detected
Carbaryl	E 0.064	Not Detected
Caffeine	E 0.043	Not Detected

Data from Winslow et al. (1965) and Eckhardt et al. (2000).

E = Estimated concentration.

than 200 m) to a surface-water body (Eckhardt et al., 2000). Wells were completed in unconfined sand and gravel aquifers, except for one well, which was completed in karstic limestone.

Atrazine and metolachlor were the parent compounds most frequently detected, and deethylatrazine (maximum concentration 0.09 µg/L) and metolachlor ESA (maximum concentration 3.58 µg/L) were the degradates most frequently detected. Atrazine was detected in water from 10 of 16 wells thought to be affected by induced infiltration at a maximum concentration of 0.019 µg/L. Deethylatrazine was detected in 11 of these 16 wells affected by induced infiltration at an estimated maximum concentration of 0.012 µg/L. Metolachlor was detected in six of the 16 wells, and metolachlor ESA was detected in seven of the 16 wells. Generally, concentrations and detection frequencies of pesticide compounds were larger in the group of wells classified as affected by induced infiltration than in the group classified as unaffected by induced infiltration.

Presence of Small Concentrations of Pesticides in Two Well Fields Along the Seine and Yonne Rivers in Paris, France

Two well fields contribute to the drinking-water supply of the City of Paris in France (Benedicte Welte, City of Paris, written communication, 2001). The first well field consists of 21 wells with a maximum depth of 25 m on the banks of the Seine River and can produce up to 50,000 m^3/d. The wells are completed in alluvial gravel, sand, and clay, and are about 100 m from the river. The travel times of water from the river to the wells are unknown.

The second well field consists of 10 wells with a mean depth of 15 m on the banks of the Yonne River, which is a tributary of the Seine River. It has a total production capacity of 50,000 m³/d. The wells are developed in an alluvial aquifer (with an average thickness of 6 m) and in chalk (with an average thickness of 9 m). The wells are 50 to 200 m from the river. The travel times of water from the river to the wells are unknown.

Water samples were analyzed for triazines, one triazine degradate, and phenylureas by HPLC, with a reporting limit of 0.05 µg/L. The maximum concentration of atrazine in river water was 0.50 µg/L. Atrazine, deethylatrazine, isoproturon, and propazine were detected at maximum concentrations of less than 0.25 µg/L, with an average concentration of less than 0.10 µg/L (Table 9-11). Deethylatrazine was the compound with the largest maximum and median concentrations, followed by atrazine. The concentrations of atrazine and its degradate were less than 0.10 µg/L, more often in water from the alluvial aquifer along the Yonne River than in water from the well field near the Seine River. It should be noted that fewer samples were collected from the well field along the Yonne River than along the Seine River. Also, seasonal variations of the concentrations of herbicides in these rivers are unknown as well as the timing of sample collection of the river presented herein. The travel times from the river to the wells have not been quantified.

Table 9-11. Pesticide Concentrations in Riverbank-Filtered Water Samples Collected from Two Well Fields Along the Seine and Yonne Rivers that Supply Water to the City of Paris in France

| | Well Field Along the Seine River | | | | | | | | |
	Atrazine	Chlortoluron	Cyanazine	DEA	Diuron	Isoproturon	Monolinuron	Propazine	Simazine
Minimum	<0.05	<0.05	<0.05	<0.05	<0.05	<0.05	<0.05	<0.05	<0.03
Maximum	0.13	<0.05	<0.05	0.19	<0.05	0.03	<0.05	0.03	<0.03
Median	0.08	<0.05	<0.05	0.1	<0.05	<0.05	<0.05	<0.05	<0.03
Number of Analyses	200	40	200	100	100	100	50	200	200
	Well Field Along the Yonne River								
	Atrazine	Chlortoluron	Cyanazine	DEA	Diuron	Isoproturon	Monolinuron	Propazine	Simazine
Minimum	<0.05	<0.05	<0.05	<0.05	<0.05	<0.05	<0.05	<0.05	<0.05
Maximum	0.19	<0.05	<0.05	0.23	<0.05	<0.05	<0.05	<0.05	<0.05
Median	0.09	<0.05	<0.05	0.1	<0.05	<0.05	<0.05	<0.05	<0.05
Number of Analyses	65	50	65	50	50	50	50	65	65

Data from Benedicte Welte, City of Paris, written communication, 2001.
DEA = Deethylatrazine.
Concentrations in µg/L.

Occurrence of the Pharmaceutical Degradate Clofibric Acid and of N-(Phenylsulfonyl)-Sarcosine in Drinking Water as a Result of Bank Filtration and Artificial Groundwater Infiltration in Berlin, Germany

In 1993, clofibric acid was the first pharmaceutically active compound that was found in drinking-water samples. It was detected at varying concentrations, with as much as 165 ng/L in more than 50 samples collected from water taps in Berlin (Stan et al., 1994). Further investigations of drinking-water samples collected from all 14 waterworks in Berlin (Heberer and Stan, 1996, 1997) demonstrated the relation between:

- The proportions of drinking water derived from bank filtration and artificial groundwater enrichment.
- The level of drinking-water contamination by clofibric acid (Table 9-12).

In this study, clofibric acid and another polar contaminant, N-(phenylsulfonyl)-sarcosine, which also originates from sewage effluents, were found at maximum concentrations of 170 and 105 ng/L, respectively (Heberer and Stan, 1996, 1997). The concentrations of these polar contaminants detected in tap-water samples collected from individual Berlin waterworks were found to correspond well with the proportions of drinking water produced by bank filtration or water from groundwater infiltration by the individual water utility; thus, the highest concentrations of N-(phenylsulfony)-sarcosine and clofibric acid were found in tap waters of the water utilities of Beelitzhof and Johannisthal. Both water supplies are characterized by high proportions of bank-filtered water in their drinking water and are located near waterways that are highly contaminated by municipal sewage effluents. On the other hand, drinking water from the water utilities of Spandau and Kaulsdorf did not contain any measurable concentrations of both compounds. The water utility in Spandau is located along the Upper Havel River, which is not contaminated by any of these pollutants. Drinking water produced by the water utility at

Table 9-12. Concentrations of Clofibric Acid and N-(phenylsulfonyl)-Sarcosine in Drinking-Water Samples from Berlin Area Waterworks in Germany, 1994

Waterworks	Site Number	BF (%)	GWE (%)	Clofibric Acid (ng/L)	NPS (ng/L)	SWC
Beelitzhof	I	66	7	60	105	+
Buch	II	0	0	<5	—	+
Friedrichshagen	III	82	0	10	25	o
Johannisthal	IV	62	0	170	105	++
Jungfernheide	V	52	43	45	—	+
Kaulsdorf	VI	0	0	—	—	++
Kladow	VII	68	0	6	17	+
Köpenick	VIII	74	0	6	8	o
Riemeisterfenn	IX	17	0	11	70	+
Spandau	X	30	48	—	—	o
Stolpe	XI	Unknown	Yes	13	—	o
Tegel	XII	54	27	25	—	+
Tiefwerder	XIII	61	0	55	35	+
Wuhlheide	XIV	58	0	75	—	++

Adapted with permission from Heberer and Stan (1997).
BF = Bank Filtration
GWE = Artificial groundwater enrichment.
NPS = N-(phenylsulfonyl)-sarcosine.
SWC = Degree of contamination of the neighboring watercourse.
— = Below the limit of detection of 1 ng/L for clofibric acid or 2 ng/L for N-(phenylsulfonyl)-sarcosine.
++ = High level of contamination by municipal sewage effluents.
+ = Moderate level of contamination.
o = Low or even very low level of contamination.

Kaulsdorf was not polluted because this utility does not use any groundwater enrichment for drinking-water production; thus, the determining factors for the concentrations of clofibric acid and N-(phenylsulfonyl)-sarcosine in drinking-water samples were the proportions of drinking water produced from bank filtration, the use of groundwater infiltration, the geographic location of the water utility, and the degree of contamination of the neighboring streams used as sources for groundwater recharge by the particular water utility (Heberer and Stan, 1997).

Pharmaceutical Residues in Groundwater at a Bank Filtration Site in Berlin, Germany

Several pharmaceutically active compounds were found at concentrations up to the microgram-per-liter level in investigations of wells from the drainage area of a drinking-water treatment plant in Berlin, Germany (Heberer et al.,1997) (Table 9-13). The drinking-water treatment plant is near a canal downstream from the sewers of a municipal sewage-treatment plant and a former pharmaceutical production plant. The drinking-water plant uses high proportions (greater than 60 percent) of bank filtrate (induced infiltration) for drinking-water production (SenStadtUm, 1997).

Table 9-13. Concentrations of Pharmaceutical Residues in Wells Located Near Contaminated Surface Water in Berlin, Germany

Residues of PhACs	Concentration Range (ng/L)
Clofibric Acid	ND to 7,300
Diclofenac	ND to 380
Fenofibrate	ND to 45
Gentisic Acid	ND to540
Gemfibrozil	ND to 340
Ibuprofen	ND to 200
Ketoprofen	ND to 30
Phenazone	ND to 1,250
Primidone	ND to 690
Propyphenazone	ND to 1,465
Salicylic Acid	ND to 1,225
Clofibric Acid Derivative	ND to E 2,900
N-methylphenacetin	ND to E 470

Data from Heberer et al. (1997) and Heberer (in press).
PhAC = Pharmaceutically active compounds.
ND = Not detected (detection level was between 1 and 10 ng/L).
E = Concentrations were estimated because standards were not available commercially.

Irrespective of the source of contamination, such as actual discharges of purified municipal sewage or former discharges of production residues from the pharmaceutical plant into the canal, the results of the two studies (see Table 9-13) demonstrate that some pharmaceutically active compounds are not eliminated on their way through the subsoil. Conditions caused by RBF allow these polar residues to leach readily into the aquifers that supply the drinking-water plant. These results show that whenever contaminated surface water is used for groundwater recharge, polar pharmaceutically active compounds also leach through the subsoil into the groundwater (Heberer et al., in press).

Fate of Pharmaceutical Residues at an Experimental Bank Filtration Site at Lake Tegel in Berlin, Germany

The fate of pharmaceutical residues was investigated at a bank filtration site at Lake Tegel in Berlin, Germany, as part of a bank filtration research project of the Free University, Berlin. This site is located on the east bank of Lake Tegel and belongs to an active Berlin waterworks water supply that uses infiltrated water from Lake Tegel (Figure 9-9).

The aquifer has a thickness of about 50 m and consists of Quaternary sand and gravel (Figures 9-10 and 9-11) with hydraulic conductivities between 10^{-4} m/s and 10^{-3} m/s. Groundwater samples were collected from 13 monitoring wells and three drinking-water wells (e.g., B13) at five distances (deeper wells) from Lake Tegel. The transect was established along a

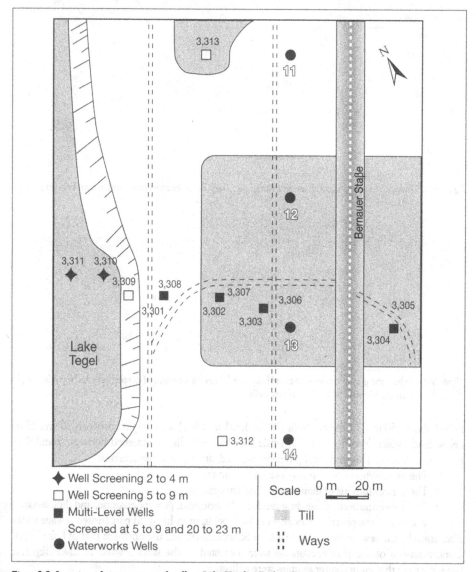

Figure 9-9. Location of piezometers and wells at Lake Tegel in Berlin, Germany.

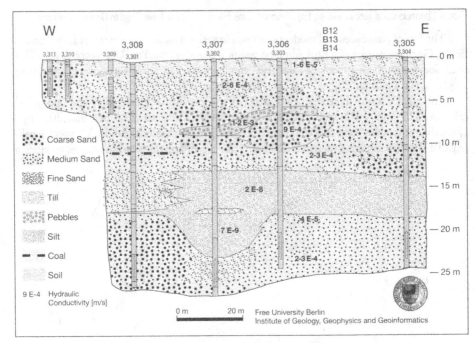

Figure 9-10. Geologic cross-section of the transect along Lake Tegel. Reproduced with permission from Fritz (in press).

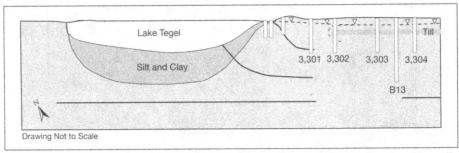

Figure 9-11. Schematic geologic cross-section at Lake Tegel showing locations of deeper wells and a public-supply well (numbered), as well as shallow wells.

line of about 150 m. The deeper wells (Wells 3,301 to 3,304) are screened between 20 and 23 m below land surface (see Figure 9-10), while the shallow wells are screened between 5 and 9 m below land surface. The analytical program included, among other parameters:

- The measurement of physico-chemical parameters.
- The concentrations of main anions and cations.
- The concentrations of the drug residues clofibric acid, propyphenazone, and diclofenac.

The source of the pharmaceuticals in Lake Tegel is, most likely, in part treated sewage water discharged by the sewage treatment plants in Schönerlinde into the northern part of Lake Tegel. Concentrations of the pharmaceuticals were measured in the surface water of Lake Tegel at the same time that groundwater samples were collected.

Clofibric acid was found in concentrations of:
- 120 ng/L (May 1998).
- 140 ng/L (June 1998).
- 190 ng/L (November 1998).

Concentrations of propyphenazone in Lake Tegel were 30 ng/L (May 1998) and 60 ng/L (June 1998), and diclofenac was below the detection limit both times.

The aquifer is divided by a till, with aerobic groundwater conditions above the till and anaerobic conditions beneath. Lake Tegel is underlain by silts and clays that are rich in organic material and form a relatively impermeable layer. Hence, the most important flow path is through the thin layer of till at the bank of Lake Tegel (see Figure 9-10). The recharge of groundwater through the silt and clay layer and groundwater flow beneath Lake Tegel is of minor importance. The portion of landward water in waterworks wells accounts for 15 to 25 percent of the total water withdrawn.

Clofibric acid appeared at concentrations as much as 290 ng/L, propyphenazone as much as 105 ng/L, and diclofenac as much as 50 ng/L (Table 9-14). A distinct decrease in the concentrations measured for propyphenazone and diclofenac could be observed between Wells 3,301 and 3,302. In waterworks wells, clofibric acid and propyphenazone were detected, although propyphenazone was not found in the wells upgradient of the waterworks wells.

Table 9-14. Distribution of Pharmaceuticals in Groundwater from Deep Wells at the Bank Filtration Site at Lake Tegel in Germany

Compound	Sampling Date	Lakeward Wells			Public-Supply Wells			Landward Well
		3,301	3,302	3,303	B 12	B 13	B 14	3,304
Clofibric Acid	May 1998	15	ND	ND	45	30	50	225
	June 1998	20	<10	ND	50	35	40	<5
	November 1998	20	240	290	70	90	10	55
Propyphenazone	May 1998	40	ND	ND	35	30	90	ND
	June 1998	55	ND	ND	45	40	105	ND
Diclofenac	May 1998	ND	ND	ND	ND	ND	ND	ND
	June 1998	50	ND	ND	ND	ND	ND	ND

Concentrations are in ng/L.
ND = Not detected.
< = Less than determination limit.

Interestingly, clofibric acid was found not only in groundwater originating from bank filtration, but also in groundwater from Well 3,304, which samples landward water. The concentration of clofibric acid was, at certain times, even higher in landward water than in bank-filtered water, indicating another possible source besides bank filtration.

The reduction in concentrations or removal of the pharmaceutical compounds may be caused by dilution, sorption, or degradation. For propyphenazone and diclofenac, the reduction in concentrations within a distance of some 10 m (Wells 3,301 to 3,302) is so great that it cannot be explained by dilution alone, when compared to chloride acting as a conservative tracer in this part of the aquifer. All the pharmaceuticals shown in Table 9-14 are water-soluble, with low to moderately-high octanol/water distribution coefficients, indicating that they are mobile and probably migrate in the aquifer. Among the three pharmaceuticals, diclofenac has the highest octanol/water distribution coefficient (log K_{ow} = 4.51), followed by clofibric acid (log K_{ow} = 2.84) and propyphenazone (log K_{ow} =2.32) (see Table 9-5). Thus, diclofenac is expected to show the strongest retardation in the aquifer. In laboratory studies, however, propyphenazone exhibited the highest retardation factor (Rd = 2.7). The retardation factor of diclofenac (Rd greater than 1.6) was lower, and clofibric acid was not retarded at all.

The concentrations of diclofenac in Lake Tegel should be much higher considering the high concentrations observed in sewage effluents. Data from degradation studies suggest that diclofenac is degraded significantly by photolytic degradation. This observation agrees with the results of Buser et al. (1998), who found that diclofenac was rapidly photodegraded in water of Swiss lakes. Furthermore, Buser et al. found that diclofenac was degraded in natural water, but not in desalted water, indicating that this degradation is microbiologically catalyzed. Most likely, the concentration of diclofenac decreases in surface water during transport from the sewage treatment plant to Lake Tegel, but is not totally removed, considering that diclofenac was found in one sample of Well 3,301. As photolytic degradation is unlikely in groundwater, the remaining amounts are either degraded or sorbed to subsurface material; however, sorption is more likely. Zwiener et al. (2000) reported very low degradation under anaerobic conditions compared to aerobic oxygen conditions.

The distribution of propyphenazone is more difficult to explain, as there are occurrences of this compound in Well 3,301 and in waterworks wells, but no further detections along the flow path (see Table 9-14). In column experiments, propyphenazone was retarded with a factor of 2.7, which explains the decrease in concentration along the flow path between Wells 3,301 and 3,302. The waterworks wells are pumped during alternating time intervals. Because of the great amount of water pumped, the flow regime is strongly influenced at the beginning of a pump interval. As already observed in column experiments (see Section 2), this may lead to a mobilization of colloids and, therefore, to a mobilization of propyphenazone that is sorbed to those colloids. In the case of the bank filtration site at Lake Tegel, sorption seems to be the main process for removing propyphenazone. Holm et al. (1995) found indications that degradation is much more important than sorption in groundwater downgradient of the Grindsted landfill in Denmark. The attenuation of propyphenazone in landfill leachate from Grindsted took place under strongly reducing conditions, whereas the redox conditions along the transect of Lake Tegel are anaerobic but not methanogenic. Clofibric acid was not eliminated during bank filtration, which agrees with the laboratory results that prove the high persistence and high mobility of this compound.

Occurrence and Fate of Pharmaceutical Residues and Other Polar Contaminants During RBF Along the Rhine River in Germany

In a recent RBF study carried out by Brauch et al. (2000), the behavior of several selected polar organic contaminants, including pharmaceutically active compounds, complexing agents (EDTA, NTA, DTPA, etc.), aromatic sulfonates, aliphatic amines, alkylphenols, and bisphenol A, was studied at two waterworks near the lower Rhine River. The Düsseldorf-Flehe Waterworks have one collector well and 76 vertical wells, which are parallel to the Rhine River at a distance of about 50 m, and six wells farther from the river. The second waterworks, Wittlaer, run by the Stadtwerke Duisburg AG, produces water from wells also parallel to the banks of the Rhine River. In the study by Brauch et al. (2000), raw-water samples from both waterworks, containing known amounts of riverbank-filtered water mixed with groundwater, and groundwater samples from an additional sampling point (M1t) were analyzed for the polar organic compounds previously mentioned. On average, raw water produced by the Düsseldorf-Flehe and Wittlaer Waterworks contains 75 and 60 percent riverbank-filtered water from the Rhine River, respectively. The well at Sampling Point M1t is between the Rhine River and the well galleries of the Wittlaer

Waterworks. Water from this well consists of only riverbank-filtered water and is not influenced by groundwater.

In their investigations, Brauch et al. (2000) observed the removal of some of the investigated compounds, which was attributed to enhanced microbial activity in the riverbed. Several compounds were not removed or poorly removed by RBF; thus, it was concluded that the decrease in concentrations was caused mainly by dilution with uncontaminated groundwater.

Among synthetic complexing agents, significant differences were observed in their RBF behavior. EDTA was found at an average concentration of 7.9 µg/L in the Rhine River. It also was found at considerable concentrations in raw water from the waterworks and at Sampling Point M1t (Figure 9-12).

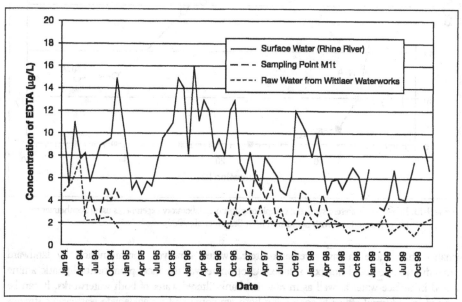

Figure 9-12. Concentrations of EDTA in the Rhine River (732.1 km) at Sampling Location M1 and in raw water from the Wittlaer Waterworks (January 1994 to October 1999). Reproduced with permission from Brauch et al. (2000).

EDTA was not found in groundwater unaffected by induced infiltration. In raw water from Düsseldorf-Flehe and Wittlaer, EDTA was detected at average concentrations of 4.8 and 2.2 µg/L, respectively. At sampling location M1t, located between the Rhine River and Wittlaer Waterworks, the average concentration for EDTA was 3.7 µg/L. Removal rates of 40 and 70 percent were calculated for EDTA during RBF used by the Düsseldorf-Flehe and Wittlaer Waterworks, respectively. The concentrations of EDTA were further reduced by using ozonation and active charcoal filtration; however, total removal was not achieved (Brauch et al., 2000).

In contrast to EDTA, NTA was detected only in surface-water samples from the Rhine River and was not detected in raw riverbank filtrate. This can be explained by the rapid microbial degradation of NTA, which was confirmed in laboratory experiments (Figure 9-13).

Brauch et al. (2000) also reported that several aliphatic amines, such as methylamine, dimethylamine, diethylamine, morpholine, and diethanolamine, were detected in raw water from both waterworks at concentrations <1 µg/L. They theorized that these compounds may be formed in the subsurface during RBF, because the concentrations found in the surface water always were

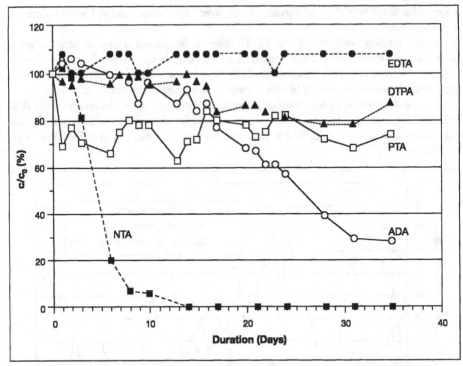

Figure 9-13. Behavior of different synthetic complexing agents in laboratory experiments using Sontheimer test filters (Reproduced with permission from Brauch et al., 2000).

smaller than those in raw riverbank-filtered water; however, contamination by landward groundwater also could not be excluded. Urotropine was the most prominent aliphatic amine found in surface water as well as in raw riverbank-filtered water of both waterworks. It can be formed in industrial effluents containing both ammonia and formaldehyde wastewater streams (Lindner et al., 1996). Urotropine was found at average concentrations of 7.1 µg/L in the Rhine River, at concentrations of 4.6, 2.5, and 2.8 µg/L in the samples from Sampling Point M1t, and in the raw water from the Wittlaer and Flehe Waterworks. A significant reduction, most probably caused by microbial degradation, was observed for urotropine during RBF. Ozonation achieved a complete removal of all aliphatic amines (Brauch et al., 2000).

Brauch et al. (2000) also detected analgesic, diclofenac, and the antiepileptic drug, carbamazepine, in the Rhine River. Diclofenac was almost completely removed during RBF. In the case of carbamazepine, the transport behavior and the changes of its concentrations observed in the subsurface were found to be much more complicated than those caused by normal mixing effects that occur during the RBF process (Brauch et al., 2000). The temporal concentration profiles were developed for both compounds in the Rhine River at Sampling Point M1t and in raw water from the Wittlaer Waterworks (Figures 9-14 and 9-15).

The transport behavior of carbamazepine was determined by subsequent adsorption and desorption processes, making a mass-balance calculation very difficult. Carbamazepine was found at average concentrations of 0.11 µg/L in the Rhine River, at concentrations of 0.11, 0.067, and 0.088 µg/L in the samples from Sampling Point M1t, and in raw water from the Wittlaer and Flehe Waterworks. In general, the removal of carbamazepine was not observed during RBF.

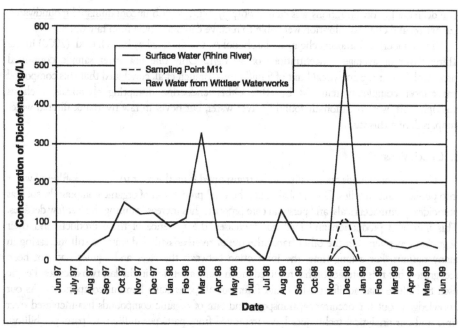

Figure 9-14. Concentrations of diclofenac in the Rhine River (732.1 km) at Sampling Point M1t and in raw water from Wittlaer Waterworks (June 1997 to June 1999). Reproduced with permission from Brauch et al. (2000).

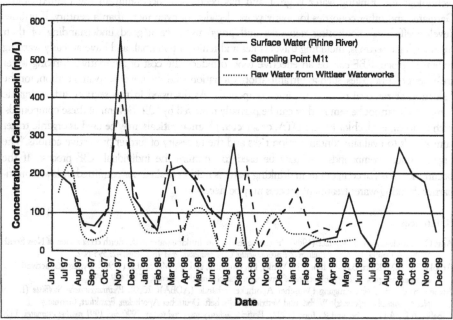

Figure 9-15. Concentrations of carbamazepine in the Rhine River (732.1 km) at Sampling Point M1t and in raw water from Wittlaer Waterworks (June 1997 to December 1999). Reproduced with permission from Brauch et al. (2000).

The decrease in concentrations was caused only by dilution with uncontaminated groundwater. Ozonation and charcoal filtration were able to remove carbamazepine from raw water.

The endocrine disrupting chemical, bisphenol A, was detected by Brauch et al. (2000) in the Rhine River at average concentrations of 30 ng/L. Only one raw-water sample contained bisphenol A near the analytical limit of 5 ng/L; thus, Brauch et al. concluded that this compound was almost completely removed by RBF. Other endocrine disrupting chemicals, such as nonylphenols, were occasionally found in river water, but never in raw riverbank-filtered water samples during this study.

5. Conclusions

Organic compounds potentially can be transported from the river to riverbank-filtered water into public-water supplies. On a global scale, the total production of organic compounds, such as pesticides, pharmaceuticals, and personal care products, has increased during the last few decades. This increased production and load has increased the presence of these products and their degradates in rivers. On the other hand, the use of riverbank-filtered water is still increasing in many nations. For a long time, the interaction between the river and aquifer had not been recognized, but this interaction has received much more attention in the last two decades. People used to believe that aquifer media could filter out all potentially harmful components. As our knowledge about the occurrence, transport, and fate of organic compounds has increased over time and our analytical techniques have improved from parts per million, to parts per billion, down to parts per trillion concentrations, concerns about the presence of organic compounds in our drinking water and their effect on public health also have increased.

Nevertheless, the potential for the development of RBF worldwide as a pretreatment technology for drinking water is great and has been used successfully in the United States, Germany, and other countries for many years, decades, or even more than a century. If utilities develop effective monitoring networks and programs, have a good understanding of their hydrogeologic settings, including travel times and removal potential, and have an early warning system in place, RBF can be an effective tool to reduce the cost of alternative drinking-water treatment technologies, such as ultrafiltration, ozonation, and the use of activated carbon, for the reduction or removal of most organic compounds. As discussed in this section, a few organic compounds cannot be removed or can be partially removed by RBF. Several of these compounds, such as atrazine, alachlor ESA, EDTA, or selected pharmaceuticals, may be used as organic tracer compounds to evaluate contamination risks and the necessity of further protective actions. The organic tracer compounds also may be used to optimize the individual RBF process. If the occurrence of these compounds in drinking water is neither desired nor permitted by law, a multi-barrier drinking-water treatment process may be necessary.

References

Ang, C., K. Meleady, and L. Wallace (1989). "Pesticide residues in drinking water in the north coast region of New South Wales, Australia, 1986-87." *Bulletin of Environmental Contamination and Toxicology*, 42: 595-602.

Arnold, S.F., D.M. Klotz, B.M. Collins, P.M. Vonier, L.J. Guillette, and J.A. McLachlan (1996). "Synergistic activation of estrogen receptor with combinations of environmental chemicals." *Science*, 272: 1489-1492.

Arzneibüro der Bundesvereinigung Deutscher Apothekerverbände (ADBA)(2000). *Pharmazeutische Stoffliste (List of pharmaceutical compounds)*. Werbe- und Vertriebsgesellscheft, Deutscher Apotheker, Frankfurt, Germany.

Aspelin, A.L., A.H. Grube, and R. Toria (1992). *Pesticide industry sales and usage: 1990 and 1991 market estimates*. U.S. Environmental Protection Agency, Washington D.C., 37 p.

Avdeef, A., K.J. Box, J.E.A. Comer, C. Hibbert, and K.Y. Tam (1998). "pH-metric logP 10: Determination of liposomal membrane-water partitioning coefficients of ionizable drugs." *Pharmaceutical Research*, 15(2): 209-215.

Barbash, J.E., and E.A. Resek (1996). "Pesticides in groundwater - Distribution, trends, and governing factors." *Pesticides in the hydrologic system, U.S. Geological Survey Water Quality Assessment,* Volume 2, R.J. Gilliom, ed., Ann Arbor Press, Inc., Ann Arbor, Michigan, 588 p.

Barber, L.B., G.K. Brown, and S.D. Zaugg (2000a). "Potential endocrine disrupting organic chemicals in treated municipal wastewater and river water." *Analysis of environmental endocrine disruptors,* L.H. Keith, T.L. Jones-Lepp, L.L. Needham, eds., American Chemical Society, Washington D.C., 97-124.

Barber, L.B., G.K. Brown, D.W. Kolpin, J.H. Writer, and S.D. Zaugg (2000b). "A reconnaissance for hormone compounds in the surface waters of the United States." *Proceedings, Effects of Animal Feeding Operations on Water Resources and the Environment,* F.D. Wilde, L.J. Britton, C.V. Miller, and D.W. Kolpin, eds., U.S. Geological Survey Open-File Report 00-204, U.S. Geological Survey, Reston, Virginia.

Beer, A.J. (1997). "Something in the water." *Biologist,* 44(2): 296.

Belgaqua and Phytofar (1999). *Groenbook (Green book),* Belgaqua and Phytofar, Belgium, 48 p.

Biologische Bundesanstalt für Land- und Forstwirtschaft (1995). *Verzeichnis der Wirkstoffe in zugelassenen Pflanzenschutzmitteln (List of active substances of permitted pesticides).* Biologische Bundesanstalt für Land- und Forstwirtschaft, Braunschweig, Germany.

Boyd, R.A. (1999). "Herbicides and herbicide degradates in shallow ground water and the Cedar River near a municipal well field, Cedar Rapids, Iowa." *Proceedings, Contamination of Hydrologic Systems and Related Ecosystems,* D.W. Morganwalp and H.T. Buxton, eds., U.S. Geological Survey Water Resources Investigative Report 99-4018B, U.S. Geological Survey, Reston, Virginia, 289-300.

Boyd, R.A. (2000). "Herbicides and herbicide degradates in shallow groundwater and the Cedar River near a municipal well field, Cedar Rapids, Iowa." *The Science of the Total Environment,* 248: 241-253.

Brauch, H.J., F. Sacher, E. Denecke, and T. Tacke (2000). "Wirksamkeit der Uferfiltration fuer die Entfernung von polaren organischen Spurenstoffen" (Efficiency of bank filtration for the removal of polar organic tracer compounds). *Gas- und Wasserfach Wasser/Abwasser* 14(1/4): 226-234.

Budavari, S. (1996). *The Merck Index,* Whitehouse Station, Rahway, New Jersey, 1741 p.

Burkart, M.R., and D.W. Kolpin (1993). "Hydrogeologic and land use factors associated with herbicide and nitrate occurrence in near-surface aquifers." *Journal of Environmental Quality,* 22: 646-656.

Burkart, M.R., D.W. Kolpin, and D.E. James (1999a). "Assessing groundwater vulnerability to agrichemical contamination in the Midwest U.S." *Water Science and Technology,* 39(3): 103-112.

Burkart, M.R., D.W. Kolpin, R.J. Jaquis, and K.J. Cole (1999b). "Groundwater quality: Agrichemicals in groundwater of the midwestern USA: Relations to soil characteristics." *Journal of Environmental Quality,* 28: 1908-1915.

Buser, H.R., T. Poiger, and M.D. Müller (1998.) "Occurrence and fate of the pharmaceutical drug Diclofenac in surface waters: Rapid photodegradation in a lake." *Environmental Science and Technology,* 32: 3449-3453.

Chaueveheid, E., C. Bertinchamps, and R. Savoir (1999). "Evolution des teneurs en micropolluants organiques dans les captages de la Cibe" (Evolution of the contents of organic micropollution in the capture zones of the Cibe). An article in "Les eaux souterraines en region wallonne" (Groundwater in the Wallonne region of Belgium). F. Edeline, ed., *Editions Cebedoc,* 600-601 (4-5): 75-83.

Chiou, C.Y., D.W. Schmedding, and M. Manes (1982). "Partitioning of organic compounds in octanol-water systems." *Environmental Science and Technology,* 16: 79-105.

Crites, R.W. (1985). "Micropollutant removal in rapid infiltration." *Artificial recharge of groundwater,* T. Asano, ed., Butterworth Publishers, Boston, Massachusetts.

Daughton, C.G., and T.A. Ternes (1999). "Pharmaceuticals and personal care products in the environment: Agents of subtle change?" *Environmental Health Perspective,* 107(6): 907-938.

Djuangsih, N. (1993). "Understanding the state of river basin management from an environmental toxicology perspective: An example from water." *The Science of the Total Environment, Supplement 1993.* Elsevier Science, Bilthoven, The Netherlands, 283-292.

Dreher, J.E., and A. Gunatilaka (1998). "Ground water management system in Vienna: An evaluation after three years of operation." *Artificial recharge of groundwater,* J.H. Peters, et al., eds., A.A. Balkema, Rotterdam, The Netherlands, 167-172.

Dünnbier, U., T. Heberer, and C. Reilich (1997). "Occurrence of bis(chlorophenyl) acetic acid (DDA) in surface and ground water in Berlin." *Fresenius' Environmental Bulletin,* 6: 753-759.

Duncan, D., T.R. Peterson, and J.D. Carr (1991). "Atrazine used as a tracer of induced recharge." *Groundwater Monitoring Reviews,* 11: 144-150.

Eckhardt, D.A.V., K.K. Hetcher, P.J. Philips, and T.S. Miller (2000). "Pesticides and their metabolites in community water-supply wells of central and western New York, August 1999." *U.S. Geological Survey Water Resources Investigative Report* 00-4128, U.S. Geological Survey, Reston, Virginia, 12 p.

Edwards, W.M., G.B. Triplett, and R.M. Kramer (1980). "A watershed study of glyphosate transport in runoff." *Journal of Environmental Quality,* 9(4): 661-665.

Eschke, H.D., D. Traud, and H.J. Dibowski (1995). "Untersuchungen zu Vorkommen polycylischer Moschus-Duftstoffe in verschiedenen Umweltkompartimenten – Nachweis und Analytik mit GC/MS in Oberflächen-, Abwässern und Fischen (1. Mitteilung)" (Investigations on the occurrence of polycyclic musk fragrances in different environment compartments – Detection and analysis in surface water, sewage, and fish applying GC-MS). *Zeitschriff fuer Umweltchemie und Oekotoxikologie*, 6(4): 183-189.

Exner, M.E. (1990). "Pesticide contamination of groundwater artificially recharged by farmland runoff." *Groundwater Monitoring Reviews*, Winter 1990: 147-159.

Fini, A., G. Fazio, and I. Rapaport (1993). "Diclofenac/N-(2-Hydroxymethyl) Pyrrolidine – A new salt for an old drug." *Drugs Under Experimental and Clinical Research*, 19(3): 81-88.

Frank, R., B.S. Clegg, C. Sherman, and N.D. Chapman (1990). "Triazine and chloroacetamide herbicides in Sydenham River water and municipal drinking water, Dresden, Ontario, Canada, 1981-1987." *Archives of Environmental Contamination and Toxicology*, 19: 319-324.

Fritz, B. (2002). "Untersuchung der Uferfiltration im Bereich der Transekte Tegeler See und Müggelsee." Ph.D. thesis, Free University, Berlin, Germany, in press.

Frycklund, C. (1998). "Long-term sustainability in artificial groundwater recharge." *Artificial recharge of groundwater*, J.H. Peters et al., eds., A.A. Balkema, Rotterdam, The Netherlands, 113-117.

Ghijsen, R.T., and W. Hoogenboezem (2000). *Endocrine disrupting compounds in the Rhine and Meuse Basin: Occurrence in surface, process, and drinking water*. Sub-project of the National Research Project on the occurrence of endocrine disrupting compounds, De Eendracht, Schiedam, Netherlands, 96 p.

Gledhill, W.E., and T.C. Feijtel (1992). "Environmental properties and safety assessment of organic phosphonates used for detergents and water treatment applications." *The handbook of environmental chemistry*, Volume 3, O. Hutzinger, ed., Springer Verlag, Berlin, Germany.

Gojemrac, T., B. Kartal, M. Žurić, and V. Zidar (1994). "Use of immunoenzyme method (ELISA) for determination of s-triazine herbicide atrazine residues in drinking water." *Veterinarska Stanica*, 25: 75-80 (in Croatian).

Grischek, T., W. Nestler, J. Dehnert, and P. Neitzel (1994). "Groundwater/river interaction in the Elbe River in Saxony." *Proceedings, Second International Conference on Groundwater Ecology*, J.A. Stanford and H.M. Valett, eds., American Water Resources Association, Middleburg, Virginia, 309-318.

Grischek, T., P. Neitzel, T. Andrusch, U. Lagois, and W. Nestler (1997). "Fate of EDTA during infiltration of Elbe River water and identification of infiltrating river water in the aquifer." *Vom Wasser*, 89: 261-282.

Groshart, C.P., and F. Balk (2000). *Biocides*. Association of River Waterworks (RIWA), Amsterdam, The Netherlands, 110 p.

Halling-Sørensen, B., N. Nielsen, P.F. Lansky, F. Ingerslev, L. Hansen, H.C. Lützhøft, and S.E. Jørgensen (1998). "Occurrence, fate and effects of pharmaceutical substances in the environment: A review." *Chemosphere*, 36: 357-394.

Hansen, R.E. (1970). *Geology and groundwater resources of Linn County, Iowa*. Iowa Geological Survey Water-Supply Bulletin 10, Des Moines, Iowa, 66p.

Hansch, C., D. Hoekman, A. Leo, L.T. Zhang, and P. Li (1995). "The expanding role of quantitative structure-activity-relationships (QSAR) in toxicology." *Toxicology Letters*, 79(1-3): 45-53.

Heberer, T. (1995). *Identifizierung und Quantifizierung von Pestizidrückständen und Umweltkontaminanten in Grund- und Oberflächenwässern mittels Kapillargaschromatographie - Massenspektrometrie (Identification and quantification of pesticide residues and environmental contaminants in ground and surface water applying capillary gas chromatography/mass spectrometry)*. Wissenschaft and Technik Verlag, Berlin, Germany, 437 p.

Heberer, T. (2002). "From municipal sewage to drinking water: Occurrence and fate of pharmaceutical residues in the aquatic system of Berlin." Article in "Attenuation of groundwater pollution by bank filtration," T. Grischek and K. Hiscock, eds., *Journal of Hydrology*, in press.

Heberer, T., and U. Dünnbier (1999). "DDT metabolite bis(chlorophenyl) acetic acid: The neglected environmental contaminant." *Environmental Science and Technology*, 33: 2346-2351.

Heberer, T., and H.J. Stan (1996). "Vorkommen von polaren organischen Kontaminanten im Berliner Trinkwasser." *Vom Wasser*, 86: 19-31.

Heberer, T., and H.J. Stan (1997). "Determination of clofibric acid and N-(Phenylsulfonyl)-sarcosine in sewage, river and drinking water." *International Journal of Environmental Analytical Chemistry*, 67: 113-124.

Heberer, T., and H.J. Stan (1998). "Arzneimittelrückstände im aquatischen System." *Wasser und Boden*, 50(4): 20-25.

Heberer, T., S. Gramer, and H.J. Stan (1999). "Occurrence and distribution of organic contaminants in the aquatic system in Berlin-Part III: Determination of synthetic musks in Berlin surface water applying solid-phase microextraction (SPME)." *Acta Hydrochimica et Hydrobiologica*, 27: 150-156.

Heberer, T., K. Schmidt-Bäumler, and H.J. Stan (1998). "Occurrence and distribution of organic contaminants in the aquatic system in Berlin-Part I: Drug residues and other polar contaminants in Berlin surface and groundwater." *Acta Hydrochimica et Hydrobiologica*, 26: 272-278.

Heberer, T., U. Dünnbier, C. Reilich, and H.J. Stan (1997). "Detection of drugs and drug metabolites in groundwater samples of a drinking water treatment plant." *Fresenius' Environmental Bulletin*, 6: 438-443.

Heberer T., B. Fuhrmann, K. Schmidt-Bäumler, D. Tsipi, V. Koutsouba, and A. Hiskia (2001). "Occurrence of pharmaceutical residues in sewage, river, ground and drinking water in Greece and Germany." *Pharmaceuticals and personal care products in the environment: Scientific and Regulatory Issues, Symposium Series 79b*, C.G. Daughton and T Jones-Lepp, eds., American Chemical Society, Washington, D.C., 70-83.

Heitkamp, M.A., W.J. Adams, and L.E. Hallas (1992). "Glyphosate degradation by immobilized bacteria: Laboratory studies showing feasibility for glyphosate removal from wastewater." *Canadian Journal of Microbiology*, 38: 921-928.

Henschel, K.P., A. Wenzel, M. Diedrich, and A. Fliedner (1997). "Environmental hazard assessment of pharmaceuticals." *Regulatory Toxicology and Pharmacology*, 25: 220-225.

Herrmann R., W. Kaa, and R. Bierl (1986). "Organic micropollutant behavior in a river water-groundwater infiltration system." *Proceedings, Conjunctive Water Use: Understanding and Managing Surface Water/Groundwater Interactions*, S.M. Gorelick, ed., International Association of Hydrological Sciences, Oxfordshire, United Kingdom, 187-197.

Hodiaumont, A., R. Cantallina, and J.M. Compere (1999). "Les eaux souterrainnes de la CILE contexte, captage et qualite" (Groundwater of the CILE, water use and quality). An article in "Les eaux souterraines en region Wallonne" (Groundwater in the Wallonie region of Belgium), F. Edeline, ed., *Editions Cebedoc*, 600-601 (4-5): 31-50.

Holm, J.V., K. Rügge, P.L. Bjerg, and T.H. Christensen (1995). "Occurrence and distribution of pharmaceutical organic compounds in the groundwater downgradient of a landfill (Grindsted, Denmark)." *Environmental Science and Technology*. 29(5): 1415-1420.

Industrieverband Agrar (2000). *Jahresbericht 1999/2000 (Annual Report 1999/2000)*. Industrieverband Agrar, Frankfurt a.m., Germany.

International Association of River Waterworks (RIWA) (1998). *Annual report 1998, Part C: The Rhine and Meuse*, International Association of River Waterworks (RIWA), Amsterdam, The Netherlands

Jani, J.P., C.V. Raiyani, J.S. Mistry, J.S. Patel, N.M. Desai, and S.K. Kashyap (1991). "Residues of organochlorine pesticides and polycyclic aromatic hydrocarbons in drinking water of Ahmedabad City, India." *Bulletin of Environmental Contamination and Toxicology*, 47: 381-385.

Jüttner, F. (1999). "Efficacy of bank filtration for the removal of fragrance compounds and aromatic hydrocarbons." *Water Science and Technology*, 40(6): 123-128.

Kalajzic, T., M. Bianchi, H. Muntau, and A. Kettrup (1998). "Polychlorinated biphenyls (PCBs) and organochlorine pesticides (OCPs) in the sediments of an Italian drinking water reservoir." *Chemosphere*, 36(7): 1615-1625.

Kalkhoff, S.J., D.W. Kolpin, E.M. Thurman, I. Ferrer, and D. Barceló (1998). "Degradation of chloroacetanilide herbicides: The prevalence of sulfonic and oxanilic acid metabolites in Iowa groundwaters and surface waters." *Environmental Science and Technology*, 32(11): 1738-1740.

Karickhoff, S.W., D.S. Brown, and T.A. Scott (1979). "Sorption of hydrophobic pollutants on natural sediments." *Water Research*, 13: 241-248.

Kolpin, D.W. (1997). "Agricultural chemicals in groundwater of the Midwestern United States: Relations to land use." *Journal of Environmental Quality*, 26: 1025-1037.

Kolpin, D.W., J.E. Barbash, and R.J. Gilliom (1998). "Occurrence of pesticides in shallow groundwater in the United States: Initial results from National Water Quality Assessment Program." *Environmental Science and Technology*, 32: 558-566.

Kolpin, D.W., E.T. Furlong, M.T. Meyer, E.M. Thurman, S.D. Zaugg, L.B. Barber, and H.T. Buxton (2002). "Pharmaceuticals, hormones, and other emerging contaminants in U.S. streams, 1999-2000: Methods, development, and national reconnaissance." *Environmental Science and Technology*, in press.

Kolpin, D.W., D.A. Goolsby, and E.M. Thurman (1995). "Pesticides in near-surface aquifers: An assessment using highly sensitive analytical methods and tritium." *Journal of Environmental Quality*, 24: 1125-1132.

Kolpin, D.W., D. Riley, M.T. Meyer, P. Weyer, and E.M. Thurman (2000). "Pharm-Chemical contamination: Reconnaissance for antibiotics in Iowa streams, 1999." *Proceedings, effects of animal feeding operations on water resources and the environment*. F.D. Wilde, L.J. Britton, C.V. Miller, and D.W. Kolpin, eds., U.S. Geological Survey, Open-File Report 00-204, U.S. Geological Survey, Reston, Virginia, 40 p.

Kolpin, D.W., E.M. Thurman, and D.A. Goolsby (1996). "Occurrence of selected pesticides and their metabolites in near-surface aquifers of the Midwestern United States." *Environmental Science and Technology*, 30: 335-340.

Kruhm-Pimpl, M. (1993). "Pesticides in surface water – Analytical results for drinking water reservoirs and bank filtrate waters." *Acta Hydrochimica et Hydrobiologica*, 21(3): 145-152.

Kühn, W., and U. Müller (2000). "River bank filtration – An overview." *Journal American Water Works Association*, December 2000: 60-69.

Kuhlmann, B., B. Kaczmarczyk and U. Schottler (1995). "Behavior of phenoxyacetic acids during underground passage with different redox zones." *International Journal of Environmental Analytical Chemistry*, 58: 199-205.

Lange, F.T., R. Furrer, and H.J. Brauch (2000). *Polar aromatic sulfonates and their relevance to waterworks*. De Eendracht, Schiedam, Netherlands, 70 p.

Lindner, K., T. Knepper, F. Karrenbrock, O. Rörden, H.J. Brauch, F.T. Lange, and F. Sacher (1996). *Erfassung und Identifizierung von trinkwassergängigen Einzelsubstanzen in Abwässern und im Rhein (Detection and identification of individual compounds relevant for drinking water supply in sewage effluents and in the Rhine River)*. IAWR-Rheinthemen 1, GEW, self-published, Köln, Germany.

Lindsey, M.E., and E.M. Thurman (2000). "Extraction of select antibiotics from agricultural wastes." *Proceedings, 220th American Chemical Society Meeting*, American Chemical Society, Washington D.C.

Marei, A.S.M., J.M.E. Quirke, G. Rinaldiz, J.A. Zoro, and G. Eglinton (1978). "The environmental fate of DDT." *Chemosphere*, 12: 993-998.

Matthess, G., E. Bedbur, H. Dunkelberg, K. Haberer, K. Hurle, F.H. Frimmel, R. Kurz, D. Klotz, U. Müller-Wegener, A. Pekdeger, W. Pestemer, and I. Scheunert (1997). Transport- und Abbauverhalten von Pflanzenschutzmitteln im Sicker- und Grundwasser (Transport and degradation behavior of pesticides in the vadose zone and in groundwater). Fischer, New York, New York.

Matthess, G., and A. Pekdeger (1985). "Survival and transport of pathogenic bacteria and viruses in ground water." *Ground Water Quality*, C.H. Ward, W. Giger, and P.I. McCarty, eds., Wiley, New York, New York.

Mathys, W. (1994). "Pestizidbelastungen von Grund- und Trinkwässern durch die Prozesse der 'künstlichen Grundwasseranreicherung' oder der Uferfiltration: Unterschätzte Kontaminationsquellen" (Pesticide pollution of ground and public drinking waters caused by artificial groundwater recharge or bank filtration: Underestimated sources for water contamination). *Zentralblatt für Hygiene und Umweltmedizin*, 196(4): 338-359.

Mazounie, P., D. D'Arras, and E. Brodard (2000). "Surveillance et contrôle des pesticides: Le point de vue d'un exploitant, Lyonnais des Eaux" (Monitoring and control of pesticides: The operation approach of Lyonnaise des Eaux). *Hydrogéologie*, 1: 7-11.

McCarthy, J.F., and J.M. Zachara (1989). "Subsurface transport of contaminants." *Environmental Science and Technology*, 23(5): 496-502.

Meitzler, L., P. Schmellenkamp, I. Schill, and F. Schredelseker (1996). "Bentazon als Uferfiltrattracer" (Bentazone as a tracer for bank filtrate). *Vom Wasser*, 87: 163-170.

Meyer, M.T., J.E. Bumgarner, J.V. Daughtridge, D.W. Kolpin, E.M. Thurman, and K.A. Hostetler (2000). "Occurrence of antibiotics in liquid waste at confined animal feeding operations and in surface and ground water." *Proceedings, effects of animal feeding operations on water resources and the environment*. F.D. Wilde, L.J. Britton, C.V. Miller, and D.W. Kolpin, eds., U.S. Geological Survey Open-File Report 00-204, U.S. Geological Survey, Reston, Virginia, 45 p.

Meylan, W.M., and P.H. Howard (1995). "Atom/fragment contribution method for estimating octanol-water partition coefficients." *Journal of Pharmaceutical Science*, 94: 83-92.

Meylan, W.M., P.H. Howard, and R.S. Boethling (1996). "Improved method for estimating water solubility from octanol water partition coefficient." *Environmental Toxicology and Chemistry*, 15(2): 100-106.

Moeder, M., S. Schrader, M. Winkler, and P. Popp (2000). "Solid-phase microextraction-gas chromatography-mass spectrometry of biologically active substances in water samples." *Journal of Chromatography A*, 873: 95-106.

Möhle, E., S. Horvath, W. Merz, and J.W. Metzger (1999). "Bestimmung von schwer abbaubaren organischen Verbindungen im Abwasser-Identifizierung von Arzneimittelrückständen – Identifizierung von Arzneimittelrückständen" (Determination of hardly degradable organic compounds in sewage water – Identification of pharmaceutical residues). *Vom Wasser*, 92: 207-223.

Mons, M.N., J. Van Genderen, and A.M. Van Dijk-Looijaard (2000). *Inventory on the presence of pharmaceuticals in Dutch water*. Kiwa N.V., Nieuwegein, The Netherlands, 39 p.

Neely, W.B., and G.E. Blau (1985). *Environmental exposure from chemicals*. CRC Press, Boca Raton, Florida.

Nondek, L., and N. Frolikova (1991). "Polychlorinated byphenyls in the hydrosphere of Czechoslovakia." *Chemosphere*, 23(3): 269-280.

Notenboom, J., A. Verschoor, A. Van der Linden, E. Van de Plassche, and C. Reuther (1999). "Pesticides in groundwater – Occurrence and ecological impacts." *National Institute of Public Health and the Environment (RIVM) Report 601506002*, Bilthoven, The Netherlands, 77 p.

Pereira, W.E., F.D. Hostettler, and J.B. Rapp (1996). "Distributions and fate of chlorinated pesticides, biomarkers, and polycyclic aromatic hydrocarbons in sediments along a contamination gradient from a point-source in San Francisco Bay, California." *Marine Environmental Resources*, 41: 299-314.

Perkow, W. (1992). *Wirksubstanzen der Pflanzenschutz- und Schädlingsbekämpfungsmittel, Second Edition (Active substances of pesticides, Second Edition)*. Blackwell Wissenschafts-Verlag/Parey Buchverlag, Berlin, Germany.

Preuss, G., N. Zullei-Seibert, J. Nolte, and B. Grass (1998). *Verhalten von Pflanzenbehandlungs- und Schaedlingsbekaempfungsmitteln in Sicker- und Grundwasser unter definierten Randbedingungen (Behavior of pesticides during infiltration and artificial groundwater recharge under defined environmental conditions)*. Umweltbundesamt Texte (German Environmental Agency), Berlin, Germany, 125 p.

Prior, J.C. (1991). *Landforms of Iowa*. University of Iowa Press, Iowa City, Iowa, 153 p.

Quensen, J.F. III, S.A. Mueller, M.K. Jain, and J.M. Tiedje (1998). "Reductive dechlorination of DDE to DDMU in marine sediment microsomes." *Science*, 280: 722-724.

Rafols, C., M. Roses, and E. Bosch (1997). "A comparison between different approaches to estimate the aqueous pKa of several non-steroidal anti-inflammatory drugs." *Analytica Chimica Acta*, 338: 127-134.

Ray, C., T. Grischek, J. Schubert, J.Z. Wang, and T.F. Speth (2002). "A perspective of riverbank filtration." *Journal of American Water Works Association*, in press.

Ray, C., T.W.D. Soong, G.S. Roadcap, and D.K. Borah (1998). "Agricultural chemicals: Effects on wells during floods." *Journal of American Water Works Association*, 90: 90-100.

Renner, R. (1998). "'Natural' remediation of DDT, PCBs debated." *Environmental Science and Technology*, 32: 60A-363A.

Roberts, P.V. (1985). "Field observations of organic contaminant behavior in the Palo Alto Baylands." *Artificial recharge of groundwater*, T. Asano, ed., Butterworth Publishers, Boston, Massachusetts.

Roger, M., and U. Fontanelle (1999). "Evolution des nitrates et des pesticides dans les captages de la S.W.D.E." (Evolution of nitrates and pesticides in water from the S.W.D.E. well fields). In an article on "Les eaux souterraines en region Wallonne (Groundwater in the Wallonne region of Belgium), F. Edeline, ed., *Editions Cebedoc*, 600-601 (4-5): 63-73.

Schaffner, C., M. Ahel, and W. Giger (1987). "Field studies on the behaviour of organic micropollutants during infiltration of river water to groundwater." *Water Science and Technology*, 19: 1195-1196.

Scheytt, T. (2002). *Arzneimittel in Grundwasser – Eintrag, Abbau, und Transport (Pharmaceuticals in ground water – Input, degradation, and transport)*. Ph.D. thesis, Technical University of Berlin, Berlin, Germany, 148 p.

Scheytt, T., S. Grams, and H. Fell (1998). "Occurrence and behaviour of drugs in ground-water." *Gambling with groundwater – Physical, chemical, and biological aspects of aquifer-stream relations*, J.V. Brahana, Y. Eckstein, L.K. Ongley, R. Schneider, and J.E. Moore, eds., IAH/AIH Proceedings Volume, St. Paul, Minnesota, 13-18.

Scheytt, T., M. Leidig, P. Mersmann, and T. Heberer (2002). "Natural attenuation of pharmaceuticals." *Ground Water*, in preparation.

Schulmeyer, P.M. (1991). "Relation of selected water-quality constituents to river stage in the Cedar River, Iowa." *Proceedings, U.S. Geological Survey Toxic Substances Hydrology Program*, U.S. Geological Survey, Reston, Virginia, 227-231.

Schulmeyer, P.M., and D.J. Schnoebelen (1998). "Hydrogeology and water quality in the Cedar Rapids Area, Iowa, 1992-96." *U.S. Geological Survey Water Resources Investigative Report* 97-4261, U.S. Geological Survey, Reston, Virginia, 77 p.

Schwarzenbach, R.P., W. Giger, E. Hoehn, and J.K. Schneider (1983). "Behavior of organic compounds during infiltration of river water to groundwater: Field studies." *Environmental Science and Technology*, 17: 472-479.

Schwarzenbach, R.P., and W. Giger (1985). "Behaviour and fate of halogenated hydrocarbons in groundwater." *Ground water quality*, C.H. Ward, W. Giger, and P.L. McCarty, eds., Wiley, New York, New York, 446-471.

Schweinsberg, F., W. Abke, K. Rieth, U. Rohmann, and N. Zullei-Seibert (1999). "Herbicide use on railway tracks for safety reasons in Germany." *Toxicology Letters*, 107: 201-205.

SenStadtUm (Senatsverwaltung für Stadtentwicklung und Umweltschutz)(1997). Berlin digital atlas, Second Edition. Kulturbuchverlag, Berlin, Germany. Free online access: www.sensut.berlin.de/sensut/umwelt/uisonline/dua96/html/edua_index.shtml.

Siegener, R., and R.F. Chen (2000). "Detection of pharmaceuticals entering Boston Harbor." *Analysis of environmental endocrine disruptors*, L.H. Keith, T.L. Jones-Lepp, L.L. Needham, eds., American Chemical Society, Washington D.C.

Seiler, R.L., S.D. Zaugg, J.M. Thomas, and D.L. Howcroft (1999). "Caffeine and pharmaceuticals as indicators of wastewater contamination in wells." *Ground Water*, 37(3): 405-410.

Skark, C., and N. Zullei-Seibert (1995). "The occurrence of pesticides in groundwater: Results of case-studies." *International Journal of Analytical Chemistry*, 58: 387-96.

Smith, N.J., R.C. Martin, and R.G. St. Croix (1996). "Levels of the herbicide glyphosate in well water." *Bulletin of Environmental Contamination and Toxicology*, 57: 759-765.

Sontheimer, H. (1980). "Experience with riverbank filtration along the Rhine River." *Journal American Water Works Association*, December 1980: 386-390.

Spliid, N.H., and B. Køppen (1998). "Occurrence of pesticides in Danish shallow groundwater." *Chemosphere*, 37(7): 1307-1316.

Stan, H.J., and T. Heberer (1997). "Pharmaceuticals in the aquatic environment." Article in a Special Issue on "Dossier water analysis," M.J.F. Suter, ed., *Analusis*, 25: M20-23.

Stan, H.J., and M. Linkerhägner (1992). "Identifizierung von 2-(4-Chlorphenoxy)-2-methyl-propionsäure im Grundwasser mittels Kapillar-Gaschromatographie mit Atomemissionsdetektion und Massenspektrometrie" (Identification of 2-[4-chlorophenoxy]-2-methyl-propionic acid in ground water using capillary-gas chromatography with atomic emission detection and mass spectrometry). *Vom Wasser*, 79: 75-88.

Stan, H.J., T. Heberer, and M. Linkerhägner (1994). "Vorkommen von Clofibrinsäure im aquatischen System: Führt die therapeutische Anwendung zu einer Belastung von Oberflächen-, Grund- und Trinkwasser?" (Occurrence of clofibric acid in the aquatic system: Is their therapeutic use responsible for the loads found in surface, ground- and drinking water?). *Vom Wasser*, 83: 57-68.

Steele, G.V., and I.M. Verstraeten (1999). "Effects of pumping collector wells on river-aquifer interaction at Platte River Island near Ashland, Nebraska, 1998." *U.S. Geological Survey Water Resources Investigative Report 99-4161*, U.S. Geological Survey, Reston, Virginia, 6 p.

Stuer-Lauridsen, F., M. Birkved, L.P. Hansen, H.C. Holten Lützhoft, and B. Halling-Sorensen (2000). "Environmental risk assessment of human pharmaceuticals in Denmark after normal therapeutic use." Article in a special issue on "Drugs in the Environment," S.E. Jorgensen and B. Halling-Sorensen, eds., *Chemosphere*, 40 (2000): 759-765.

Stuyfzand, P.J. (1989). "Hydrology and water quality aspects of Rhine bank groundwater in the Netherlands." *Journal of Hydrology*, 106(3/4): 341-363.

Stuyfzand, P.J. (1998a). "Fate of pollutants during artificial recharge and bank filtration." *Artificial recharge of groundwater*, J.H. Peters et al., eds., A.A. Balkema, Rotterdam, The Netherlands.

Stuyfzand, P.J. (1998b). "Simple models for reactive transport of pollutants and main constituents during artificial recharge and bank filtration." *Artificial recharge of groundwater*, J.H. Peters et al., eds., A.A. Balkema, Rotterdam, The Netherlands.

Syracuse Science Center (2000). *Database of experimental octanol-water partition coefficients (Log P)*. Syracuse Research Corporation, Syracuse, New York, http://www.syrres.com

Szabo Z., D.E. Rice, T. Ivahnenko, and E.F. Vowinkel (1994). "Delineation of the distribution of pesticides and nitrate in an unconfined aquifer in the New Jersey coastal plain by flow-path analysis." *Proceedings, Fourth National Conference on Pesticides: New Directions in Pesticide Research, Development, Management, and Policy*, Virginia Polytechnic Institute and State University Water Resources Center, Blacksburg, Virginia.

Taets, C., S. Aref, and A.L. Raybum (1998). "The clastogenic potential of triazine herbicide combinations found in potable water supplies." *Environmental Health Perspectives*, 106(4): 197-201.

Taylor, A.W., and W.F. Spencer (1990). "Volatilization and vapor transport processes." *Pesticides in the soil environment: Processes, impacts, and modeling*, H.H. Cheng, ed., SSSA Book Series 2, Madison, Wisconsin, 213-269.

Taylor, A.G. (1994). "Pesticides in the Illinois public water supplies: A year of compliance monitoring." *Proceedings, Illinois Agriculture Pesticides Conference 1994*, Philips Brothers, Urbana, Illinois, 94-99.

Ternes, T.A. (1998). "Occurrence of drugs in German sewage treatment plants and rivers." *Water Resources*, 32: 3245-3260.

Ternes, T.A., P. Kreckel, and J. Mueller (1999b). "Behavior and occurrence of estrogens in municipal sewage treatment plants - II. Aerobic batch experiments with activated sludge." *The Science of the Total Environment*, 225: 91-99.

Thurman, E.M., D.A. Goolsby, M.T. Meyer, and D.W. Kolpin (1991). "Herbicides in surface waters of the Midwestern United States: The effect of spring flush." *Environmental Science and Technology*, 25: 1784-1796.

Thurman, E.M., and K.A. Hostetler (2000). "Analysis of tetracycline and sulfamethazine antibiotics in ground water and animal-feedlot wastewater by high-performance liquid chromatography/mass spectrometry using positive-ion electrospray." *Proceedings, Effects of animal feeding operations on water resources and the environment*, F.D. Wilde, L.J. Britton, C.V. Miller, and D.W. Kolpin, eds., U.S. Geological Survey Open-File Report 00-204, U.S. Geological Survey, Reston, Virginia, 47 p.

Thurman, E.M., and M.E. Lindsey (2000). *Transport of antibiotics in soil and their potential for ground-water contamination*. Society of Environmental Toxicology and Chemistry meeting, Brussels, Belgium.

Tsipi, D., and A. Hiskia (1996). "Organochlorine pesticides and triazines in the drinking water of Athens." *Bulletin of Environmental Contamination and Toxicology*, 57: 250-257.

Umweltbundesamt (UBA) (1996). *Sachstandsbericht zu Auswirkungen der Anwendung von Clofibrinsäure und anderen Arzneimitteln auf die Umwelt und die Trinkwasserversorgung (Report on the the effects of the application of clofibric acid and other pharmaceuticals on the environment and drinking water supplies)*. Umweltbundesamt (German Environmental Protection Agency), Berlin, Germany.

Umweltbundesamt (UBA) (2000). *Annual Report 1999*. Umweltbundesamt (German Environmental Protection Agency), Berlin, Germany.

Upham, B.L., B. Boddy, X. Xing, J.E. Trosko, and S.J. Masten (1997). "Non-genotoxic effects of selected pesticides and their disinfection by-products on gap junctional intercellular communication." *Ozone Science and Engineering*, 19: 351-369.

U.S. Department of Agriculture (1976). *Iowa-Cedar Rivers basin study*. U.S. Department of Agriculture, Des Moines, Iowa.

U.S. Environmental Protection Agency (1989). *Guidance manual for compliance with the filtration and disinfection requirements for public water systems using surface water sources*. U.S. Environmental Protection Agency, Washington D.C.

U.S. Environmental Protection Agency (1991). *Maximum contaminant levels (subpart B of part 141, National primary drinking water regulations)*: U.S. CFR, Title 40, parts 100-149, U.S. Environmental Protection Agency, Washington, D.C., 659-660.

U.S. Environmental Protection Agency (1994). *Drinking water regulations and health advisories*. U.S. Environmental Protection Agency, Washington D.C.

U.S. Environmental Protection Agency (1996). *CT Compliance regulations*: U.S. CFR, Title 40, parts 100-149, U.S. Environmental Protection Agency, Washington, D.C.

Van Genderen, J., M.N. Mons, and J.A. Van Leerdam (1999). *Inventory and toxicological evaluation of organic micropollutants — Revision 1999*. De Eendracht, Schiedam, Netherlands, 160 p.

Van Hoorick, M., J. Feyen, and L. Pussemier (1998). "Experimental and numerical analysis of atrazine behaviour in an artificial recharge site." *Artificial recharge of groundwater*, J.H. Peters et al., eds., A.A. Balkema, Rotterdam, The Netherlands.

Vasconcelos, J., and S. Harris (1992). *Consensus method for determining groundwaters under the direct influence of surface water using microscopic particulate analysis*. U.S. Environmental Protection Agency, Washington, D.C.

Verstraeten, I.M., and M.J. Ellis (1994). "Reconnaissance of the ground-water quality in the Papio-Missouri River Natural Resources District, northeastern Nebraska, June-September 1992." *U.S. Geological Survey Water Resources Investigative Report*, 94-4197, U.S. Geological Survey, Reston, Virginia, 90 p.

Verstraeten, I.M., and J.G. Miriovsky (1999). "Herbicide removal through bank filtration using horizontal collector wells at Ashland, Nebraska, United States." *Proceedings, International Riverbank Filtration Conference*, National Water Research Institute, Fountain Valley, California.

Verstraeten, I.M., M.J. Soenksen, G.B. Engel, and L.D. Miller (1999a). "Determining travel time and stream mixing using tracers and empirical equations." *Journal of Environmental Quality*, 28(5): 1387-1395.

Verstraeten, I.M., E.M. Thurman, E.C. Lee, and R.D. Smith (2002). "Degradation of triazines and acetamide herbicides by bank filtration, ozonation, and chlorination in a public water supply." *Journal of Hydrology*, Special issue on bank filtration, in press.

Verstraeten, I.M., D.T. Lewis, D.L. McCallister, A. Parkhurst, and E.M. Thurman (1996). "Influence of landscape position and irrigation on alachlor, atrazine, and selected degradates in selected upper regolith and associated shallow aquifers in northeastern Nebraska." American Chemical Society Symposium Series 630: Herbicide metabolites in surface water and groundwater. M.T. Meyer and E.M. Thurman, eds., American Chemical Society, Washington, D.C., 178-197.

Verstraeten, I.M., J.D. Carr, G.V. Steele, E.M. Thurman, and D.F. Dormedy (1999b). "Surface-water/ground-water interaction: Herbicide transport into municipal collector wells." Journal of Environmental Quality, 28(5): 1396-1405.

Wahl, K., and B.J. Bunker (1986). "Hydrology of carbonate aquifers in southwestern Linn County and adjacent parts of Benton, Iowa, and Johnson Counties, Iowa." Bulletin 15, Iowa Geological Survey Water Supply, Des Moines, Iowa, 56 p.

Wang, W., and P. Squillace (1994). "Herbicide interchange between a stream and the adjacent alluvial aquifer." Environmental Science and Technology, 28: 2336-2344.

Ware, G.W., D.G. Crosby, and J.W. Giles (1980). "Photodecomposition of DDA." Archives of Environmental Contamination and Toxicology, 9: 135-146.

World Health Organization (1993). Guidelines for Drinking Water Quality, Second Edition, Volume 1: Recommendations. World Health Organization, Geneva, Switzerland.

Wilderer, P.A., U. Forstner, and O.R. Kunschik (1985). "The role of riverbank filtration along the Rhine for municipal and industrial water supply." Artificial recharge of groundwater, T. Asano, ed., Butterworth Publishers, Boston, Massachusetts.

Wilken, R.D., T.A. Ternes, and T. Heberer (2000). "Pharmaceuticals in sewage, surface and drinking water in Germany." Security of public water supplies, R.A. Deininger et al., eds., Kluwer Academic Publishers, Dordrecht, The Netherlands.

Winslow J.D., H.G. Stewart, R.H. Johnson, and L.J. Crain (1965). Ground-water resources of eastern Schenectady County. Bulletin 57, New York State Water Resources Commission, New York, New York, 148 p.

Yalkowsky, S.H., and R.M. Dannenfelser (1992). Aquasol database of aqueous solubility, Version 5, College of Pharmacy, University of Arizona, Tucson, Arizona.

Zullei-Siebert, N. (1996a). "Pesticides and artificial recharge of groundwater via slow sand filtration – Elimination potential and limitations." Proceedings, International Symposium on Artificial Recharge of Groundwater, A.L. Kivimäki and T. Suokko, eds., The Nordic Coordination Committee for Hydrology (KOHYNO), Helsinki, Finland, 247-253.

Zullei-Siebert, N. (1996b). "Belastung von Gewässern mit Pflanzenschutzmitteln: Ergebnisse einer Bestandsaufnahme in Deutschland und der Europäischen Union" (Loads of pesticides in watercourses: Results from a review in Germany and the EU). Proceedings of the Groundwater Colloquium of the TU Dresden, TU Dresden, Dresden, Germany, 179-201.

Zullei-Siebert, N. (1998). "Your daily 'drugs' in drinking water? State of the art for artificial groundwater recharge." Artificial recharge of groundwater, J.H. Peters et al., eds., A.A. Balkema, Rotterdam, The Netherlands.

Zwiener, C., T. Glauner, and F.H. Frimmel (2000). "Biodegradation of pharmaceutical residues investigated by SPE-GC/ITD-MS and on-line derivatization." Journal of High Resolution Chromatography, 23(7-8): 474-478.

Chapter 10. Effectiveness of Riverbank Filtration Sites to Mitigate Shock Loads

Hans-Joachim Mälzer, Ph.D.
IWW Rheinisch-Westfälisches Institut für Wasserforschung
Institut an der Gerhard-Mercator-Universität Duisburg
Mülheim a.d. Ruhr, Germany

Jürgen Schubert, M.Sc.
Stadtwerke Düsseldorf AG
Düsseldorf, Germany

Rolf Gimbel, Ph.D.
IWW Rheinisch-Westfälisches Institut für Wasserforschung
Institut an der Gerhard-Mercator-Universität Duisburg
Mülheim a.d. Ruhr, Germany

Chittaranjan Ray, Ph.D., P.E.
University of Hawaii at Mānoa
Honolulu, Hawaii, United States

1. Introduction

Some of the busiest waterways in the United States are the Ohio, Illinois, Mississippi, Missouri, Columbia, and Hudson Rivers. Many riparian communities draw their water from these rivers, either through direct intake or RBF systems. This is also true for the Rhine River, which flows north from the Swiss Alps to the North Sea, and for the Danube River, which flows through the heart of Europe, traversing through the countries of Germany, Austria, Slovakia, Hungary, Yugoslavia, Bulgaria, and Romania. Many large cities along the Rhine and Danube Rivers use RBF as the source of drinking-water production.

Accidental releases of chemicals, fertilizers, and other hazardous materials are routine in navigable waterways like these. Unless the magnitudes of these releases are severe, many go unreported. Minor fuel releases from barge or ferry traffic have negligible impacts on RBF systems; however, conventional treatment plants with direct intake systems may have to make some adjustments in treatment operations to remove most contaminants prior to delivery into the distribution system. For large accidental releases, the direct intake systems are, generally, shut down and some sort of emergency management is activated at the intake point or in the water treatment plant. In contrast, the impacts of releases of a comparable magnitude on RBF systems are much lower. The contaminant travel time in an RBF system is significantly greater than in a direct intake system because of such phenomena as sorption and degradation. Another advantage is that there is a time lag between the peaks observed between pumped water and surface water. This lag time in RBF allows water utilities to respond to an emergency faster within an RBF system than within a direct intake system.

C. Ray et al. (eds.), Riverbank Filtration, 229–259.
© 2002 *Kluwer Academic Publishers. Printed in the Netherlands.*

A release can have devastating impacts. In January 1998, the Monongahela River in the northeastern United States was contaminated with 4-million gallons of diesel oil when an Ashland Oil Company, Inc. storage tank ruptured. The oil release overtopped the containment dikes, entered an adjacent property, and then entered the Monongahela River through an uncapped storm drain. The Monongahela River serves as a tributary to the Ohio River, which ultimately drains to the Mississippi River. Ultimately, the oil release affected a million people along the waterways by temporarily contaminating water-supply systems as well as affected the ecology of the waterways, resulting in the death of fish and wildlife. Several important lessons were learned from this release, such as the magnitude of the disaster could have been reduced if:

- Water utilities and industries on the riverbank communicated efficiently.
- Adequate containment and monitoring devices were available.
- A rapid response team was mobilized.

If in place, such systems may have also reduced the negative impacts of a more recent release. On January 30, 2000, there was a massive cyanide release from a holding pond in the City of Baia Mare in northwest Romania. The cyanide traveled through the Sasar, Lapus, Somes, Tisza, and Danube Rivers before reaching the Black Sea about 4 weeks later. In Romania, the release negatively affected the operation of 24 municipalities and caused massive fish kills.

To manage such emergencies, a number of interstate or inter-country organizations have been formed. In Europe, the Rhine Warning and Alarm service provides such functions for Germany and The Netherlands. In the United States, the Ohio River Valley Water Sanitation Commission, established in 1948 and located in Cincinnati, Ohio, also coordinates emergency response activities for releases or accidental discharges into the river in addition to coordinating its regularly mandated water-quality functions for eight member states.

This chapter will examine the role of RBF systems in attenuating these releases, known as "shock loads." In addition, this chapter will examine the role of modeling in managing emergencies.

2. Experience Gained from the Sandoz Accident on the Rhine River in Europe

The Sandoz Accident

On the night of Saturday, November 1, 1986, a fire erupted in an agrochemical store of the Sandoz Chemical Plant in Basel, Switzerland. Insecticides, herbicides, and fungicides were carried out into the Rhine River with fire-fighting water. The effects of this accident on the Rhine River were serious. The report of the German Commission for the Control of Pollution in the Rhine River (DRK, 1986) contained, interalia, the following observations: "On the stretch of the Rhine up to the Loreley Rock, the entire stock of eels has been destroyed. In addition, other species of fish have also been affected in Baden-Württemberg. Damaging effects can be determined on fish food organisms up to the mouth of the Mosel River."

Through the media, the accident and its consequences were observed by the general public; each day, the eels killed by the uncheckable, ever-advancing "wave of poison" were shown on television. The almost unanimous opinion of the experts was that the pollutants would not only destroy, but also prevent any life in the Rhine River for years to come and would simultaneously destroy the basis for water catchment within the scope of influence of the Rhine River.

Initially, due to the poor flow of information, those responsible at the Rhine waterworks and cooperating institutions had difficulty in estimating the effects of the accident on RBF because:

- Vital information on the type and quantity of the substances that had been carried into the Rhine River with the fire-fighting water as well as reliable data on concentrations were limited.

- The models available for a prognosis of the flow time and concentration gradient along the Rhine River were inadequate, particularly as far as the dam-controlled section of the Upper Rhine was concerned.
- Critical data on the properties of the substances, their biodegradability during passage through the soil, and their adsorbability on activated carbon were not available. These data had to be determined by means of tests and, consequently, became available only after some delay.
- Particularly for the case of peak loads, no systematic analyses had been carried out up to that time; therefore, no models existed for simulating RBF behavior in the case of actual accidents. The knowledge acquired by the operators of RBF plants from observations and tests on the time required for soil passage and concentration equalization processes was available with respect to RBF.

In this situation, the experience gained by waterworks along the Rhine River from earlier accidents was an important aid in mitigating the crisis. Because of this experience and the timely experiments performed to determine the behavior of the substances during passage through the soil and in treatment plants, the waterworks on the Lower Rhine were assured that the pollutants would not be transported into potable water. Nevertheless, authorities ordered that wells on the Rhine River be shut down, at least during passage of the wave of contaminated water in front of the adjacent well fields.

Consequences of the Sandoz Accident

The Sandoz Accident gave fresh impetus to improve pollution control on the Rhine River. Worthy of particular note were *in situ* measures (e.g., retention tanks) to provide protection in the case of accidents during the production and storage of water-hazardous substances. In addition, the riparian states of the Rhine River initiated the "Rhine Action Program," the objective of which was a distinct improvement of the quality of Rhine River water by the year 2000. Remarkable progress has been made concerning the improvement of water quality and the restoration of the continuity of the Rhine River and its tributaries for fish migration. The discharge of hazardous substances in municipal and industrial wastewaters has been considerably reduced. Meanwhile, a new program for the sustainable development of the Rhine, "Rhine 2020," was started, which combines ecological requirements with those of flood prevention and the protection of surface water with the protection of groundwater, including ecological, economic, and social aspects.

The deficiencies that still existed at the time of the accident have, in the meantime, been eliminated to a far-reaching degree:

- The reporting system of the "International Rhine Warning and Alarm Service" has been gradually improved.
- Flow-time and concentration curve models have been developed, which allow the waterworks and authorities to make reliable prognoses with respect to mass transport in the Rhine River.
- RBF hydraulic processes have been systematically analyzed in a comprehensive research project (Sontheimer, 1991). From this, among other things, new knowledge was gained with regard to mitigating accident-related peak loads. Much of what the waterworks had "experienced" could be confirmed and traced back on a scientific basis.

- Unsteady groundwater flow and transport models were developed in several waterworks by means of which accidental releases into the Rhine River can be simulated and their effects observed up to the wells (König and Schubert, 1992).

The reporting system, models, and additional tools from the research project for monitoring shock loads has improved safety as far as drinking-water supply from riverbank filtrate is concerned.

3. Mechanisms of Pollutant Transport and an Assessment of the Effects of Shock Loads on RBF

Pollutants may arrive at the abstraction wells with some delay and with attenuated concentrations, as schematically shown in Figure 10-1. The main processes responsible for the temporal delay and attenuation of the shock-load concentration are:

- Dilution caused by transport on different path lines.
- Hydrodynamic dispersion.
- Biodegradation.
- Adsorption.

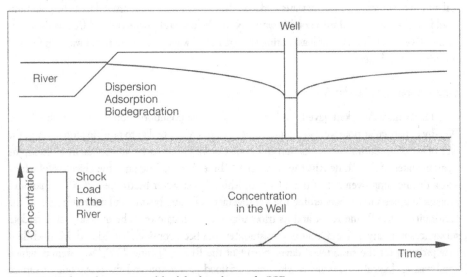

Figure 10-1. Delay and attenuation of shock loads in the river by RBF.

The effects of dispersion, adsorption, and biodegradation are considered in Equation 1, which describes the one-dimensional transport of a pollutant in a homogeneous parallel flow through a porous medium

$$R\frac{\partial c}{\partial t} + u\frac{\partial c}{\partial x} - u\alpha\frac{\partial^2 c}{\partial x^2} + R\lambda c = 0 \qquad (1)$$

where c is the solution concentration (kg/m^3); t is the time (s); x is the distance in flow direction (m); α is the dispersivity (m); and λ is the first-order degradation coefficient (1/s).

The mean pore water velocity u is given by $u=v_F/\varepsilon$, where v_F is the filtration rate (m/s) and ε is the porosity. The retardation factor R includes instantaneous adsorption equilibrium in combination with a linear isotherm. For an instantaneous intake of the mass m (kg) of a pollutant into an aquifer with the thickness M (m) and the width B (m) at x = 0 and t = 0, the analytical solution of Equation 1 is:

$$c(x,t) = \frac{m}{2BM\epsilon R \sqrt{\alpha \pi u \frac{t}{R}}} \; exp\left(-\frac{(x-ut/R)^2}{4\alpha ut/R}\right) exp\left(-\lambda_t\right) \tag{2}$$

Further analytical solutions for the one-, two-, and three-dimensional transport of a pollutant and for various boundary conditions can be found in literature (Van Genuchten and Wierenga, 1976; Van Genuchten and Wierenga, 1977; Van Genuchten, 1981; Bear, 1972; Bear, 1979; Kinzelbach, 1987; Luckner and Schestakow, 1991; Segol, 1994).

Beside the effects mentioned above, Fick's diffusion and the mixing of pollutants with immobile pore water are also responsible for a delayed arrival of the pollutant at the well and the attenuation of the pollutant concentration during subsoil passage. In most practical cases, these effects will be of minor importance or will only be observed coupled with other dominating effects; therefore, the influence of Fick's diffusion and the presence of immobile pore water on the transport of pollutants in RBF will be not discussed in this chapter.

Effect of Transport on Different Path Lines

In RBF, the pollutant infiltrates into the aquifer and follows the flow paths of the water during subsurface flow. The pre-condition for the infiltration of water from the river into the aquifer is a hydraulic gradient caused by pumping wells, which lowers the water table in the aquifer under the water level in the river, or is caused by a water level in the river that is currently higher than the water table in the aquifer (e.g., during floods). Darcy's law can describe the seepage velocity of a flow through a porous medium. For one-dimensional flow, it is obtained that

$$v_F = k_f \frac{\Delta H}{L} \tag{3}$$

where k_f (m/s) is the permeability; ΔH (m) is the hydraulic head difference between two observation points; and L (m) is the distance between the observation points.

The positions and size of the areas of infiltration in the riverbed and the positions of the wells, as well as the geometry of the aquifer/river system, are decisive for pollutant transport on different flow paths through the aquifer. Gölz et al. (1991) showed that the permeability of the Rhine riverbed can vary temporally as well as spatially and, as model calculations of Gotthardt (1992) indicate, infiltration at the Rhine River at Düsseldorf, Germany, really occurs over an area in the riverbed and not only at the riverbank. Pollutants infiltrating at different places in the riverbed will take different pathways through the aquifer and will finally converge at the well. As a result, the shock-load concentration is attenuated by RBF. The effect is demonstrated by simplified model calculations (Figures 10-2 to 10-5).

Because of the wastewater discharge of the French potash mines on working days (from Monday to Friday), the Rhine River showed an oscillation of the chloride concentration with an oscillation time of about 1 week (Figure 10-2), which is still detectable in the Rhine River at Wittlaer (near the City of Düsseldorf) after a distance of several hundred kilometers. These oscillations may be considered to be a series of shock loads of an ideal tracer, which is not affected by adsorption or biodegradation.

Figure 10-3 shows a vertical section of the aquifer at Wittlaer, where the Department of Works for the City of Duisburg operates several pumping and monitoring wells.

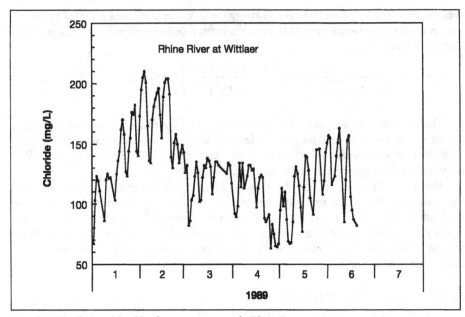

Figure 10-2. Oscillation of the chloride concentration in the Rhine River.

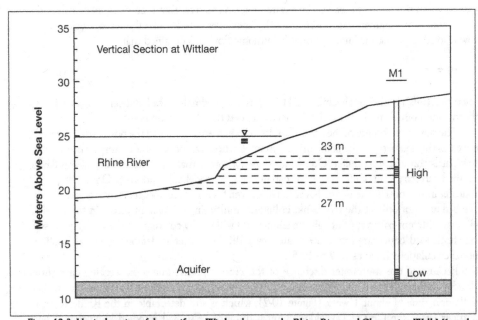

Figure 10-3. Vertical section of the aquifer at Wittlaer between the Rhine River and Observation Well M1, and a simplified model assumption of five parallel tubes. Observation Well M1 can be sampled in both a high (M1h) and low (M1l) vertical position.

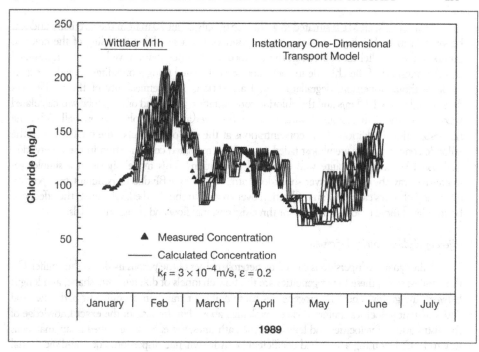

Figure 10-4. Chloride concentration at the end of each tube.

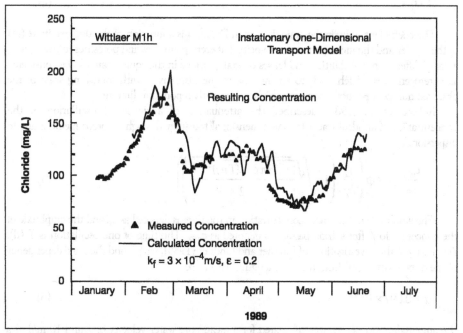

Figure 10-5. Mean chloride concentration at Observation Well M1.

Monitoring Well M1 is situated in a distance of only about 20 m from the riverbank and can be sampled in both a high and low vertical position. For a simplified modeling of the chloride transport, five parallel tubes between the riverbed and the observation well have been assumed and the transport of the chloride in each tube was calculated using a one-dimensional transport model without dispersion, degradation, and adsorption. The permeability of the aquifer was determined as 3×10^{-4} m/s and the chloride concentration at the end of each tube was calculated according to the daily measured water levels in the river and in the observation well. After the flow-proportional addition of the concentrations at the end of each tube, the oscillations of the chloride concentrations are almost faded out and the measured concentration in the observation well could be described quite well by model calculations. This model shows in a simple, yet impressive way the effect of river shock-load attenuation by RBF due to pollutant transport on different path lines caused by an infiltration over an area in the riverbed. Of course, the effect can be calculated more exactly with two- or three-dimensional flow and transport models.

Effect of Hydrodynamic Dispersion

Hydrodynamic dispersion is caused by heterogeneities of the porous media in the aquifer. On the smallest scale, these heterogeneities are the flow channels of different size, shape, and length between the grains. These heterogeneities cause pollutant transport on different path lines and with different velocities similar to those explained above but, for lack of the exact knowledge of the distribution of velocities and lengths of the path lines, the effect is described in an analogous way to Fick's law using a so-called coefficient of hydrodynamic dispersion. For one-dimensional transport conditions, the coefficient of hydrodynamic dispersion can be described as:

$$D = \alpha u \tag{4}$$

The coefficient of hydrodynamic dispersion D (m^2/s) is a function of the dispersivity α (m) of the aquifer and the mean pore water velocity. Different grain sizes and porosities in the aquifer, areas of different permeabilities, and lenses or stratifications in the aquifer can also be considered as heterogeneities, which leads to an increase of the dispersivity with increasing scale of the observed transport phenomenon and which is frequently reported in literature.

Roberts et al. (1982) described the attenuation of a sinusoidal oscillation of the concentration of an ideal tracer by one-dimensional transport through a porous medium due to dispersion.

$$\frac{c_{max}}{c_0} = exp\left(\frac{L}{2\alpha}\left(1 - \sqrt{\frac{\sqrt{1+((8\pi\alpha)/(Tu))^2}+1}{2}}\right)\right) \tag{5}$$

The amplitude of the tracer concentration in the river is denoted as c_0 and the amplitude of the concentration after subsoil passage is denoted as c_{max}. The time of one oscillation is T (d). According to the investigations of Lindner and Lindner (1986), who found the scale dependency of the dispersivity α (m) from the scale length L (m) to be

$$\alpha = 0.097 \times L^{0.825} \tag{6}$$

the amplitude ratio c_{max}/c_0 was calculated for a mean pore water velocity of 1 and 10 m/d as a function of the passed distance L. The results of calculations for an oscillation frequency of 7 days are shown in Figure 10-6.

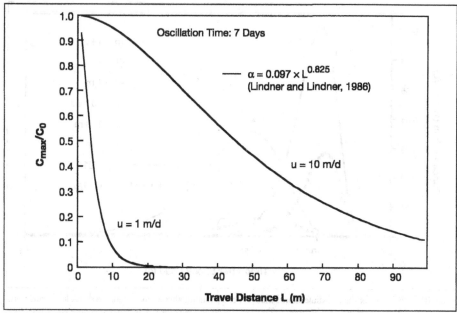

Figure 10-6. Attenuation of a sinusoidal oscillating concentration by dispersion in one-dimensional transport for different mean pore velocities.

Especially at lower mean pore velocities, oscillations in the river will be completely faded out by RBF after very short distances; therefore, the effect of dispersion will also be responsible for attenuating the oscillation of a chloride concentration by RBF (see Figure 10-5).

Effect of Adsorption and Biodegradation

In most cases, the effect of the adsorption of pollutants on soil is described by a linear isotherm and, usually, an instantaneous equilibrium is assumed in transport modeling. For these assumptions, the effect of adsorption on pollutant transport can be described by retardation factor R, as already used in Equation 1.

$$R = 1 + K_A \rho_b \frac{(1 - \varepsilon)}{\varepsilon} \tag{7}$$

The retardation factor is a function of the soil bulk density ρ_b (kg/m³), the porosity ε, and the adsorption coefficient K_A (m³/kg). The adsorption coefficient is frequently found to be dependent of the organic carbon content of the soil and the octanol-water partition coefficient (Briggs, 1981; Chiou et al., 1983; Karikoff et al., 1979; Schwarzenbach and Westall, 1981). For a pollutant that is not affected by adsorption, the value of the retardation factor is R = 1.

In most cases, the biodegradation of pollutants during transport through the aquifer is described by first-order kinetics with the degradation rate λ (1/s).

$$\frac{dc}{dt} = -\lambda c \tag{8}$$

Figure 10-7 shows an example for the influence of the adsorption and biodegradation effect on the one-dimensional transport of a pollutant after an instantaneous intake into homogeneous parallel flow.

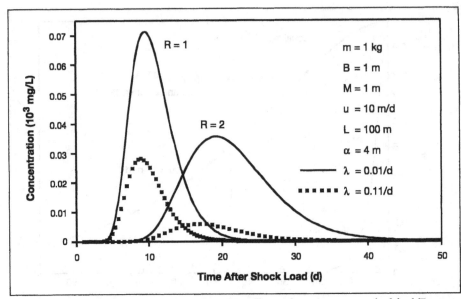

Figure 10-7. Example for the one-dimensional transport of a pollutant after an instantaneous shock load (Equation 2). Assumed distance between infiltration and detection point L = 100 m, mean pore water velocity u = 10 m/d, porosity of the aquifer material ε = 0.2, dispersivity α = 4 m, and there are two different values of the retardation factor R and the first order degradation rate λ.

Due to adsorption of the pollutant, which is assumed by a value of R=2 in Figure 10-7, the maximum concentration of the pollutant arrives approximately after twice the traveling time of water to the well. Because of hydrodynamic dispersion, the maximum concentration is lower for an adsorbing pollutant and the time range in which the pollutant may be detected in the well is much longer than in the non-adsorbing case. If no degradation occurs (λ=0), the amount of the pollutant that was carried into the aquifer stays constant during transport, and the areas under the concentration curves for R=1 and R=2 are constant. When degradation is assumed to occur (see Figure 10-7), it is estimated to be 0.01 1/day. If the pollutant is additionally retarded due to adsorption, the time for biodegradation is extended and the pollution concentration would be less than that without adsorption.

Simulation and Assessment of Shock Loads by Control Filters

To assess biodegradation processes in RBF at the Rhine River, control filters have been developed in which DOCs are degraded by sessile microorganisms under conditions similar to RBF, but under exactly known and definite flow conditions (Mälzer et al., 1993). The control filters consist of a feed vessel, pre-filter, storage vessel, and four columns in sequence with a total length of 8.4 m filled with granular pumice. This material has been proven to be a suitable carrier for microorganisms in previous investigations (Gimbel and Mälzer, 1987). A schematic drawing of a control filter is shown in Figure 10-8. The feed vessel was charged once per week with Rhine River water, which was pre-filtered by a sand filter with a high filtration rate to remove suspended solids. During a residence time of 1 week in the storage vessel, only slight biodegradation was observed. The filter columns had been charged with the pre-filtered river water using filtration rates between 1.67 × 10⁻⁵ and 6.67 × 10⁻⁵ m/s (0.06 and 0.24 m/h). Water samples had been taken in filter depths of 2.1, 4.2, 6.4, and 8.4 m.

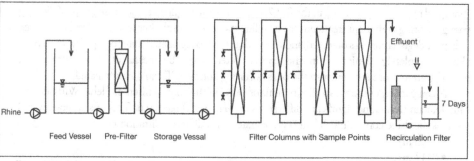

Figure 10-8. Schematic drawing of a control filter and the recirculation filter.

Control filters had been operated at the waterworks in Mündelheim, Flehe, and Wittlaer at the Rhine River in Germany, respectively. These waterworks are located between the Cities of Duisburg and Düsseldorf, over a distance of about only 30 km. To achieve an almost complete degradation of organic compounds, samples of the effluent of the control filter had been additionally treated in a recirculation filter (see Figure 10-8). The column of the recirculation filter was filled with granular pumice, and the residence time of water in the recirculation filter was 7 days.

Before starting the simulation of the influence of shock loads on RBF with the control filters, the normal undisturbed case had to be studied. This was done by measuring the DOC concentration. After the control filters were operated for about 2 months, a nearly constant removal of DOC was observed.

A comparison between elimination in a control filter and by RBF is given in Figure 10-9, showing the arithmetic means of DOC in the period between March 1990 and March 1991. The two bars at the left side represent DOC in the Rhine River at Mündelheim and in an observation well about 10 m away from the riverbank. The bars at the right side represent the DOC at several sampling points of the control filter in Mündelheim. The effluent collected at 8.4 m from RBF and the control filter treatment had the same concentrations.

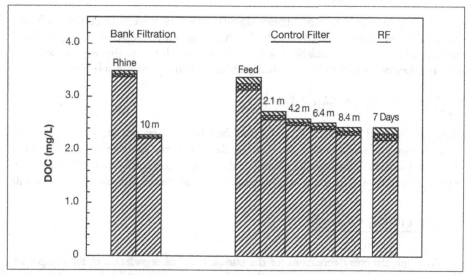

Figure 10-9. Elimination of DOC by RBF and by control filter treatment at Mündelheim (arithmetic means and confidence limits for 90-percent probability). RF = Recirculation filter.

Biodegradation mainly occurred in the first 2.1 m of the control filter and decreased in the following passage. During additional treatment of the effluent of the control filter in the recirculation filter, only poor degradation was observed, indicating that the biodegradable water compounds had been almost completely eliminated in the control filter. Thus, by means of the control filter, biodegradation processes similar to the ones occurring during RBF could be simulated.

A comparison of the mean eliminations of DOC in the control filters at Flehe, Wittlaer, and Mündelheim showed no significant differences between the three filters. This was an unexpected result as the three control filters had been operated with different filtration rates:

- 0.06 m/h.
- 0.12 m/h.
- 0.24 m/h.

Obviously, the elimination was not dependent on the residence time in the control filters. Furthermore, the assumption of only one degradable substance, according to Equation 1, proved to be inappropriate because it was not possible to determine the first-quarter decay coefficient (λ) for the measured degradation of DOC; therefore, one well-degradable, one poorly-degradable, and one non-degradable fraction have been assumed to describe the biodegradation of DOC in the control filters (Mälzer et al., 1992). The first-order decay coefficients of the degradable DOC fractions have been found to be dependent on the rate of filtration. Further investigations showed that the reason for this behavior must be biological activity or the amount of active biomass in the control filters, which is dependent on the filtration rate or, more exactly, on the feed of degradable DOC. Furthermore, there must exist a steady state for the biomass distribution in the filter columns because neither blocking of the columns caused by bacteria nor an exceeding elimination with rising operating time was observed. This led to the assumption that degradation must be dependent on local biomass in the filter (Gimbel et al., 1992). In this case, the transport and biodegradation of one constituent in the control filters can be described by the following equation:

$$R\frac{\partial c}{\partial t} + u\frac{\partial c}{\partial x} - u\alpha\frac{\partial^2 c}{\partial x^2} + Rkbc = 0 \tag{9}$$

where b is the active biomass concentration (kg/m^3) and k is the biodegradation-rate coefficient (cubic meters per kilogram per second [$m^3/kg/s$]). Assuming again two different degradable DOC fractions and one non-degradable fraction, the relation between the biomass and the concentrations of the well and the poorly degradable fractions c_1 and c_2 can be described as:

$$\frac{\partial b}{\partial t} = Y\varepsilon b(R_1 k_1 c_1 + R_2 k_2 c_2) - k_D b^n \tag{10}$$

where Y is the yield coefficient; R_1 and R_2 are the retardation factors; and k_1 and k_2 are the biodegradation-rate coefficients of the two degradable fractions. Furthermore, a decay of the biomass was assumed in Equation 10, where k_D is the biomass decay coefficient ($m^{3(n-1)}/kg^{(n-1)}/s$). A steady state ($\partial b/\partial t = 0$) for the biomass will result for n=2. In this case, the solution of Equation 10 is:

$$b = \frac{Y\varepsilon(R_1 k_1 c_1 + R_2 k_2 c_2)}{k_D} \tag{11}$$

Assuming stationary conditions for the transport and biodegradation of the compounds ($\partial c/\partial t = 0$) and neglecting dispersion ($\alpha=0$), it follows from Equation 9 for the sum of the three fractions (c_1 is well degradable; c_2 is poor degradable; c_3 is non degradable):

$$u \frac{\partial(c_1 + c_2 + c_3)}{\partial x} + b\,(R_1\,k_1\,c_1 + R_2\,k_2\,c_2) = 0 \tag{12}$$

By inserting Equation 11 in Equation 12, a second-order kinetics for the biodegradation of dissolved and adsorbed water constituents is received:

$$v_F \frac{d\,(c_1 + c_2 + c_3)}{dx} + P_F\,(k_{E1}\,c_1\,k_{E2}\,c_2)^2 = 0 \tag{13}$$

where $P_F = Y/k_D$ will subsequently be called population factor (kg × s/m); $k_{E1} = R_1 \times \varepsilon \times k_1$ and $k_{E2} = R_2 \times \varepsilon \times k_2$ are the effective biodegradation rate coefficients.

The concentrations of the three fractions have to be determined by fitting the parameters of Equation 13 to the measured concentrations in the control filters. As the concentrations and their effective biodegradation rate coefficients could not be determined independently from each other, the values of the effective biodegradation-rate coefficients had to be chosen arbitrarily as $k_{E1} = 13.9$ m³/s/kg (= 50 m³/h/g) and $k_{E2} = 1.39$ m³/s/kg (= 5 m³/h/g). It was found that the ratio between population factor and filtration rate was nearly constant for all control filters with a mean value of $P_F/v_F = 8.3 \times 10^{-10}$ kg × s²/m⁴ (=3 × 10⁻³ g × h²/m⁴). Furthermore, the feed DOC-concentrations of the differently degradable fractions have been found to be nearly constant for all control filters:

- 0.3×10^{-3} kg/m³ well degradable.
- 1.0×10^{-3} kg/m³ poorly degradable.
- 1.8×10^{-3} kg/m³ non-degradable.

Using Equation 11, the biomass concentration in the control filters as a function of the filter length was calculated and is shown in Figure 10-10. The active biomass concentration decreases rapidly in the first few meters. For higher filtration rates, a higher amount of biomass will be found

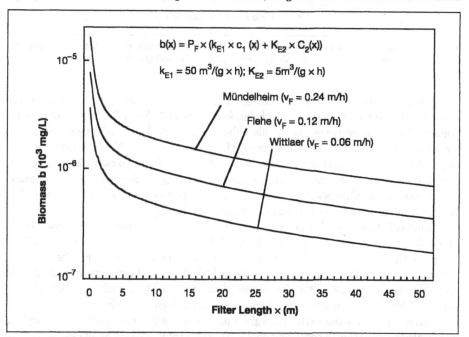

Figure 10-10. Calculated active biomass concentration in the control filters as a function of filter length.

in the filters. An extrapolation to the range of RBF (more than 10 m) shows only a low and slightly decreasing amount of biomass at higher filter lengths.

The biodegradation model explained above was used to describe the transport and degradation of pollutants. To simulate shock loads in the river, pollutants were added to the influent of the control filters. For example, Figure 10-11 shows the breakthrough curve of benzene in the effluent of the control filter in Wittlaer.

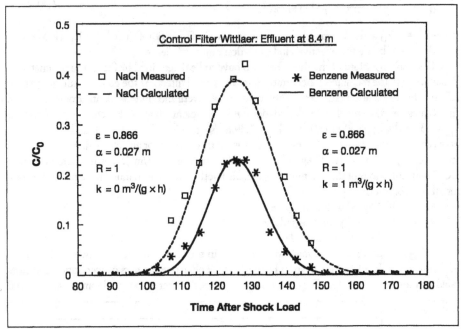

Figure 10-11. Shock-load simulation at the control filter in Wittlaer using benzene as the pollutant and determined parameters for porosity, dispersivity, retardation factor, and biodegradation rate (time in hours).

A shock load of 1.1×10^{-3} kg/m³ benzene and 0.5 kg/m³ sodium chloride (NaCl) as a tracer were added to the influent of the filter columns for a period of 10 hours (h). Because of the high porosity of the granular pumice in the filter, the maximum of the breakthrough curve finally arrived after about 125 h at the end of the total filter length of 8.4 m. A retardation of benzene caused by adsorption could not be observed. The values of the porosity, dispersivity, and biodegradation-rate coefficient had been determined by fitting the parameters to the measured concentrations, using the NaCl tracer for the determination of porosity and dispersivity. The calculated curves describe the measured values quite exactly.

Some pollutants caused a decrease of DOC elimination in comparison to the normal undisturbed case. An example is shown in Figure 10-12 for 2,4,6-trichlorophenole (2,4,6-TCP) as the pollutant, which was added to the influent of the control filter in Mündelheim with a concentration of 9×10^{-3} kg/m³ and for a period of 3.5 h; however, this decrease turned out to be reversible because approximately the normal elimination of DOC could be observed after 2,4,6-TCP had completely passed the filter. Similar effects have been observed using benzene, anthracene (initial concentration 6×10^{-4} kg/m³), naphthene (initial concentration 8×10^{-4} kg/m³), and dichloropropene (initial concentration 2×10^{-3} kg/m³) as shock-load pollutants.

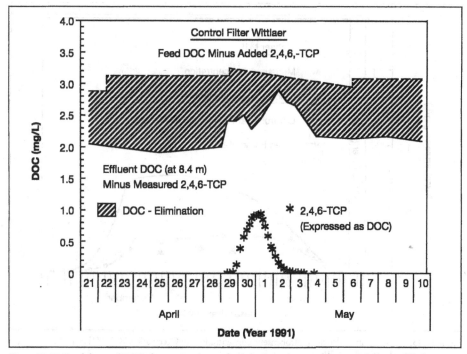

Figure 10-12. Breakdown of DOC elimination during shock-load simulation at the control filter in Wittlaer using 2,4,6-trichlorophenole (2,4,6-TCP) as the pollutant.

Application of Control Filter Results to Predict Pollutant Behavior During RBF

The application of the results of the control filters to the Wittlaer RBF site was demonstrated at the water catchment site of the City's Department of Works at Duisburg (Mälzer, 1993; Mälzer et al., 1993). A vertical section of the aquifer between the Rhine River and Observation Well M1 in Wittlaer was already shown in Figure 10-3. Between November 7 to November 8, 1989, nitrobenzene with a maximum concentration of 1.9×10^{-5} kg/m^3 was detected in the Rhine River at Wittlaer. The effect of this shock load on RBF was calculated using a simplified, one-dimensional transient transport model. As the simulation of this shock-load event by control filter tests was not possible in advance, the values for the biodegradation-rate coefficient and the retardation factor determined by the control filter experiment (using benzene as the pollutant) have been taken as a first approximation. The biomass concentration was determined using Equation 11, with a mean value of the RBF rate and with mean concentrations of the different degradable fractions, as determined by control filter investigations. Transient flow conditions have been considered. Figure 10-13 compares the measured and calculated concentrations at Observation Well M11. The maximum measured concentration in the well was only about 2 percent of the maximum concentration in the river. The nitrobenzene concentrations measured at M11 could be only approximately described by the calculations using the retardation factor and the biodegradation-rate coefficient for benzene, which have been determined by control-filter simulations. A better description will be possible if a higher value for the biodegradation-rate coefficient is assumed. In Figure 10-13, the calculated nitrobenzene concentration using a biodegradation rate coefficient of k=5 m^3/(g × h) is shown. This was determined by fitting the biodegradation-rate coefficient to the measured concentrations.

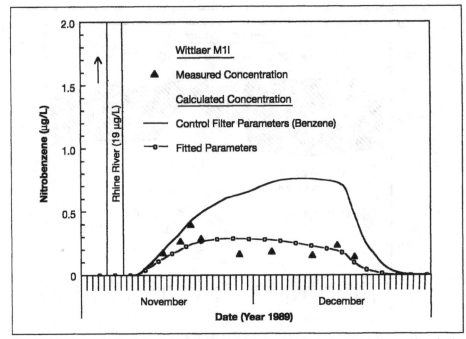

Figure 10-13. Measured and calculated nitrobenzene concentrations in Observation Well M11 caused by a shock
load in the Rhine River.

The use of control filters for monitoring river-water quality leads to a procedure shown in
Figure 10-14. In the case of a shock load in the river, data describing subsurface adsorption and
biodegradation can be obtained by the control filters, which makes it possible to predict the
approximate effect of the shock load on RBF. Thus, the affected waterworks will have the
opportunity to take necessary precautions before the shock-load wave arrives at the well.

4. Transient Three-Dimensional Simulation of RBF Systems to Attenuate Shock Loads

Section 3 describes the mechanisms of contaminant attenuation in a RBF system. In reality,
most RBF systems operate under transient conditions. Two types of chemical emergencies can be
expected:
- High concentrations of dissolved synthetic chemicals and nitrate during periods of high
  river flow.
- An sudden release of a chemical during normal flow.

In the first case, runoff from agricultural watersheds is expected to carry large loads of land-applied
chemicals that will eventually enter the river. Also, the hydraulic gradient from the river to the
well is high during flood conditions, since the head in the river is high. Further, the suspension
and subsequent washoff of low permeability sediments from the river/aquifer sediment can
enhance hydraulic conductivity, thus allowing the entry of more water from the river to the wells;
therefore, the hydrodynamics of the river/aquifer system is mostly in a transient state. In the
second case, most releases occur when river flows are nearly constant (most barge traffic is stopped
during peak flow conditions). In navigable rivers, such releases typically occur at loading or
unloading facilities and, at such times, river levels are nearly constant.

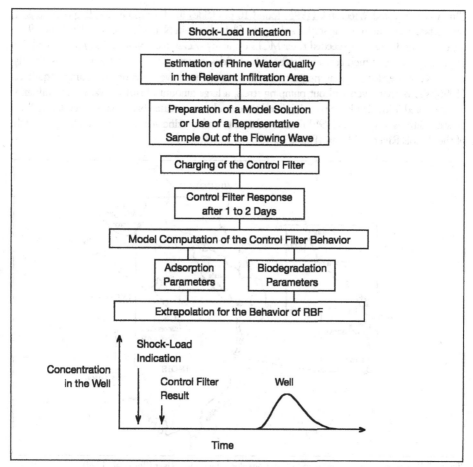

Figure 10-14. Use of control filters in the case of a shock load in the Rhine River.

The purpose of this section is to show that RBF systems have the ability to attenuate contaminant peaks under transient conditions. Also considered are:

- The impact of riverbed hydraulic conductivity variation during flood conditions on the transport of contaminants towards RBF wells.
- The effect of the distance between wells and the river on contaminant peak concentrations in well water.

In addition, simulations were conducted for vertical and horizontal collector wells to estimate peak concentrations in pumped water for various pumping and bed hydraulic conductivity scenarios.

In the midcontinental United States, many rivers traverse through agricultural watersheds. Typically, farmers apply pre-emergent herbicides just prior to the onset of warm temperatures in late spring. In addition, anhydrous ammonia, applied during the fall or early winter, also converts to nitrate with the warming of soil. Soil moisture is close to saturation during this period. Large flushing storms in late spring often contribute the the loss of significant amounts of nitrate and spring-applied pesticides. Coupe and Johnson (1991) and Demissie et al. (1996) have reported the matching of peak contaminant concentrations with peak flows in various rivers and streams in

Illinois. Stamer and Wieczorek (1996) found 10 pesticides in 10 or more sampling occasions at two surface-water monitoring stations along the Platte River in Nebraska between 1992 and 1994. Cyanazine and atrazine exceeded their MCLs on many occasions. Stamer (1996) also raised the implications of such high concentrations of herbicides in surface water for water utilities directly using river water for drinking purposes or for utilities that have wells on riverbanks. Squillace (1996) shows that, even without pumping stress, a large amount of surface-water contaminants can be stored in the bank areas during flood times and then released back to the river during low-flow conditions. Ray et al. (1998) monitored nitrate and atrazine at four locations along a stretch of the Illinois River in Illinois (Figure 10-15).

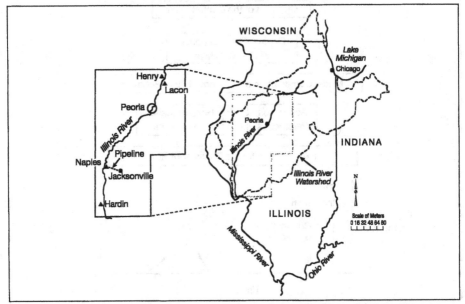

Figure 10-15. Locations of the four sampling locations along the Illinois River (Ray et al., 1998).

At Naples, Illinois (the source-water site for the horizontal collector well for the City of Jacksonville), Ray et al. found that atrazine in river water reached a concentration peak of 11.2 μg/L, whereas the concentration in the well was about 1.3 μg/L, indicating about 1-log attenuation (Figure 10-16). Typical flooding events in the river lasted between 4 to 6 weeks. For rivers traversing agricultural watersheds, such concentrations are not unusual. For the Platte River in Nebraska, Verstraeten et al. (1999) reported peak concentrations of atrazine to be 13 and 26 μg/L during the flood seasons of 1995 and 1996, respectively. Concentrations of DOC and several other contaminants (such as volatile organic compounds) can be high during flood seasons. On the contrary, the concentrations of TDSs and other dissolved inorganics can be low in flood seasons; however, exceptions can be found for rivers that are used for the disposal of treated wastewater and that are located in less intensively cultivated watersheds. In such river systems, nitrate loads could be high during low-flow periods. These simulations are primarily focused on flood-induced enhancement in chemical loads.

Chemical releases are common in navigable rivers. Such releases occur during the loading or unloading activities of fuel or other organic chemicals, the rupture of tanks (that do not have enough containment) located on riverbanks, and from minor releases of transported chemicals or

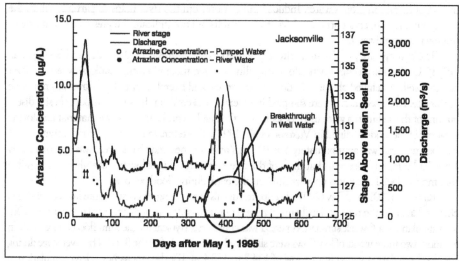

Figure 10-16. Concentration of atrazine in river water and pumped water for the collector well of Jacksonville, Illinois, along with the stage and discharge of the Illinois River (Ray et al., 2001). The maximum contaminant limit for atrazine is 3 μg/L. The arrows indicate that the actual concentrations of the two samples were higher than those reported. The detection limit was 0.1 μg/L.

fuel from barges or tug boats. The release quantities are expected to vary widely. What is important after a release is the mixing process in the flowing river and the overall distribution of the contaminants in the water column; however, in emergency situations, many utilities and government agencies lack the manpower for such precise monitoring efforts. Most monitoring data are taken from selected depths or locations of the river cross-section. The Ohio River Valley Water Sanitation Commission in Cincinnati, Ohio, keeps information on chemical releases along the Ohio River. For example, the concentration of ethylene dibromide from a release in the Ohio River in 1996 ranged from 7,200 μg/L at River Mile 188 (300 km) from the origin to 0.95 μg/L at River Mile 660.6 (1,057 km). As the contaminants distribute in the flowing water stream, a portion of the contaminants are sorbed to bottom sediments, a portion is lost to the atmosphere (if the contaminant is volatile), and a portion could be degraded by microbial action. Without the benefit of a complex river transport model and its validation and testing, we have resorted to simulations using chemical hydrographs measured near the water intakes of specific water utilities.

A transient simulation of the effect of flooding on the transport of atrazine and nitrate from river water to a vertical well and a simulated horizontal collector wells was conducted at the town of Henry, located on the west bank of the Illinois River in Illinois. The town has two vertical wells (Well 3 and Well 4) on the bank of the Illinois River. Each well has a pumping capacity of 0.0312 m³/s, and the depths of these wells range between 19 and 23 m (each with a 4.2-m screen). The distance between Well 3 and the normal pool level of the Illinois River is close to 50 m. Well 4 is 55 m from the river. A series of lock and dams control the river flow between Chicago, Illinois, and its confluence with the Mississippi River near Grafton, Illinois. The town also has another backup well (Well 5) approximately 2-km upgradient from the river. Well 5 has a higher background nitrate than Wells 3 and 4. The town prefers to operate either Well 3 or 4 and the system pressure activates pump operation in these wells. The background nitrate level in groundwater in the area often exceeds 10 milligrams as nitrogen per liter (mg-N/L) (which is the health standard in the United States). Normally, river nitrate levels are lower than the concentration of nitrate in groundwater. Induced infiltration from the river helps to partially dilute the background nitrate concentration. Monitoring results did not indicate the presence of atrazine in the town's groundwater.

The hydrogeology of the study site is described in detail in Ray et al. (2001) and Soong et al. (1998). Unlike most other towns along the Illinois River, there is a sandy bank between the river and the well. A detailed characterization of the vertical and lateral extent of this sand was not done at this site. River sediments were sampled in several transects from the site. The particle size distributions of the sediments were used for estimating hydraulic conductivity of the material employing a variety of methods outlined in Vukovic et al. (1992). The estimated hydraulic conductivity of the riverbed material was approximately 5×10^{-8} m/s and, for the sandy bank area, was approximately 1.3×10^{-3} m/s. It is apparent that most of the riverbed material is of low permeability; it is possible that most induced infiltration occurs through the bank during flood periods.

Ray et al. (2001) and Ray (2001) describe the details of modeling efforts at this site for flood-induced chemical loads in the river and a chemical release during low-flow periods, respectively. For the flood-load simulation, a 6-week simulation period was considered (typical for medium flooding events). In addition, two more weeks of low flows were simulated prior to and after the flood. The river stage during the flood simulation varied from a low of 134.7 to 138.4 m. For the emergency release simulation, a constant river stage of 134.7 m was considered. Water flow simulation was conducted by using the U.S. Geological Survey model, MODFLOW (Harbaugh and McDonald, 1996). The model accounted for transient saturated flow and used the "river package" to account for the exchange of water between the river and aquifer. First, a model was created for existing vertical wells and calibrated (Ray et al., 2001). Later, simulations were carried out for a hypothetical medium-capacity horizontal collector well at the same site. There is no horizontal collector well at Henry and at the four sites studied by Ray et al. (1998). Jacksonville has a horizontal collector well at Naples, Illinois (see Figure 10-15). No modeling of the performance of this collector well has been made due to lack of data and site access. In essence, the hypothetical collector well at Henry was assumed to represent conditions elsewhere. The transport of atrazine and nitrate was simulated using the solute transport model, MT3D (Zheng, 1992). The model solves the advection-dispersion equation along with sorption and degradation reactions to estimate liquid-phase concentrations in the model domain. Once the flow equation is solved using MODFLOW, MT3D uses the same grid for solute transport.

A schematic of the study location is presented in Figure 10-17. River stage and concentrations of nitrate and atrazine varied depending upon flood conditions and are presented in Table 10-1.

The rate of groundwater recharge was assumed to vary between 0.013 and 0.635 m/yr for each of the 2-week simulation periods. The river stage was considered to inundate a portion of the land between the normal pool level of the river and the well. Appropriate hydraulic conductivity, based upon measurements, were assigned to areas that are above the normal pool level of the river. The background concentration of nitrate is normally higher than river nitrate, except during peak-flow periods in the spring. River water is expected to dilute the concentration of nitrate in pumped water.

A controlled aquifer test was conducted onsite to estimate the transmissivity of the aquifer. Geologic logs and grain-size distribution data were used to calculate the hydraulic conductivity of aquifer material. In essence, three distinct zones were found with differing hydraulic conductivity: the top 20 m belonged to Henry sand; the upper 15 m had a hydraulic conductivity of 122 m/d; and the bottom 5 m had a hydraulic conductivity of 5 m/d. The bottom 10 m of the aquifer was Sankoty sand with a hydraulic conductivity of approximately 213 m/d. A relatively low permeability Pennsylvanian Shale was present at the bottom of the Sankoty sand and formed the bottom boundary of the aquifer.

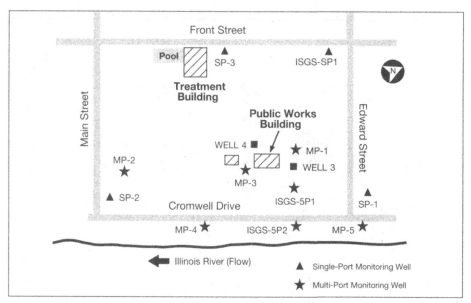

Figure 10-17. Schematic of the site layout at the Henry well site (Ray, 2001). ISGS stands for Illinois State Geological Survey.

Table 10-1. River Stage and Concentrations of Nitrate and Atrazine in Illinois River Water and at the Specified Head Boundary Used in Simulations

Days	River Stage (m above MSL)	River Water Nitrate (mg/L)	River Water Atrazine (μg/L)	Nitrate at the Western Boundary (mg/L)	Head (m above MSL) at the Western Boundary
0 to 14	134.7	4	0.2	9.5	135.1
14 to 28	136.9	8	3	9.5	136.0
28 to 42	138.4	10	4	9.0	136.9
42 to 56	136.9	8	3	9.0	137.8
56 to 70	134.7	3	1	9.5	135.1

MSL = Mean sea level.

The model area covered 424 m along the flow direction of the river and extended 394-m westward. The model area that is on the other side of the center line of the river was not considered to contribute to flow (i.e., the center line of the river was treated as the no-flow boundary). The horizontal discretization included 130 equally spaced rows and 140 equally spaced columns. The model was divided into seven layers for vertical well simulation and eight layers for horizontal collector well simulation. The sixth layer was spilt into two layers, and the top layer was used for representing the laterals of the collector well. A cross-section of the model area passing through Well 3 is presented in Figure 10-18. The collector well simulated at the site has five laterals with lengths ranging from 77 to 86 m, all directed towards the river. Figure 10-18 provides an orientation of these laterals. Each lateral was approximately 0.6- to 0.9-m thick and 3-m wide, and some extended more than 30-m underneath the riverbed. The hydraulic conductivity of the laterals were assigned large values to account for the slots on the pipes and to minimize headloss. A collector caisson normally connects to all laterals. For our simulation, a

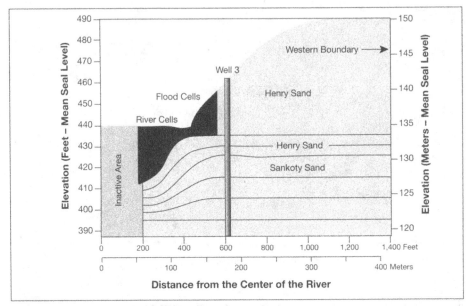

Figure 10-18. Cross-section of the aquifer through the pumping well.

vertical well was placed with screen openings in the new sixth layers, where all laterals converge. In essence, pumping stress from Layer 6 allowed free passage of water from the laterals to the well. Average pumping from the collector well was 0.0875 m³/s.

For low-flow and selected high-flow periods, the flow model was calibrated against measured water levels. The bed and bank hydraulic conductivity were allowed to vary during flood simulations. Ray et al. (2001) conducted several sets of simulations for the vertical and horizontal collector wells for atrazine and nitrate transport by varying the hydraulic conductivity of the riverbed and bank. It is expected that the bed and bank material can be scoured or silted, depending upon the flow regime of the river. For flood simulation, only two cases are presented here: one with a highly conductive bank and the other with a less conductive bank.

The accidental release data for ethylene dibromide, presented in Figure 10-19, was used as the source concentration in the river. This hydrograph was measured after nearly 400 km of travel of the contaminant plume. The chemical pulse, as shown in this figure, lasts about 10 days when the river level is at the normal navigation pool. As observed in Figure 10-19, the concentration hydrograph is somewhat asymmetric, with a peak appearing quickly and a tail end lasting for more than a week. For the emergency release simulation, the hydraulic conductivity of riverbed material (the bank is exposed during low flow time; hence, no flow) was assigned a value of 5×10^{-5} cm/s or 5×10^{-4} cm/s or 5×10^{-3} cm/s. While sorption and degradation are expected as this water travels to the well, for worst-case conditions, sorption and degradation reactions were ignored.

The simulation results indicated that for a highly conductive bank, the concentration of atrazine at the collector well could almost reach near peak value (4 µg/L) in the water; however, we do not have accurate information about the hydraulic conductivity of bank material beyond a depth of 1.0 m. The simulation conducted for a moderately conductive bank (5×10^{-7} to 5×10^{-6} m/s) and less conductive riverbed (5×10^{-8} m/s) showed that the peak concentration of atrazine would be about 0.035 µg/L without considering sorption and decay.

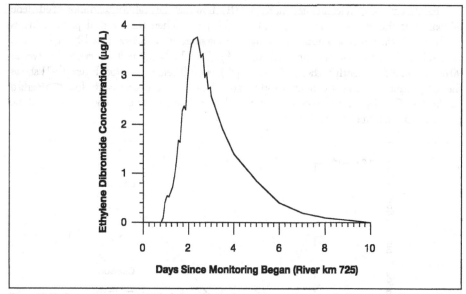

Figure 10-19. Source concentration of ethylene dibromide used for simulation at the horizontal collector well (Ray, 2001).

If equilibrium sorption and literature-reported degradation values are used in the simulation, the expected peak concentration would be lower than 0.01 µg/L (Figure 10-20). The lag time for the contaminant peak to reach the well in a highly permeable bank was between 1 and 2 weeks. From this figure, it is evident that lag times for contaminant peaks increase with lower sediment/ bank material conductivity. It is expected that local conditions, well placement, and other hydrologic, geologic, and biochemical factors will affect the peak concentrations of chemicals in river water.

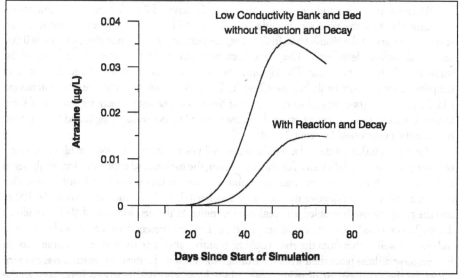

Figure 10-20. Atrazine concentration in a horizontal collector well for a less conductive bank (Ray, 2001).

For the emergency release situation, Ray (2001) show that the peak concentration of ethylene dibromide at the well site (Figure 10-19) was 3.8 µg/L. When the riverbed permeability is 5×10^{-5} cm/s, the peak concentration of ethylene dibromide at the caisson (5×10^{-5} µg/L) would be significantly below detection limit (about 10^{-2} µg/L). The lateral that intersects the river at 90 degrees would be slightly higher (1.2×10^{-4} µg/L), but still below detection. Figure 10-21 shows the concentration peaks (without sorption and decay) for the case with lowest riverbed conductivity. The lag time between the surface and groundwater peaks is around 4 weeks, and the peak attenuation factor is more than 76,000.

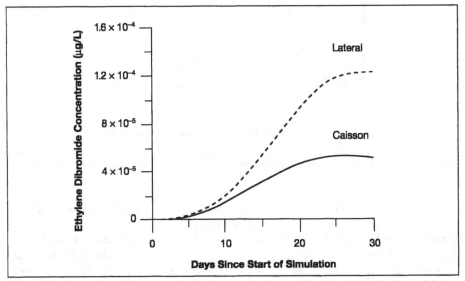

Figure 10-21. Ethylene dibromide concentration in a horizontal collector well for less conductive bed material during low flow (Ray, 2001).

Riverbed hydraulic conductivity was increased 100 times (5.0×10^{-3} cm/s) and simulations were carried out for the same set of conditions. Figure 10-22 shows the expected concentrations at the caisson and at the river lateral. In this case, the peak concentration in the caisson was 0.007 µg/L (still below detection). The peak concentration in the shortest lateral would be approximately twice that value. The lag time between the concentration peaks in the river and the pumped water would be slightly over 2 weeks. Ray (2001) showed that if five other laterals are added to this well (two parallel to the river and three in a landward direction) that are of same length, the expected concentration in the caisson would be below 0.001 µg/L and the lag time between the peaks would be about 3 weeks.

Ray et al. (2001) conducted a simulation of well placement as a function of nitrate content of water pumped from RBF wells. For this simulation, the background concentration was lowered to 3.0 mg/L, typical for many riparian communities along the Illinois River. A highly conductive riverbank and a low conductive riverbed were simulated. If Well 3 is moved upstream by 100 m and the pumping rate is doubled, the peak concentration at the new location of Well 3 would be 3.5 mg/L, compared to a concentration of 7.8 mg/L at the present location. It is obvious that moving the well farther from the river could help further attenuate river-water contaminants. In Europe, most utilities place their wells some distance away from the river. Increased travel distance allows for the development of redox zones, where biogeochemical reactions contribute to the

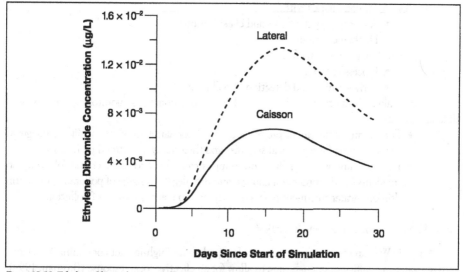

Figure 10-22. Ethylene dibromide concentration in a horizontal collector well for highly conductive bed material during low flow (Ray, 2001).

degradation of certain contaminants. Ray et al. (2001) also presented some scenarios where small-capacity pumping wells could be placed between the river and production wells to create a hydraulic barrier during flood times; however, for the present scenario involving a highly conductive bank, such wells do not help in creating a hydraulic divide. For low conductive banks, such mechanisms may be feasible.

From the results of simulations and field observations, RBF wells have the potential to attenuate the magnitude of concentration peaks and delay their arrival in the pumped wells. In ideal conditions, sorption and degradation reactions could further attenuate the contaminants. Water utilities that employ RBF could expect a higher quality of water prior to treatment compared to direct intake systems. RBF systems may have some advantage over direct intake systems in terms of mitigating intentional chemical releases into rivers or lakes (a form of terrorism). An adequate response time and significant peak attenuation would benefit utilities in mitigating such emergencies. To evaluate the effectiveness of RBF systems, water utilities must have monitoring systems in place, calibrated models for the site, a test filter setup (see Section 3), and an early warning system to warn utilities about potential releases.

5. Early Warning System in the Rhine River Basin — Emergency Management

Water-Quality Monitoring

The main task of monitoring systems is to provide the right information to make management decisions. Information needs depend upon the users, and the required information can change in time; therefore, dynamic elements are a part of monitoring strategies. Research, political decisions, events, or accidents may change monitoring activities.

Different objectives of monitoring water quality can be distinguished:

Strategic monitoring for identifying:

- Status and trends (concentrations).
- Mass flow (loads).

Testing for compliance with:
  - Water-quality standards and classification.
  - Discharge permits.

Operational monitoring for:
  - Process control.
  - Early warning and detection of pollution.

Water-quality monitoring in the Rhine River basin combines monitoring systems with different objectives:

  - Long-term measurement programs to give information on water quality to safeguard natural waters (as ecological systems) and their many uses (trend monitoring).
  - Early warning water-quality monitoring is carried out to discover and follow-up on incidents of environmental damage and the illegal discharge of pollutants, to inform drinking-water treatment plants quickly, and to identify sources of pollution.

Early Warning System in the Rhine River Basin

An Early Warning System has to identify low-probability/high-impact contamination events in river water in a sufficient enough time to allow for an effective local response (Brosnan, 1999). The system should cover all potential threats and work year-round.

The following information will focus on the structure and components of the Early Warning System in the Rhine River Basin from the point of view of waterworks using RBF.

After the Sandoz Accident in November 1986, the riparian states of the Rhine River initiated the "Rhine Action Program." As a result, an improved International Quality Monitoring Program and the Rhine Warning and Alarm Service were installed. The waterworks developed additional tools for monitoring and assessing the effects of shock loads following accidents on RBF and water supply.

a. Information Needs

The design of such a system has to regard the general information needs of the users (authorities, waterworks, others). In the event of contamination, the main questions (from a waterworks' perspective) include:

  I   Time and site of the release.
      Quantity, type, and characteristics of the pollutants.
      Effects on the ecosystem of the river.
  II Propagation of the pollution along the river.
      Time of arrival and concentration of a pollutant wave at any observation point.
  III Can the pollutant survive the subsoil region during RBF?
      Will drinking-water quality be affected?

Monitoring stations along the river and an efficient reporting system will provide the data for Numbers I and II. Additional waterworks tools (e.g., test filters, RBF flow and transport models, monitoring wells for depth-oriented sampling) are needed to answer the questions in Number III.

b. Information Strategy

To meet the information needs of users, a tailor-made monitoring concept must be established and a sampling network must be prepared (Lowis and von Danwitz, 1997); however, before proceeding, several questions must be answered:

- Which parameters should be measured at how many sampling sites, and how reliable should this data be?
- Which criteria are relevant for a warning or alarm message?
- How much of a lag time should be accepted between sampling and warning?
- How can sources of pollution be detected in a short time frame?

As a result of answering such questions, a basic monitoring concept was introduced in the Rhine region, which consists of the following:

- Dynamic biological tests (continuous-flow fish test, dynamic Daphnia assay, Dreissena-monitor).
- Physio-chemical parameters, online (pH value, electrical conductivity, temperature, oxygen content) and daily laboratory analyses (e.g., heavy metals, nutrients).
- Organic micropollutants: adsorber units (such as XAD) for chromatographic screening analysis, online; and daily employment of GC and HPLC analyses for screening purposes to find any deviations from the "normal status."
- Daily water samples that are not analyzed periodically, but stored over about 10 days to refer to at a later date in case of emergency.

The network of monitoring stations has an international backbone with stations on each national border and, in Germany, on each federal state border where the Rhine River or its main tributaries enter or leave a region. The basic network includes about 25 monitoring stations in the Rhine River area. Moreover, there are additional sampling stations under operation of the water and environmental authorities in the federal states at points of interest (e.g., downstream of industrial plants) along the river and the main tributaries.

This network is supplemented by more than 20 monitoring stations from waterworks that have monitoring programs coordinated by the International Association of the Waterworks in the Rhine Region. The objective of the monitoring program is the early warning and detection of pollution, but an important point is to receive timely information about "new" trace substances that can concern the quality of riverbank-filtered water or even of drinking water (e.g., pharmaceuticals, antibiotics, phenols, methyl tertiary butyl ether [MTBE]).

c. Data Collection

Only a few parameters can be monitored continuously in real time. Many pollutants cannot even be detected by biological tests. The only way to discover and follow-up pollutants of interest (e.g., non-biodegradable contaminants) during shock loads is through a laboratory analysis of samples; therefore, decisions must be made in advance regarding the strategy of data collection (composite versus discrete samples) and the frequency of monitoring.

An early warning system has to provide warning in a sufficient enough time for action. One of the advantages of RBF is that it provides time for an emergency response. The mean travel time between the river and the wells in water-supply systems should never be less than 2 weeks because of issues like microorganism removal, biodegradation, and the equalization of fluctuating concentrations. On the contrary, the travel time in the river between the site of pollution and the infiltration areas of waterworks can be very short — at least for waterworks just downstream of the pollution site.

The rule for data collection is:

- Continuous (whenever possible) online analysis.
- Periodic (daily) analyses of composite samples (laboratory).
- Frequent analyses of discrete samples (laboratory).

d. Data Analysis and Reporting

Interpreting the collected data for early warning may be particularly problematic. It requires a clearly established baseline for contaminants or events and a clearly established deviation from the baseline. Baseline development will be contaminant-specific and influenced by the following:

- Characteristics of the compound (e.g., toxicity, degradability, volatility, adsorption).
- Ability to detect the compound.
- Perceived risk associated with exposure to the compound.

There are six alert stations along the Rhine River and two at the Mosel River. In the case of an accidental release, one of these stations collects data from direct information and affected monitoring stations about site, time, type, and the amount of pollutants. Furthermore, data about the effects on the ecosystem, sudden measures, and actual data of the flow regime are added. When the contaminants exceed threshold levels, an alarm signal will be triggered. The affected alert station transmits information or a warning to the other alert stations, environmental authorities, and waterworks along the Rhine River by facsimile.

e. Information Utilization

Waterworks receive the alert report and must assess the possible effects on their water supplies. Generally, they must answer the following questions about the behavior of the pollutants:

- Time of arrival of the pollutant wave at the site.
- Concentration of the contaminants in the river water at the site.
- Biodegradability of the compounds during RBF.
- Time of arrival and concentration of contaminants at the wells.
- Behavior of the contaminants in the treatment plant.

Several tools have been developed and tested under real conditions to answer the above-mentioned questions:

Rhine Alarm Model to calculate the fate and transport of pollutants along the Rhine River.

A key component is the availability of a method for calculating the fate and transport of contaminants along the river and its main tributaries. For this reason, the Rhine Alarm Model was developed to predict the course of a pollutant wave. The model has been tested by means of tracer experiments and has been improved upon. It is available as an operator friendly PC-Program.

RBF flow and mass transport model to calculate the fate and transport of pollutants during RBF. An important objective within the framework of a former waterworks research project was the development of groundwater models to simulate flow and transport in RBF, particularly during shock loads. As the models are rather complex (three-dimensional, dynamic), to keep from activating calculations each time an accident occurs on the Rhine River, a series of 1- to 3-day shock loads was simulated, taking as an example a relatively extreme case. The simulated cases were based not only on unfavorable hydraulic conditions (flood wave), but also on the assumption that substance behavior would be persistent. The simulation results include a wide range of possible cases, which can be supplemented and extended as required (Schubert and Gotthardt, 1994).

Test filters to simulate biodegradability during subsoil passage in quick motion. To allow the waterworks to take measures that may limit the effects of accident-related pollution on raw water, it is important to be familiar with the degradation behavior of the substances during passage through the soil. The substance data sheets do not, as a rule, provide adequate information in this regard; therefore, test filters, with which degradation processes during passage through the soil can be simulated in quick motion, were also tested in the course of the waterworks' research project. Since then, test filters have always been held in readiness at the larger Rhine waterworks and are an important component of the monitoring system.

Additional testing facilities to test behavior in the treatment plant. In cases where the equalization of fluctuating concentration and degradation processes during passage through the soil cannot prevent the quality of raw water from the wells near the Rhine River from being impaired, the behavior of the substances during water treatment has to be evaluated. Here, conventional laboratory tests are performed to determine the adsorption of the compounds and their reaction under oxidizing conditions.

Monitoring wells between the river and production wells to check-up the tool results. The subsoil passage of the infiltrated water from the river to the wells takes several weeks. Consequently, this is also the time frame for the monitoring activities mentioned above. In addition, pollutant diffusion is monitored at several multi-level sampling points between the riverbank and the wells to check the validity of the model-based predictions.

f. Emergency Management

Experience has shown that the major portion of substances emanating from shock loads in the Rhine River is removed by the self-purifying capacity during subsoil passage, or their residual concentrations are reduced to a non-critical level. But, as substance behavior during subsoil passage is, generally, not well known and the question of sufficient microbial adaptation in the infiltration areas cannot be predicted in advance, parts of the above mentioned tools (e.g., samples from monitoring wells) have to be used every time. This is a good way to validate predictions, to accumulate experience and, if necessary, to improve tools.

Decision about the operation mode of the waterworks (emergency response). Based on the information of the Warning and Alarm Service, additional information (e.g., from upstream waterworks and personal investigations) create a rough picture about the nature, magnitude, and effects of the contamination event, including:

- If there is no indication that raw water will be concerned, the normal operation mode of the wells can be maintained.
- If there are any doubts regarding contamination of raw water, precautionary measures (e.g., reducing of the depression cone to extend retention time) should be taken.
- If raw-water quality is of concern, then the behavior of breakthrough compounds in the treatment plant must be tested in advance. If those compounds cannot be removed by treatment, then it may be necessary to shut off the wells.

Informing authorities and consumers (communication). It is not only a question of existing regulations, but also of the confidence in the management of waterworks to communicate the choice of emergency response with the relevant authorities and to inform customers.

Activities for further prevention measures. Some of the former accidents in the Rhine region (e.g., the Sandoz Accident of 1986 and others) clearly revealed deficiencies in water-protection measures. To avoid severe water pollution, prevention measures should be discussed with the authorities and responsible parties and implemented, if possible.

Other information needs? It is recommended to keep a journal of accidental releases with all relevant data. From time to time, one should ask if there are any changes in future information needs; if so, please return to (a) Information Needs.

References

Bear, J. (1972). *Dynamics of fluids in porous media.* American Elsevier, New York, New York.

Bear, J. (1979). *Hydraulics of groundwater.* McGraw-Hill, New York, New York.

Briggs, G.G. (1981). "Theoretical and experimental relationship between soil adsorption, octanol-water partition coefficient, water solubilities, bioconcentration factors, and the parachor." *Journal of Agricultural Food Chemistry,* 29: 1050-1059.

Brosnan, T. (1999). *Early warning monitoring to detect hazardous events in water supply.* ILSI Risk Science Institute, Washington, D.C.

Chiou, C.T., P.E. Porter, and D.W. Schmedding (1983). "Partition equilibria of organic compounds between soil organic matter and water." *Environmental Science and Technology,* 17: 227-231.

Coupe, R.H., and G.P. Johnson (1991). "Triazine herbicides in selected streams in Illinois during storm events, Spring 1990." *Water Resources Investigation Report 91-4034,* U.S. Geological Survey, Reston, Virginia.

Demissie, M., L. Keefer, D. Borah, V. Knapp, S. Shaw, K. Nichols, and D. Mayer (1996). "Watershed monitoring and land use evaluation for the Lake Decatur watershed." *Miscellaneous Publication 169,* Illinois State Water Survey, Champaign, Illinois.

Deutsche Kommission zur Reinhaltung des Rheins (DKR) (1986). *Deutscher Bericht zum Sandoz-Unfall mit Meßprogramm (German report on the Sandoz accident including the monitoring program).* German Commission for the Control of Pollution in the Rhine River (DKR), Koblenz, Germany.

Gimbel, R., M. Gerlach, and H.J. Mälzer (1992). *Die Testfiltermethode als Simulationsgrundlage des Störstoffverhaltens bei Untergrundpassagen (The test filter method for simulation of pollutants behavior during subsoil passage).* Erfassung und Bewertung von Xenobiotika aus der Sicht der Trinkwasserversorgung, Berichte aus dem Rheinisch-Westfälisches Institut für Wasserchemie und Wassertechnologie GmbH, Ruhr, Germany, 6: 126-163.

Gimbel, R., and H.J. Mälzer (1987). "Testfilter zur Beurteilung der Trinkwasserrelevanz organischer Inhaltsstoffe von Fließgewässern." *Vom Wasser,* 69: 139-153.

Gölz, E., J. Schubert, and D. Liebich (1991). "Sohlenkolmation und Uferfiltration im Bereich des Wasserwerks Flehe (Düsseldorf)." *Gas- und Wasserfach Wasser/Abwasser,* 132(2): 69-76.

Gotthard, J. (1992). "Dreidimensionales Strömungs- und Transportmodell für das Wasserwerk Flehe." *Grundwassermodelle für die Uferfiltration - Vorträge zum Kolloquium am 29. September 1992 an der Ruhr-Universität Bochum (Riverbank filtration flow and transport models. Papers presented at the colloquium on flow and transport models for riverbank filtration, Ruhr University Bochum, September 29, 1982),* C. König and J. Schubert, eds., Ruhr Universität, Bochum, Germany, 87-106.

Harbaugh, A.W., and M.G. McDonald (1996). "User's documentation for MODFLOW-96: An update to the U.S. Geological Survey Modular Finite-difference Groundwater Flow Model." *Open File Report 96-485,* U.S. Geological Survey, Reston, Virginia.

Karickhoff, S.W., D.S. Brown, and T.S. Scott (1979). "Sorption of hydrophobic pollutants on natural sediments." *Water Research,* 13: 241-248.

Kinzelbach, W. (1987). *Numerische Methoden zur Modellierung des Stofftransports von Schadstoffen im Grundwasser (Numerical methods for contamination transport modeling in groundwater).* Schriftenreihe gwf Wasser/Abwasser Bd. 21, R. Oldenbourg Verlag, München, Germany.

König, C., and J. Schubert. (1992). Grundwassermodelle für die Uferfiltration. Vorträge zum Kolloquium am 29. September 1992 an der Ruhr-Universität Bochum *(Riverbank filtration flow and transport models. Papers presented at the colloquium on flow and transport models for riverbank filtration, Ruhr University Bochum, September 29, 1982),* C. König und J. Schubert, eds., Ruhr-Universität, Bochum, Germany.

Lindner, K., and W. Lindner (1986). "Bemessung der Schutzzone II des Wasserwerks Köln-Weiler unter Berücksichtigung der Längsdispersion." *Wasserwirtschaft,* 76: 384-387.

Lowis, J., and B. Von Danwitz (1997). "Early warning monitoring in North Rhine Westfalia." *Proceedings, Monitoring Tailor-made II: An International Workshop on Information Strategies in Water Management,* ELMA Edities B.V., Noord-Scharwoude, The Netherlands.

Luckner, L., and W.M. Schestakow (1991). *Migration processes in the soil and groundwater zone,* Lewis Publishers, Chelsea, Michigan.

Mälzer, H.J. (1993). "Untersuchungen zu Transport- und Abbauvorgängen bei der Uferfiltration im Hinblick auf die Auswirkungen von Stoßbelastungen." Ph.D. Dissertation, Universität Duisburg, Duisburg, Germany.

Mälzer, H.J., M. Gerlach, and R. Gimbel (1992). "Investigations on transport and degradation processes in riverbank filtration." *Vom Wasser,* 78: 343-353.

Mälzer, H.J., M. Gerlach, and R. Gimbel (1993). "Effects of shock loads on bank filtration and their prediction by control filters." *Water Supply,* 11: 165-176.

Ray, C., T.W. Soong, G.S. Roadcap, and D.K. Borah (1998). "Agricultural chemicals: Effects on wells during floods." *Journal American Water Works Association,* 90: 90-100.

Ray, C., T.W. Soong, Y. Lian, and G.S. Roadcap (2001). "Effect of flood-induced chemical load on filtrate quality at bank filtration sites." *Journal of Hydrology,* accepted.

Roberts, P.V., M. Reinhard, and J. Valocchi (1982). "Movement of organic contaminants in ground water." *Journal American Water Works Association*, 74: 408-413.

Schubert, J., and J. Gotthardt (1994). *Weiterentwicklung eines Modells zur Simulation des Stofftransports bei der Uferfiltration (Development of a model to simulate mass transport in riverbank filtration.)* Arbeitsgemeinschaft der Rhein-Wasserwerke e.v. (ARW), Köln, Germany, 123-144.

Schwarzenbach, R.P., and J. Westall (1981). "Transport of nonpolar organic compounds from surface water to ground water. Laboratory sorption studies." *Environmental Science and Technology*, 15: 1360-1367.

Segol, G. (1994). *Classic Groundwater simulations.* PTR Prentice Hall, Englewood Cliffs, New Jersey.

Soong, T.W., G.S. Roadcap, D.K. Borah, and C. Ray (1998). *Flood induced loading of agricultural chemicals to public water supply wells in selected reaches of the Illinois River.* Final Report, submitted to the Illinois Ground Water Consortium, Southern Illinois University, Carbondale, Illinois.

Sontheimer, H. (1991). *Trinkwasser aus dem Rhein? Bericht über ein Verbundforschungs-vorhaben zur Sicherheit der Trinkwassergewinnung aus Rheinuferfiltrat.* Academia-Verlag, Sankt Augustin, Germany.

Squillace, P. (1996). "Observed and simulated movement of bank-storage water." *Ground Water*, 34(1): 121-134.

Stamer, J.K. (1996). "Water supply implications of herbicide sampling." *Journal American Water Works Association*, 88(2): 76-85.

Stamer, J.K., and Wieczorek, M.E. (1996). "Pesticide distribution in surface water." *Journal American Water Works Association*, 88(11): 79-87.

Van Genuchten, M.T. (1981). "Analytical solutions for chemical transport with simultaneous adsorption, zero order production, and first order decay." *Journal of Hydrology*, 49: 213-233.

Van Genuchten, M.T., and P.J. Wierenga (1976). "Mass transfer studies in sorbing porous media: I. Analytical solutions." *Soil Science Society of America Journal*, 40: 473-480.

Van Genuchten, M.T., and P.J. Wierenga (1977). "Mass transfer studies in sorbing porous media: II. Experimental evaluations with Tritium." *Soil Science Society of America Journal*, 41: 272-278.

Verstraeten, I.M., J.D. Carr, G.V. Steele, E.M. Thurman, K.C. Bastian, and D.F. Dormedy (1999). "Surface water/ground water interaction: Herbicide transport into municipal collector wells." *Journal of Environmental Quality*, 28(5): 396-405.

Vukovic, M., A. Soro, and D. Miladinov (1992). *Determining hydraulic conductivity of porous media from grain-size composition.* Water Resources Publications, Littleton, Colorado.

Zheng, C. (1992). *MT3D Version 1.8 Documentation and User's Guide.* S.S. Papadopulos & Associates, Inc., Bethesda, Maryland.

Chapter 11. Riverbank Filtration as a Pretreatment for Nanofiltration Membranes

Thomas F. Speth, Ph.D., P.E.
United States Environmental Protection Agency
Cincinnati, Ohio, United States

Till Merkel, M.Sc.
DVGW Water Technology Center
Karlsruhe, Germany

Alison M. Gusses, M.S.
University of Cincinnati
Cincinnati, Ohio, United States

1. Introduction

The loss of membrane efficiency due to fouling is one of the main impediments to developing membrane processes for use in drinking-water treatment. The extent of membrane fouling is dependent upon feed water quality and the membrane's properties and construction. While many utilities that have brackish water or extremely high levels of organic matter in their water currently operate nanofiltration or reverse osmosis membranes, most of these utilities use groundwaters that are stable over time, and much is known about their fouling tendencies. Less is known about the membrane treatment of surface waters, where seasonal changes in water quality can cause fouling problems. Surface waters, in general, have a greater proclivity towards fouling as compared to groundwaters.

RBF changes surface water into a water with characteristics close to that of a groundwater; therefore, because groundwaters are more amenable to nanofiltration treatment, RBF should be an effective pretreatment for nanofiltration or reverse osmosis membrane technologies. Although it is well documented that RBF can remove sizeable portions of particulates and organics from river waters (Sontheimer, 1980; Wang et al., 1995), there has been little work comparing RBF to conventional treatment as a pretreatment for nanofiltration membrane systems. Merkel et al. (1998) showed that membranes fed riverbank-filtered Ohio River water at Louisville, Kentucky, had less flux decline than those fed conventionally treated Ohio River water. The goal of this work is to further evaluate RBF versus conventional treatment as a pretreatment for nanofiltration membranes. Two riverbank-filtered waters will be compared to three related surface waters that were conventionally treated.

This work presents nanofiltration membrane data from riverbank-filtered water systems in Louisville (influenced by the Ohio River) and in southwestern Ohio (influenced by the Little Miami River). The Louisville riverbank-filtered well was 20 m from the Ohio River. More information regarding this well can be found in Wang et al. (1995). At the southwestern Ohio site, there were 10 wells at various distances from the Little Miami River. The waters from these 10 wells were combined before they were fed to the nanofiltration membrane system. Based on

C. Ray et al. (eds.), Riverbank Filtration, 261–265.

temperature data, the average induced aquifer travel time from the surface water to the wells was estimated to be 2 months.

2. Methods and Procedures

Three different conventionally treated waters were studied:
- Ohio River water treated at the Louisville Water Company in Kentucky.
- Ohio River water treated at the Cincinnati Water Works in Ohio.
- Harsha Lake water treated at the U.S. Environmental Protection Agency pilot plant in southwestern Ohio.

The Louisville Water Company's facility used flocculation with iron salts followed by sedimentation and dual-media filtration to treat Ohio River water (Merkel et al., 1998). Chloramines (2.5 mg/L) were formed before filtration. The facility at the Cincinnati Water Works used coagulation and biological filtration without a disinfectant. This plant also treated Ohio River water approximately 200 km upstream from Louisville. The two facilities are compared to the Louisville RBF well to demonstrate that specific operational issues, such as chloramination or iron coagulation, were not the reason for the increased flux decline for the conventionally treated water, as shown by Merkel et al. (1998). The U.S. Environmental Protection Agency pilot plant in Cincinnati, Ohio, treated Harsha Lake water, which was coagulated with aluminum salts, settled, and cartridge filtered. Although not directly comparable to Little Miami River water (site of the RBF wells in southwestern Ohio), it is within the same watershed and, therefore, may be similar. In any event, the authors present these data to show general differences in flux decline rates between membranes treating both conventionally treated surface water and riverbank-filtered water.

TOC was measured by U.S. Environmental Protection Agency Method 415.1. Conductivity was determined with a hand-held Oakton probe (WD-35607-20), according to Standard Method 2150B. Finally, turbidity was determined with a Hach 2100N instrument, according to Standard Method 2130 (Standard Methods, 1992).

Spiral-wound nanofiltration membrane elements were used in each study. The Ohio River water membranes were all 10×100-cm units, whereas the Little Miami River and Harsha Lake membranes were 5×30-cm units. The units were operated at manufacturer's recommended flux rates, cross-flow velocities, and pHs (to control for inorganic scaling). All the systems were operated at approximately 50-percent recovery. During the operation of each membrane system, the membranes rejected the vast majority of inorganic, organic, and particulate species present in feed waters. The percent rejections were in line with manufacturer's specifications, indicating that the membranes were operated satisfactorily. Additional data and operating conditions for the Louisville, Cincinnati, Little Miami River, and Harsha Lake systems can be found in Merkel et al. (1998), Speth et al. (1998), Speth et al. (2001), and Speth (1998), respectively.

3. Results and Discussion

Table 11-1 briefly summarizes the feed-water qualities of the five waters studied. For the Louisville-Cincinnati comparisons, there is nothing to suggest that riverbank-filtered water was a higher quality membrane feed water than conventionally treated waters. TOC in Louisville riverbank-filtered water was lower than in the conventionally treated Louisville water, but it was comparable to conventionally treated Cincinnati water. The difference between the Louisville and Cincinnati conventionally treated waters could have been due to river location, plant operation (pretreatment efficiency), and seasonal effects. The Little Miami riverbank-filtered water was lower in TOC and higher in conductivity than the Harsha Lake conventionally treated water.

Table 11-1. Abridged Water Qualities for Waters Studied

Location	Water	Mean TOC (mg/L)	Mean Conductivity (µS/cm)	Mean Turbidity (ntu)
Louisville Bank Filtered	Ohio River	2.0	579	0.10
Louisville Conv. Treated	Ohio River	4.0	587	0.05
Cincinnati Conv. Treated	Ohio River	2.0	388	0.23
Southwestern Ohio – Riverbank Filtered	Little Miami River	2.3	1,020	Not Taken
Southwestern Ohio – Conv. Treated	Harsha Lake	3.2	263	0.50

mg/L = Milligrams per liter.   µS/cm = Microsiemens per centimeter.   ntu = Nephelometric turbidity unit.

Although part of the same watershed, the differences between the Little Miami riverbank-filtered water and Harsha Lake conventionally treated water were likely due to different water sources, the influence of groundwater, and the difference in pretreatment effectiveness between RBF and conventional treatment. The four-fold difference in conductivity between riverbank-filtered water and Harsha Lake water was also seen for calcium, magnesium, and sodium.

Table 11-2 contains the length of time that each system was operated, and the number of cleaning/flux cycles during operation. A cleaning/flux cycle is defined as the specific flux data between startup and the first cleaning, or between cleanings. The specific fluxes were determined by normalizing the permeate fluxes by temperature, pressure, and osmotic pressure. Table 11-2 also lists the mean calculated cleaning frequency for each system. The membranes were chemically cleaned when the specific flux was reduced by 15 percent, as compared to the stable specific flux after startup, or after the previous cleaning. For operational ease, the membranes were often put on cleaning schedules based on the expected fouling rates. Because the membranes often fouled

Table 11-2. Operation and Flux Results for Membrane Systems

Location	Water	Length of Operation (days)	Number of Cleaning/Flux Cycles[1]	Mean Calculated Cleaning Frequency (days)	% Flux Lost After Approximately 62 Days
Louisville Bank Filtered	Ohio River	62	7	75[2]	24
Louisville Conv. Treated	Ohio River	79	8	36[2]	46
Cincinnati Conv. Treated	Ohio River	460	51	8	36
Southwestern Ohio Riverbank Filtered	Little Miami River	79	1	62	12
Southwestern Ohio	Harsha Lake	70	12	8	50

[1] Includes initial cycle.   [2] Not arithmetic mean; projected from slope of entire run.

at unexpected rates, the time to achieve 15-percent specific flux decline for each cleaning/flux cycle was calculated to facilitate comparisons among membranes (calculated cleaning frequency). The results indicate that the conventionally pretreated membranes had much shorter calculated cleaning frequencies as compared to RBF pre-filtered membranes. The shorter calculated cleaning frequencies demonstrate that conventionally pretreated membranes had higher fouling rates.

The calculated cleaning frequencies for the Louisville studies were calculated from the slope of the entire run, as opposed to calculating the arithmetic mean of the slopes between cleanings. This was done because the membranes were cleaned once a week, and there were a number of weeks during which the flux increased slightly over the course of the week rather than decreased. This resulted in a negative cleaning frequency that could not be incorporated into the arithmetic mean without falsely representing the resulting mean. By taking the slope of the entire flux decline curve, the calculated cleaning frequencies were elevated because the increase in flux due to cleaning was incorporated into the number; however, the specific flux increases due to acid cleanings were not large in these runs and, therefore, the final flux of the membrane may have been similar to a system that was never cleaned. As a further indication of how the systems behaved, the southwestern Ohio and Louisville bank pre-filtered systems lost 12 and 24 percent of their initial fluxes over 62 days of operation, respectively, whereas the conventionally pretreated systems lost between 36 and 50 percent of their initial fluxes over the same length of time. This further shows the advantages of RBF over conventional treatment for pretreating nanofiltration feed waters.

Foulant autopsies were completed on the membranes; however, no specific foulant was identified on the conventionally pretreated membranes that was not also found on the RBF-pretreated membranes. The autopsy results suggested that the differences might have been due to increased organic/colloidal/particulate fouling for the conventionally pretreated membranes, as compared to RBF-pretreated membranes. Neither inorganic precipitation nor microbial growth on the membrane surface appeared to be the causative agent for the greater fouling observed with the conventionally pretreated membranes.

Table 11-2 also shows why conventional treatment is considered the bare minimum pretreatment for nanofiltration elements. It would be difficult to justify chemical cleaning on the order of 8 days, as that needed for the Cincinnati and Harsha Lake conventionally pretreated membranes; therefore, additional (or alternative) pretreatments would be needed for these surface waters. Microfiltration or ultrafiltration membranes have been shown to be successful surface-water pretreatments for nanofiltration elements. RBF can also be an effective pretreatment for nanofiltration membrane elements.

Disclaimer

This discussion should not be interpreted as U.S. Environmental Protection Agency policy or guidance. Mention of trade names or commercial products in this article does not constitute endorsement or recommendation for use by the United States Government.

References

Merkel, T.H., T.F. Speth, J.Z. Wang, and R.S. Summers (1998). "The performance of nanofiltration in the treatment of bank-filtered and conventionally treated surface waters." *Proceedings, Water Quality Technology Conference*, American Water Works Association, Denver, Colorado.

Sontheimer, H. (1980). "Experience with riverbank filtration along the Rhine River." *Journal of American Water Works Association*, 72(7): 386-390.

Speth, T.F., R.S. Summers, and A.M. Gusses (1998). "Nanofiltration foulants from a treated surface water." *Environmental Science and Technology*, 32: 3612-3617.

Speth, T.F. (1998) "Evaluation of nanofiltration foulants from treated surface waters." Ph.D. dissertation, University of Cincinnati, Cincinnati, Ohio.

Speth, T.F., C.J. Parrett, S.M. Harmon, and K.C. Kelty (2001). "Effect of chloramination and seasonal water changes on nanofiltration fouling." *Proceedings, Membrane Technology Conference*, American Water Works Association, Denver, Colorado.

Standard Methods (1992). *Standard methods for the examination of water and wastewater*, 18th edition, A.E. Greenberg, L.S. Clesceri, and A.D. Eaton, eds., American Public Health Association, American Water Works Association, and Water Environment Federation, Baltimore, Maryland.

Wang, J., J. Smith, and L. Dooley (1995). "Evaluation of riverbank infiltration as a process for removing particles and DBP precursors." *Proceedings, Water Quality Technology Conference*, American Water Works Association, Denver, Colorado.

Chapter 12. Water-Quality Improvements with Riverbank Filtration at Düsseldorf Waterworks in Germany

Jürgen Schubert, M.Sc.
Stadtwerke Düsseldorf AG
Düsseldorf, Germany

1. Introduction

RBF improves water quality during subsoil passage. The benefits of the underground passage of river water includes the removal of particles, bacteria, viruses, and protozoa, the removal of biodegradable compounds, and the compensation of fluctuating concentrations and temperatures; however, this advantage may be obscured when quantity aspects are only considered and when wells are located too close to the river. The biological degradation of DOC and, more so, the removal of microorganisms limit the minimum retention time of infiltrated river water. This chapter discusses the results obtained at the Flehe Waterworks, one of three RBF well fields (Flehe, Staad, and Auf dem Grind) that make up the Düsseldorf Waterworks in Düsseldorf, Germany.

2. Removal of Particles and Turbidity

The concentration of suspended solids in river water depends upon flow dynamics. The highest values appear when water levels increase during flood waves. The concentration of suspended solids in the Rhine River varies between 10 g/m^3 and more than 400 g/m^3; the mean value is less than 40 g/m^3; however, raw water in the wells is clear; the turbidity of the well water is 0.05 formazine nephelometric units (fnu). This value cannot be improved by further treatment. Particle counts in Flehe Waterworks well water were investigated in 1996 and 1998 (Schubert, 2001). The total count is very low — between 70 and 250 particles per milliliter (1- to 100-μm diameter).

3. Removal of Biodegradable Compounds

Figure 12-1 compares DOC concentrations in the Rhine River and in corresponding RBF wells at Flehe Waterworks (731.5 km) during the past 25 years. The average annual DOC concentration in river water decreased during this period from 6 mg/L to approximately 2.5 mg/L. In well water, the average DOC concentration began at 2.7 mg/L and varied between 1.2 and 1.0 mg/L over the past 8 years. The reduction rate under aerobic conditions in the subsoil seems to be independent from the DOC concentration in river water and varies between 50 and 60 percent. To estimate the real removal efficiency of DOC, the relationship between riverbank-filtered water (75 percent) and groundwater (25 percent) in well water must be considered. The DOC concentration in groundwater is approximately 1 mg/L.

Studies in both 1987 and 1989 investigated the assumption that biodegradation during RBF is similar to biodegradation during slow sand filtration and that biodegradation will occur in the upper layer of infiltration areas. In 1987, water samples were taken from a diving bell with special equipment at 0.6-m beneath the riverbed in front of the Flehe Waterworks. Table 12-1 shows the

C. Ray et al. (eds.), Riverbank Filtration, 267–277.

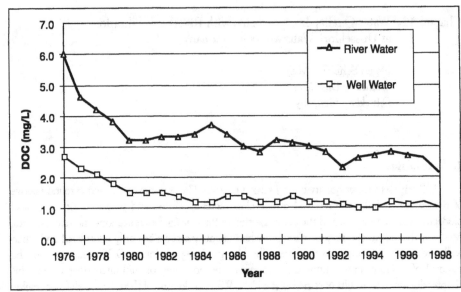

Figure 12-1. DOC concentration in the Rhine River and in well water at 731.5 km.

Table 12-1. Selected Water-Quality Parameters Measured in the Rhine River and the Leftover Fractions at 0.6-m Beneath the Riverbed and in Two Nearby Monitoring Wells

Parameter	Rhine River	0.6 m Below the Riverbed	Monitoring Well A	Monitoring Well B
DOC	1 (4.3 mg/L)	0.62	0.57	0.57
Ultraviolet Absorbance (254 nm)	1 (9.6 1/m)	0.57	0.56	0.56
Adsorbable Organic Halogen	1 (0.06 mg/L)	0.48	0.48	0.48

The "1" in the Rhine River column represents actual concentrations in the river (which are given in parenthesis). The fractions in the remaining three columns represent the fractions of actual river-water concentrations present at those locations.

leftover fractions of selected water-quality parameters monitored in the river water (0.6 m below the riverbed and in Monitoring Well A [near the river] and Monitoring Well B [between Monitoring Well A and the well gallery]).

The results of this investigation verify the expected degradation behavior: the observed real removal efficiency in the organic-rich and biologically active layer on and in the riverbed under aerobic conditions is 40 percent for DOC, 43 percent for ultraviolet absorbance, and approximately 50 percent for adsorbable organic halogen.

Figure 12-2 presents a cross-section of the aquifer in the Düsseldorf, Germany, region at 731.5 km (Flehe Waterworks). In 1989, water samples were taken at three different depths in this region via:

- Two rows of sampling wells between the river and the well gallery (Rows A and B).
- One row of sampling wells on the landside (Row C).
- The collector caisson at the gallery (Production Well 5).

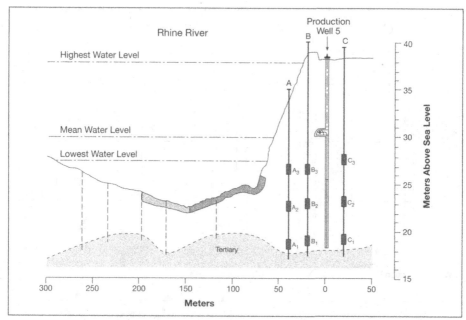

Figure 12-2. Cross-section through the aquifer in the Düsseldorf, Germany, region at 731.5 km. A, B, and C indicate rows with three monitoring and sampling wells each. Indices $_{1, 2, 3}$ indicate the different depths of the screen pipe.

The distance of the sampling wells along Cross-Section 731.5 km related to the production well is shown in Table 12-2.

DOC removal from Rhine River water at select monitoring points is presented in Figure 12-3. Similarly, the removal results for ultraviolet absorbance and adsorbable organic halogen are presented in Figures 12-4 and 12-5, respectively.

The results of both investigations confirm that the biodegradation of DOC under aerobic conditions is nearly complete after a short flow distance of a few decimeters. Between Row A and Row B of the sampling wells, there is additional NOM removal of only 1 to 2 percent. Similar results were found in other waterworks; therefore, this data can be regarded as representative for the Lower Rhine region.

Table 12-2. Locations of the River and Sampling Wells to Production Well 5

Parameter	Distance (m)
Production Well 5	0
Water Line of Rhine River	−50
Sampling Well A with Filter Screens at Depths A_1, A_2, A_3[a]	−40.2
Sampling Well B with Filter Screens at Depths B_1, B_2, B_3[b]	−20.4
Sampling Well C with Filter Screens at Depths C_1, C_2, C_3[c]	+18.1

− = Towards the river. + = Away from the river.

[a] 1-m long filter screens for A_1, B_1, and C_1 are approximately 20-m below ground.

[b] 1-m long filter screens for A_2, B_2, and C_2 are 15-m below ground.

[c] 1-m long filter screens for A_3, B_3, and C_3 are 10-m below ground.

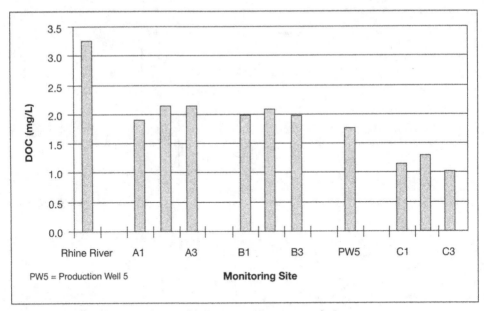

Figure 12-3. DOC level at Cross-Section 731.5 km for the Rhine River and select monitoring sites.

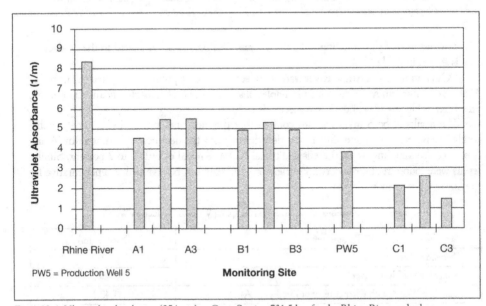

Figure 12-4. Ultraviolet absorbance (254 nm) at Cross-Section 731.5 km for the Rhine River and select monitoring sites.

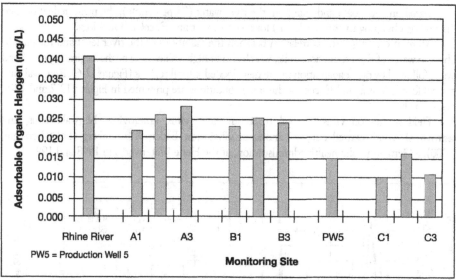

Figure 12-5. Adsorbable organic halogen level at Cross-Section 731.5 km for the Rhine River and select monitoring sites.

Removal of Bacteria, Viruses, and Parasites

Under steady-state conditions, RBF (with an average residence time of 2 weeks) shows a 5-log removal efficiency for pathogenic microorganisms, such as bacteria, viruses, and parasites. There may be variations of ±1-log removal efficiency, depending on the microorganism (Medema et al., 2001).

But steady-state conditions are not standard conditions in RBF. The river level varies and, when it increases, infiltrated river water enters subsoil regions that were previously unsaturated and not adapted. Sterilized filter media do not significantly remove microorganisms initially. This shows that there is a need for adaptation (maybe even for the formation of a food chain) to remove microorganisms.

Statistical data on indicator organisms may be used to understand microorganism removal during RBF dynamic conditions. Table 12-3 shows the results of well-water data over a 14-month period for the Flehe Waterworks (1998 to 1999).

Table 12-3. Microbial Concentrations in the Well Water of Production Well 5

Range of Data	Colony Count at 20°C per mL	Colony Count at 36°C per mL	Coliform per 100 mL	E. coli per 100 mL
Maximum	228	1,000	2,000	20
95 Percentile	11	21	409	1
90 Percentile	3	7	125	1
75 Percentile	1	2	52	0
50 Percentile	0	0	22	0
Minimum	0	0	0	0

In the above table, 95 percentile means that the concentration of 95 percent of all samples were at or below the presented values (i.e., 11 colonies/mL at 20°C, etc.). This is also true of 90, 75, and 50 percentile.

Based on these data and data from the river water (75 percentile), the mean reduction rate (including changing river-water levels) for colony count and E. *coli* is 3 to 4 logs.

More interesting is the correlation between the variations of the river level and the data in the raw water of the production wells, which shows high values only at the beginning of a flood wave followed by quick adaptation in the new flooded subsoil regions (Figure 12-6). Removal data for coliform bacteria and E. *coli* for the same conditions are presented in Figures 12-7 and 12-8, respectively.

Field studies on virus removal (rotavirus, adenovirus, enterovirus, Norwalk virus, and astrovirus) where carried out by Landesgesundheitsamt Baden-Württemberg (LGA-Berichte, 2000), a German public health administration, at the Flehe Waterworks in 1998 and 1999.

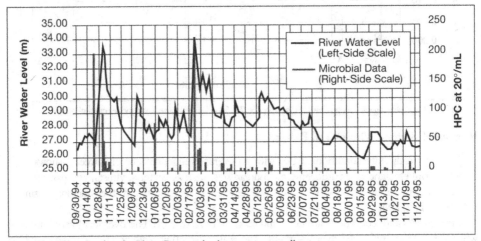

Figure 12-6. Water level in the Rhine River and colony counts in well water.

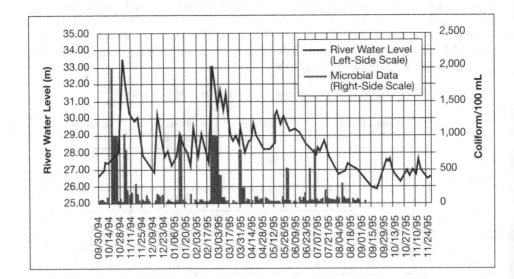

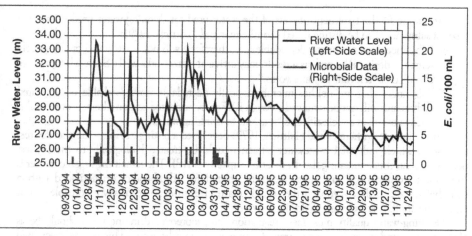

Figure 12-8. Water level in the Rhine River and E. *coli* counts in well water.

In 1998, rotavirus, adenovirus, and enterovirus (12 samples) were detected eight to 10 times in river water, but never in Flehe Waterworks well water. In 1999, rotavirus, adenovirus, enterovirus, and Norwalk virus (18 samples) were detected in river water two to 14 times. Rotavirus and enterovirus were both positive only once in well water (rising river level). This shows that the unavoidable, temporary passage of riverbank-filtered water through normally unsaturated regions of the subsoil during flood events influences the removal efficiency of microorganisms. It confirms that RBF *does not* replace the necessity of disinfection.

Limited data are available about parasite removal by RBF. During the last 5 years, studies were conducted in river water, monitoring wells near the river, well water, and backwash water of the filters in treatment plants. Parasites (*Cryptosporidium sp.* and *Giardia lamblia*) in 500-L samples could only be found in river water, but never in the monitoring wells, raw water of the production wells, or backwash water. Table 12-4 shows removal data taken during a heavy flood event in March 1997.

Table 12-4. Removal Data for *Cryptosporidium* and *Giardia* for the Rhine River and Monitoring and Production Wells

	Cryptosporidium/100 L	*Giardia*/100 L
River Water	6.2	1.1
Sampling Well A3	Not Detected	Not Detected
Sampling Well B2	Not Detected	Not Detected
Production Well 5	Not Detected	Not Detected

5. Removal of Pharmaceutical Wastes and Other Specialty Organics

Surface waters like the Rhine River are recipients of wastewater. As some major chemical companies are situated in the basin of the Rhine River, wastewater from chemical production plants have always been of special concern to waterworks (Lindner et al., 2001). Sontheimer and Völker (1987) conducted several research projects to classify industrial wastewaters with respect to their importance for waterworks.

According to the treatment process of the water used for drinking-water production, two types of substances can be distinguished (Sontheimer and Haltrich, 1979):

- Substances that are poorly degradable and, therefore, will not be removed or completely removed during RBF and, as a result, may arrive at waterworks wells. Such substances are classified as "relevant to waterworks."
- Substances that are neither microbiologically degradable nor adsorbable on activated carbon and will not be removed through RBF or by activated carbon filters in waterworks. Such substances are classified as "relevant to drinking water."

Sontheimer (Sontheimer and Haltrich, 1979) developed a testing device — the classical test filter — in which industrial wastewater effluents could be treated biologically to the highest possible extent, thus simulating the degradation capacity of RBF. The remaining organic matter is "relevant to waterworks" and must be further tested for adsorbability on activated carbon (Sontheimer and Völker, 1983). If the organic matter turns out to be poorly adsorbable or not adsorbable at all, it is "relevant to drinking water."

Drinking-water quality must meet regulations but, moreover, drinking water should be as clean as protected groundwater and free of any man-made substances. This is why specialized monitoring and screening programs for surface water, groundwater, and well water are needed.

Micropollutants have been monitored in the Rhine River by the Association of Rhine Waterworks (Association of Rhine Waterworks, 1998/1999). Micropollutants that could be detected in river water and/or are classified as "relevant to waterworks" or "relevant to drinking water" are listed in Table 12-5. The concentration in river water is characterized by the 90 percentile value in 1998 and 1999 at the Flehe Waterworks (731.5 km). Table 12-5 provides a summary of these parameters for the Rhine River.

6. The Gasoline Oxygenate MTBE

There is only limited data available from some preliminary studies on the effects of MTBE on water, but these data show that due to its physical and chemical properties, MTBE has a high potential to contaminate even groundwater. MTBE is water soluble, mobile, and very difficult to biodegrade. Ninety-seven percent of the samples from the Rhine, Main, Neckar, Elbe, and Danube Rivers in Germany tested positive for MTBE, with concentrations ranging from 0.08 and 9.8 µg/L in river water samples and from 0.05 µg/L (detection limit) and 0.13 µg/L in RBF well-water samples. MTBE can even be found in drinking water.

7. Decline of Mutagenic Activity By RBF

The presence of mutagenic compounds (chemicals showing the signs of mutation) in Rhine River water has been investigated since 1981. Significant mutagenic activity has been observed, with higher levels in the winter than in the summer (Association of Rhine River Waterworks, 1993). While mutagenic activity in the Rhine River has decreased since 1990, it is still present (Veenendal and Van Genderen, 1998). One means to assess mutagenic activity in water is by using the Ames Test method (also called a *Salmonella typhimurium* Microsomal Mutagenicy Assay). The Ames Test, which uses a series of genetically engineered strains of *Salmonella*, is conducted in samples that are concentrated using XAD resin at pH 2 and pH 7 and is frequently used in combination with drug-metabolizing enzymes isolated from rodent livers, usually in the form of supernatant liquor from a $9,000 \times g$ centrifugation of liver homogenates (referred to as the S9 fraction). The *Salmonella* bacteria are modified so that they cannot produce Histidin, which is an amino acid necessary for growth. Mutagenic active compounds, however, "switch on" the ability of the bacteria to produce Histidin; therefore, the modified bacteria revert to their natural state and form colonies. These are called "revertants." Figure 12-9 presents Ames Test data for Lobith, a small city at the German/Dutch border of the Rhine River, from 1981 to 1998.

Table 12-5. Concentrations of Pharmaceutically Active Substances, Chelating Agents, Aromatic Sulfonates, and Aliphatic Amines in the Rhine River Near the Flehe Waterworks in Düsseldorf, Germany

Compound Class/Compounds	Concentration
Pharmaceuticals	(ng/L)
*Bezafibrat**	71
Carbamacepine	256
*Clofibricacid**	<10
*Diclofenac**	72
*Ibuprofen**	<5
Chelating Agents	(µg/L)
Nitrilotriacetic Acid (NTA)	1.5
Ethylenediaminetetraacetic acid (EDTA)	8.0
Diethylenetrinitrilopentaacedic Acid (DTPA)	1.8
Aromatic Sulfonates	(µg/L)
Naphthalin-1,5-disulfonate	1.7
Naphthalin-2,7-disulfonate	0.20
Naphthalin-1,6-disulfonate	0.37
Naphthalin-1,3,6-trisulfonate	0.11
2-Aminonaphthalin-1,5,-disulfonate	0.39
2-Aminonaphthalin -4,8-disulfonate	0.50
Cis-4,4'-Dinitrostilben-2,2'-disulfonate	4.2
Aliphatic Amines	(µg/L)
Methylamine	0.33
Dimethylamine	0.18
Morpholine	0.97
Ethanolamine	0.55
Diethanolamine	0.65
Urotropine	9.3

Italics = Relevant to waterworks.   Underlined = Relevant to drinking water.

*Based upon test filter studies.

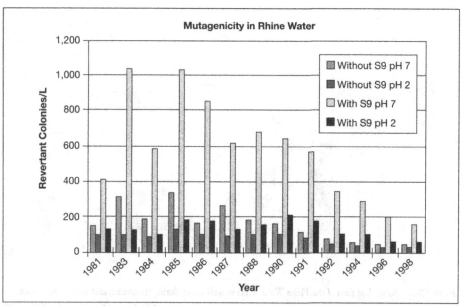

Figure 12-9. Ames Test data in the Rhine River at Lobith. Mean values of induced revertants (1981 to 1998)

During a research project sponsored by the International Association of the Waterworks in the Rhine Region in 1994 (Van Genderen and Veenendaal, 1995), the mutagenic activity in the Rhine River was investigated in three time series between 149 km (the City of Rheinfelden) and 947 km (the City of Hagestein). The results (induced revertants per liter) for locations in Düsseldorf and Lobith are shown in Table 12-6.

Table 12-6. Mutagenic Activity of the Rhine River at Düsseldorf and Lobith in 1994

Time	Without S9 pH 7	Without S9 pH 2	With S9 pH 7	With S9 pH 2
Spring 1994	49 (39)	34 (2)	427 (252)	104 (70)
Summer 1994	37 (30)	35 (25)	545 (434)	132 (104)
Autumn 1994	72 (50)	33 (30)	692 (606)	329 (204)

Lobith data is in parentheses.

To verify the effectiveness of subsoil passage and the different treatment steps on the decline of mutagenic activity, KIWA N.V. Research & Consultancy of Nieuwegein, The Netherlands, was responsible for relevant studies in the Flehe Waterworks. Samples were taken in March 1998 in the well water, after ozonation, after granular activated carbon adsorption, and in the drinking water after disinfection with 0.05 mg/L chlorine dioxide (ClO_2). For the investigations, the bacterium *Salmonella typhimurium* T98 and T100 were used. The results are shown in Figure 12-10.

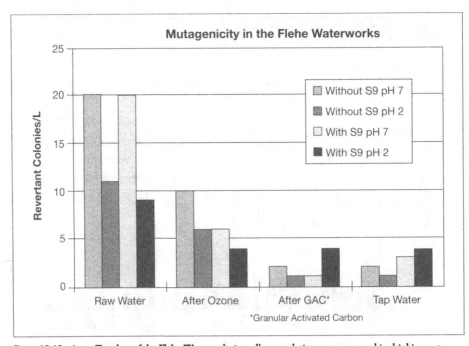

Figure 12-10. Ames Test data of the Flehe Waterworks in well water, during treatment, and in drinking water.

The results can be summarized as follows (Mons et al., 1998):

- No sample tested positive related to Ames Test criteria.
- The number of induced revertants in well water (20/L) was much lower than in river water (about 250/L).
- During treatment, the low values of the raw water (20/L) are further reduced to values of <5/L.

8. Conclusion

The Düsseldorf Waterworks supply drinking water to approximately 600,000 inhabitants. Since the Flehe Waterworks began operation in 1870, the City of Düsseldorf has depended on RBF as the main source of drinking water. For this reason, investigations and research on RBF and treatment are important in designing, constructing, operating, and maintaining the waterworks. Data show that RBF is an efficient treatment step in improving drinking-water quality. In addition, RBF is a stable process that not only does not produce residuals, but also helps decrease treatment costs.

References

Association of River Waterworks (RIWA) (1993). *Water quality of the river Rhine in 1990 and 1991*. Association of River Waterworks, Amsterdam, The Netherlands (in Dutch).

Association Rhine Waterworks (ARW) (1998/1999). *Arbeitsgemeinschaft der Rhein-Wasserwerke e.V.* Association of Rhine Waterworks (ARW), Köln, Germany.

LGA-Berichte (2000). *Untersuchungen zur Belastung von Rohwässern mit enteropathogenen Viren und deren Verhalten bei der Trinkwasseraufbereitung*. Landesgesundheitsamt Baden-Württemberg, Stuttgart, Germany.

Lindner, K., F. Karrenbrock, T. Knepper, and F. Sacher (2001). "Experimental approaches to bank filtration." *Proceedings, International Riverbank Filtration Conference*, W. Jülich and J. Schubert, eds, International Association of the Rhine Waterworks (IAWR) Amsterdam, The Netherlands, 81-93.

Medema, G.J., M.H.A. Juhasz-Holterman, and J.A. Luijten, (2001). "Removal of microorganisms by bank filtration in a gravel-sand soil." *Proceedings, International Riverbank Filtration Conference*, W. Jülich and J. Schubert, eds., International Association of the Rhine Waterworks (IAWR) Amsterdam, The Netherlands, 161-168.

Mons, M.N., J. Van Genderen, and D. Van der Kooij (1998). *Mutagenicity testing in bank filtrate and in drinking water of Stadtwerke Düsseldorf*. Kiwa N.V., Research & Consultancy, Nieuwegein, The Netherlands.

Schubert, J. (2001). "How does it work? Field Studies on Riverbank Filtration." *Proceedings, International Riverbank Filtration Conference*, W. Jülich and J. Schubert, eds., International Association of the Rhine Waterworks (IAWR) Amsterdam, The Netherlands, 41-55.

Sontheimer, H., and W. Haltrich (1979). "Untersuchungen zur Beurteilung der Qualität biologisch gereinigter Abwässer." *Vom Wasser*, 53(S): 121-132.

Sontheimer, H., and E. Völker (1983). *Charakterisierung und Beurteilung von Kläranlagenabläufen aus Sicht der Trinkwasserversorgung*. Arbeitsgemeinschaft Rhein-Wasserwerke e. V. (Association of Rhine Waterworks) (ARW), Köln, Germany, S.79 - 137.

Sontheimer, H., and E. Völker (1987). *Charakterisierung von Abwassereinleitungen aus Sicht der Trinkwasserversorgung*. Heft 31, University of Karlsruhe, Karlsruhe, Germany.

Van Genderen, J., and H.R. Veenendaal (1995). *IAWR – Rheinkampagne 1994, Toxikologische und ökologische Untersuchung des Rheins*. International Association of the Rhine Waterworks (IAWR), Amsterdam, The Netherlands.

Veenendaal, H.R., and J. Van Genderen (1998). *Mutageniteit in Rijn en Maas in 1998*. Association of the River Waterworks (RIWA), Amsterdam, The Netherlands.

Part III:
Research Needs

Chapter 13. Infiltration Rate Variability and Research Needs

William D. Gollnitz
Greater Cincinnati Water Works
Cincinnati, Ohio, United States

1. Introduction

One aspect of RBF that should be considered by future users is the potential for significant variability in the rate of infiltration over time. It is believed that RBF is similar to engineered filtration systems in that RBF follows the same basic principals; that is, the efficiency of the filtration process is controlled in part by the hydraulic gradient between the river stage elevation and the groundwater elevation. For example, slow sand filtration systems are designed and operated within a set standard, typically up to 1.5 m, that limits the maximum amount of hydraulic gradient, or "head," applied to the filter (Letterman, 1999). A significant increase in the head on a filter can force particulate material to penetrate, or "breakthrough," the bed and leach into the filter effluent. Under the right conditions, it is believed that a similar situation can occur with RBF facilities. A primary argument against pathogen removal has been the fact that the water-supply operator cannot control the rate of infiltration. RBF facilities may be located next to rivers that experience a significant fluctuation in the difference between the river stage and groundwater elevation over time. The hydraulic gradient resulting from prolonged pumping, followed by a high stage (flood) event, can significantly increase the rate of infiltration, as will be discussed in more detail later in this chapter.

2. Factors Influencing the Rate of Infiltration

In an unconfined aquifer, the initial pumping of a well causes the water level to drawdown; it also removes water from storage within the volume of aquifer immediately surrounding the well casing. The lowering of the water table creates a cone-shaped hydraulic gradient. As the cone expands outward, it draws in water from further distances within the aquifer. The volume of the aquifer that contributes water to the well is referred to as the zone of contribution. Drawdown in the well will cease and the zone of contribution will stop expanding as soon as the amount of recharge equals the amount of pumpage (e.g., the system reaches equilibrium). The size and shape of the zone of contribution is determined by:

- Aquifer characteristics (permeability, porosity, and thickness).
- Configuration of the well (depth, screen length, pumping rate, and period).
- Location of a source of recharge.

Recharge may come from precipitation, a nearby surface-water body or, in most cases, a combination of both. Surface water will flow into an aquifer along a river reach where the zone of contribution has lowered the groundwater level below the stream stage through a phenomenon called "induced infiltration."

C. Ray et al. (eds.), Riverbank Filtration, 281–290.

The rate of induced infiltration of river water depends on a multitude of factors that can change significantly over time. These variables include:

- River-stage elevation.
- Groundwater elevation (related to number of pumping wells, pumping rates, and periods).
- Water viscosity (temperature-related).
- Streambed thickness.
- Streambed permeability.
- Wetted streambed area.

Typically, the rate of infiltration is slow under certain conditions of low-river stage (Figure 13-1). The slow stream flow velocity allows for the deposition of fine-grained sediments on the streambed. During the summer months, algal growths can accumulate and be deposited on the bottom of the streambed. These sediments, along with microbial material, can increase the streambed thickness and, typically, cause the permeability to be significantly lower than the underlying aquifer matrix. At low stage, the water is also in contact with a limited amount of river channel surface area (wetted streambed area). If the water temperature is cold, the higher viscosity will slow the flow of river water into the aquifer. During periods of minimal well field pumpage, the groundwater elevation may be at or near the river stage. The hydraulic gradient may not be sufficient to provide enough hydraulic head pressure for large quantities of water to flow through the streambed.

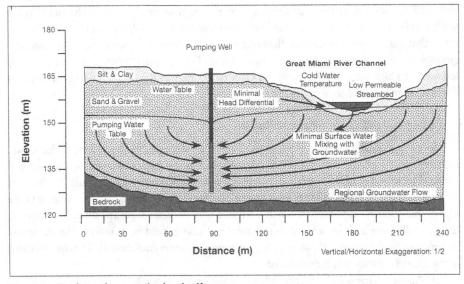

Figure 13-1. Conditions for minimal induced infiltration.

At higher river stages (Figure 13-2), the increased flow velocity may scour the streambed and re-suspend sediment. Typically, scour occurs simultaneously along long reaches, both on pools and bars (Leopold et al., 1992). The scouring effect may decrease bed thickness and increase bed permeability. Depending upon channel configuration, the increased volume of water comes into contact with a much larger channel surface area. The increased stage provides a much higher hydraulic gradient, especially in narrow river channels. If a high river stage event occurs after a

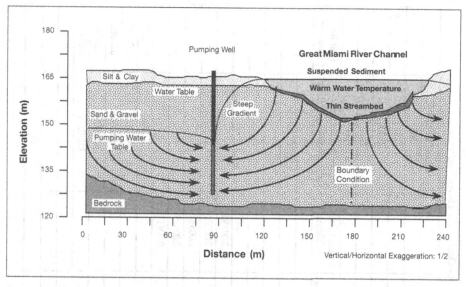

Figure 13-2. Conditions for maximum induced infiltration.

period of pumping and low infiltration, the groundwater elevation maybe at or below the streambed.

Walton et al. (1967) studied the variability of induced infiltration on the Great Miami River, located in southwestern Ohio. Walton found that the rate of infiltration was not consistent along the entire reach of the river. His work demonstrated that the rate of infiltration could increase by a factor of three within 24 hours. Under certain hydrologic conditions, the rate can change by over an order of magnitude. During the year of Walton's study, there were four events where the rate more than doubled within 48 hours.

3. Rate Evaluation

To further demonstrate this concept, a spreadsheet model was used to estimate the potential range of infiltration rates for two alluvial aquifers in the United States. These include the Great Miami River Buried Valley Aquifer in southwest Ohio (Tables 13-1 and 13-2) and the Hunt River Aquifer in eastern Rhode Island (Tables 13-3 and 13-4). The spreadsheet uses Dary's law to calculate the rate of flow through the streambed (Fetter, 1994). Hydraulic information collected from each site was used as input. Input data includes:

- River-stage elevation.
- Streambed width.
- Length of influenced reach.
- Streambed wetted area.
- Streambed thickness.
- Streambed permeability.
- Viscosity, as determined by surface-water temperature.

Three of the parameters (stage, width, and wetted area) were determined from available stream channel profile information. The other data were available from field surveys, pump test analyses, or published sources. The model uses a range of data determined for each parameter at each site.

Table 13-1. Estimates of Induced Infiltration from the Great Miami River, Southwest Ohio — Warm River Temperature

River Stage (m)	Low (0.5 to 1.5)		Moderate (1.5 to 3.0)		High (3.5 to 4.5)	
Streambed Width (m)	100		150		200	
Length of Influenced Reach (m)	300		275		250	
Streambed Wetted Area (m²)	30,000		41,250		50,000	
Streambed Thickness (m)	0.60		0.30		0.15	
Streambed Permeability (m/day)	0.5		0.6		0.7	
Available Head (m)	Infiltration Rate (m³/day/influenced reach)	Unit Rate (m³/day/m²)	Infiltration Rate (m³/day/influenced reach)	Unit Rate (m³/day/m²)	Infiltration Rate (m³/day/influenced reach)	Unit Rate (m³/day/m²)
0.3	7,500	0.00017				
0.6	15,000	0.00035				
0.9	22,500	0.00052				
1.2	30,000	0.00069				
1.5	37,500	0.00087				
1.8			148,500	0.00250		
2.1			173,250	0.00292		
2.4			198,000	0.00333		
2.7			222,750	0.00375		
3			247,500	0.00417		
3.3					770,000	0.01069
3.6					840,000	0.01167
3.9					910,000	0.01264
4.2					980,000	0.01361
4.5					1,050,000	0.01458
Surface-Water Temperature: °C	27					
Surface-Water Temperature: °F	80.7					
Surface-Water Viscosity: Poise	0.85					

Table 13-2. Estimates of Induced Infiltration from the Great Miami River, Southwest Ohio — Cold River Temperature

River Stage (m)	Low (0.5 to 1.5)		Moderate (1.5 to 3.0)		High (3.5 to 4.5)	
Streambed Width (m)	100		150		200	
Length of Influenced Reach (m)	300		275		250	
Streambed Wetted Area (m²)	30,000		41,250		50,000	
Streambed Thickness (m)	0.60		0.30		0.15	
Streambed Permeability (m/day)	0.5		0.6		0.7	
Available Head (m)	Induced Infiltration (m³/day/influenced reach)	Unit Rate (m³/day/m²)	Induced Infiltration (m³/day/influenced reach)	Unit Rate (m³/day/m²)	Induced Infiltration (m³/day/influenced reach)	Unit Rate (m³/day/m²)
0.3	3,545	0.00008				
0.6	7,091	0.00016				
0.9	10,636	0.00025				
1.2	14,182	0.00033				
1.5	17,727	0.00041				
1.8			70,200	0.00118		
2.1			81,899	0.00138		
2.4			93,599	0.00157		
2.7			105,299	0.00177		
3			116,999	0.00197		
3.3					363,998	0.00506
3.6					397,088	0.00552
3.9					430,179	0.00597
4.2					463,270	0.00643
4.5					496,361	0.00689
Surface-Water Temperature: °C	1.1					
Surface-Water Temperature: °F	34					
Surface-Water Viscosity: Poise	1.80					

Table 13-3. Estimates of Induced Infiltration from the Hunt River, Eastern Rhode Island — Warm River Temperature

River Stage (m)	Low (0.3 to 0.9)		Moderate (1.0 to 1.8)		High (2.0 to 3.0)	
Streambed Width (m)	6		15		30	
Length of Influenced Reach (m)	500		300		250	
Streambed Wetted Area (m²)	3,000		4,500		7,500	
Streambed Thickness (m)	0.60		0.30		0.15	
Streambed Permeability (m/day)	0.3		0.6		1.0	
Available Head (m)	Infiltration Rate (m³/day/influenced reach)	Unit Rate (m³/day/m²)	Infiltration Rate (m³/day/influenced reach)	Unit Rate (m³/day/m²)	Infiltration Rate (m³/day/influenced reach)	Unit Rate (m³/day/m²)
0.3	450	0.0001				
0.6	900	0.0002				
0.9	1,350	0.0003				
1.2			10,800	0.0017		
1.5			13,500	0.0021		
1.8			16,200	0.0025		
2.1					105,000	0.0097
2.4					120,000	0.0111
2.7					135,000	0.0125
3					150,000	0.0139
Surface-Water Temperature: °C	27					
Surface-Water Temperature: °F	80.7					
Surface-Water Viscosity: Poise	0.85					

Table 13.4. Estimates of Induced Infiltration from the Hunt River, Eastern Rhode Island — Cold River Temperature

River Stage (m)	Low (0.3 to 0.9)		Moderate (1.0 to 1.8)		High (2.0 to 3.0)	
Streambed Width (m)	6		15		30	
Length of Influenced Reach (m)	500		300		250	
Streambed Wetted Area (m²)	3,000		4,500		7,500	
Streambed Thickness (m)	0.60		0.30		0.15	
Streambed Permeability (m/day)	0.3		0.6		1.0	
Available Head (m)	Induced Infiltration (m³/day/influenced reach)	Unit Rate (m³/day/m²)	Induced Infiltration (m³/day/influenced reach)	Unit Rate (m³/day/m²)	Induced Infiltration (m³/day/influenced reach)	Unit Rate (m³/day/m²)
0.3	213	0.0000				
0.6	425	0.0001				
0.9	638	0.0001				
1.2			5,105	0.0008		
1.5			6,382	0.0010		
1.8			7,658	0.0012		
2.1					49,636	0.0046
2.4					56,727	0.0053
2.7					63,818	0.0059
3					70,909	0.0066
Surface-Water Temperature: °C	1.1					
Surface-Water Temperature: °F	34					
Surface-Water Viscosity: Poise	1.80					

The model calculates a rate (m^3/d) per estimated area of influence as well as a unit rate (cubic meters per day per square meter [$m^3/d/m^2$]). The primary assumptions of the model are:

- The individual parameters will change accordingly as the flow and stage in the river increase.
- There is sufficient capacity (void space) for large volumes of water to move into the aquifer during high river stage events.

In other words, prior to the event, the groundwater table below the river had been drawn down significantly to allow water to move quickly into the aquifer.

As can be seen by Tables 13-1 to 13-4, the range of potential infiltration rates at both sites can be dramatic. Throughout a 1-year period at the Great Miami River Buried Valley Aquifer, the potential infiltration rate can range from 7,500 m^3/d (0.00017 m/d) to over 1-million m^3/d (0.01458 m/d) (under certain conditions). At the Hunt River Aquifer, the potential rates range from 213 m^3/d (0.00005 m/d) to 150,000 m^3/d (0.0139 m/d). Although the extremely high rates are unlikely due to the required hydrologic conditions, it is important to note that a rate change of an order of magnitude or more is probable throughout the year. Interestingly, except for the upper end of the model, the unit rate of infiltration for both sites is still less than the unit rate of a slow sand filter (1.08×10^{-5} $m^3/s/m^2$ to 1.08×10^{-4} $m^3/s/m^2$). This poses the question as to whether or not a porous sand and gravel aquifer still has the ability to remove pathogens even under high river stage conditions.

4. Infiltration Evaluation Using Hydrographs

Surface-water and groundwater hydrograph data is valuable in providing information concerning the occurrence and rate of induced infiltration. Figure 13-3 is a hydrograph for a river stage gauging station and two groundwater-level monitoring wells in the Great Miami River Buried Valley Aquifer. The configuration of the station and wells is provided in Figure 13-4.

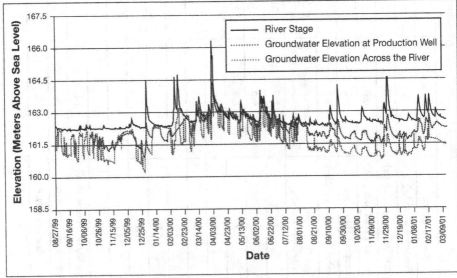

Figure 13-3. River stage and groundwater elevation data (August 1999 to March 2001).

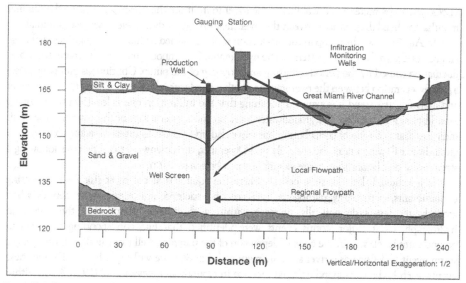

Figure 13-4. River stage and groundwater elevation points.

The period of record for this hydrograph is approximately 2 years from July 1999 to July 2001, thereby including two full hydrologic cycles.

The hydrograph starts in the late summer. During this period, extending through the fall and into early winter, the river stage is low. The groundwater level in Well A oscillates within a 1.2- to 1.5-m range, with the starting and stopping of the pump in the production well. Under these low river stage conditions, the amount of recharge via induced infiltration is lower than the amount of water pumped, causing a downward trend in the groundwater hydrographs. In late November, the production well was shut off for a period of approximately 1 month. During this down period, the groundwater levels recovered to within 0.3 m of the river stage. Pumping resumed in mid-December and continued for approximately 2 weeks. During this short time period, the groundwater levels continued to decline to the area's lowest elevation of the year, indicating that the rate of infiltration was still lower than the rate of pumping. Groundwater temperature at this time of the year had declined to just above freezing, significantly slowing the rate of recharge.

At the beginning of January, a major storm occurred in the watershed. Runoff significantly raised the river stage by 2.4 m within 24 hours. The groundwater level at Well A responded with a 2.3-m increase within 2 days. A larger volume of water entered the aquifer at a significantly faster rate than had been previously experienced. As a result of this event, groundwater levels recovered to near mid-summer levels. By the latter part of the month, surface-water stage returned to its pre-storm elevation. For the remainder of this 1-month period, the groundwater elevations approached river stage elevation due to the pump not being in operation.

Two more storms occurred in February 2000. The first had a stage increase of more than 2.1 m in 3 days. Before the hydrograph could return to pre-storm stage, a second storm increased the stage by 1.7 m within another 24-hour period. This resulted in a 2.1-m groundwater elevation increase over an 8-day period. For the remainder of March and throughout the spring, the groundwater level followed an increasing trend. The temperature of the surface water during this period decreased the viscosity, thereby increasing the ability of the water to infiltrate the streambed. During the summer months, the head differential between the river stage and groundwater elevation was minimal.

In August 2000, the pump in the production well was turned on fulltime. A significant change can be noticed in the overall pattern of the hydrograph. Continuous pumping eliminates the 1.5-m fluctuation caused by the periodic starting and stopping of the pump. Continuous pumping causes a greater separation between the river stage and both groundwater elevations. During the late summer and early fall, the trend is downward, meaning that the infiltration rate is less than the pumping rate. Pumping also causes the groundwater level on the opposite side of the river to drop. This indicates that the zone of contribution has expanded up-gradient to intercept additional regional groundwater flowing under the river. By mid-December, the downward trend reverses following a storm event and begins to increase as surface water begins to warm.

It is believed that the time periods where the aquifer is at the most risk for transmitting contaminants, particularly pathogenic protozoa and bacteria, are during those periods when groundwater levels dramatically increase within a short time period. For the illustrated hydrograph, eight major storms events were identified that had corresponding increases in groundwater levels within the cone of depression of the pumping well (for this discussion, a major storm event is defined as a river stage increase of 1.5 m or more within 48 hours). During these events, each had a corresponding rapid increase in groundwater elevation of 0.9 to 2.4 m within 2 to 3 days.

5. Research Needs

Further research is needed to investigate the overall impacts of an increased infiltration rate on groundwater quality. Studies should be conducted to answer several important questions:

- Does streambed permeability increase during the storm event?
- Does streambed thickness decrease?
- How are these two parameters measured during an event?
- Do concentrations of pathogens or their surrogates increase during and immediately after an identified period of rapid filling of groundwater?
- If pathogen breakthrough only occurs during events, how important is it if the frequency of these events is low?

These studies will necessitate an understanding of site-specific travel time considerations between the river and the well being monitored and, therefore, will be expensive; however, this type of research will provide important insight to whether or not RBF is a reliable process, even under considerable hydrologic variability.

References

Fetter, C.W. (1994). *Applied hydrogeology*, Third Edition. Prentice Hall, Inc, New York, New York.

Leopold, L.B., M.G. Wolman, and J.P Miller (1992). *Fluvial processes in geomorphology*. Dover Publications, New York, New York.

Letterman, R.L. (1999). *Water quality and treatment*, Fifth Edition. American Water Works Association, Denver, Colorado.

Walton, W.C., D.L. Hills, and D.L. Grundeen (1967). "Recharge from induced streambed infiltration under varying groundwater-level and stream-stage conditions." *Bulletin* 6, Water Resources Research Center, University of Minnesota, St. Paul, Minnesota.

Chapter 14. Siting and Design Issues for Riverbank Filtration Schemes

Thomas Grischek, M.Sc.
Institute of Water Chemistry
Dresden University of Technology
Dresden, Germany

Dagmar Schoenheinz, M.Sc.
Institute of Water Chemistry
Dresden University of Technology
Dresden, Germany

Chittaranjan Ray, Ph.D., P.E.
University of Hawaii at Mānoa
Honolulu, Hawaii, United States

1. Introduction

Often, waterworks prefer to use RBF to produce drinking water when:

- Groundwater resources are insufficient.
- The cost for treating direct intake is higher than for treating pumped riverbank filtrate with better water quality after storage passage due to natural attenuation processes.
- Water-quality fluctuations require increased efforts in water treatment to reach the desired quality.

As a treatment process, RBF not only removes contaminants, such as organics, microbiological pathogens, and particles, from surface-water sources (see Chapters 6 through 12), but it is also cost-effective as it can balance fluctuations in both water temperature and ion concentration (e.g., nitrate and ammonia), thereby eliminating the need for further treatment. Another advantage of RBF includes mixing riverbank-filtered water with native groundwater in the aquifer to increase the groundwater supply and dilute contaminants.

A relatively small number of RBF systems are used in the United States, most of which are designed and built by a handful of companies. Though RBF, as an engineering technique, is widespread throughout Europe, the design and construction are based upon personal experience. Before the publication of this book, there were no guidelines or handbooks available on where and how to install RBF systems; however, there is a demand for developing more such tools for stakeholders to ease the application of this highly effective and relatively inexpensive technique.

2. Aspects of Design

The siting and design of RBF systems depend upon river hydrology, hydrogeological site conditions, and the aims of water withdrawal; however, for effective and sustainable use of RBF, the river should be in hydraulic contact with the adjacent aquifer.

C. Ray et al. (eds.), Riverbank Filtration, 291–302.

Most RBF sites are located in alluvial sand and gravel aquifers having hydraulic conductivities higher than 1×10^{-4} m/s. The thickness of exploited aquifers ranges from 5 to 60 m. Table 14-1 provides the thickness and hydraulic conductivity of selected RBF systems in the United States and Europe for which data were readily available.

Table 14-1. Aquifer Thickness and Hydraulic Conductivity of the Aquifer Material for Selected Alluvial Sand and Gravel Aquifers in the United States and Europe

RBF System	River System	Aquifer Thickness (m)	Hydraulic Conductivity (m/s)
Henry, Illinois, United States	Illinois	15 to 20	2×10^{-3} to 3×10^{-3}
Jacksonville, Illinois, United States	Illinois	25 to 27	2×10^{-3} to 3×10^{-3}
Lincoln, Nebraska, United States	Platte	23 to 25	1.4×10^{-3}
Boardman, Oregon, United States	Columbia	13	3.7×10^{-3}
Casper, Wyoming, United States	North Platte	3 to 12	9×10^{-4} to 3×10^{-3}
Cedar Rapids, Iowa, United States	Cedar	12 to 18	7.5×10^{-5} to 1×10^{-3}
Cincinnati, Ohio, United States	Great Miami	~30	8.8×10^{-4} to 1.5×10^{-3}
Louisville, Kentucky, United States	Ohio	21	6×10^{-4}
Dresden-Tolkewitz, Germany	Elbe	10 to 13	1×10^{-3} to 2×10^{-3}
Meissen-Siebeneichen, Germany	Elbe	15 to 20	1×10^{-3} to 2×10^{-3}
Torgau-Ost, Germany	Elbe	40 to 55	6×10^{-4} to 2×10^{-3}
Auf dem Grind, Düsseldorf, Germany	Rhine	25 to 30	1×10^{-3} to 1×10^{-2}
Flehe, Düsseldorf, Germany	Rhine	10 to 12	3×10^{-3} to 6×10^{-3}
Böckingen, Germany	Neckar	3 to 5	1×10^{-2}
Maribor, Slovenia	Drava	14	2×10^{-3} to 4×10^{-3}
Karany, Czech Republic	Jizera	8 to 12	4×10^{-4}

Table 14-2 is a compilation of selected hydrogeologic information for six RBF sites in the United States and three in Germany. As can be seen, the conditions vary mainly for capacity, travel time, and distance between the river and wells. At most sites in Europe, the distance between the riverbank and production wells is >50 m and travel times are >50 days. In the United States, travel times are <50 days at the sites reported here.

Along with the hydraulic conductivity of the riverbed, the thickness determines the possible pumping rate along a river if wells are installed. Otherwise, thickness is less important if the laterals of collector wells are installed directly beneath the riverbed. Two types of RBF settings can be distinguished:

- Water extracted beneath a riverbed.
- Water extracted along a riverbed.

In terms of well construction, two types of wells can be distinguished:

- Vertical wells.
- Collector wells with laterals.

Vertical well construction has been used since the early beginnings of RBF. Nowadays, vertical wells are preferred for the extraction of low water quantities. Collector wells were developed later and are preferably installed at sites with high water extraction rates.

Table 14-2. Selected Site Data for Nine RBF Systems in the United States and Germany

Items	SCWA, California	Boardman, Oregon	Lincoln, Nebraska	Cedar Rapids, Iowa	Louisville, Kentucky	Somersworth, New Hampshire	Dresden-Tolkewitz, Germany	Meissen-Siebeneichen, Germany	Torgau-Ost, Germany
Wells									
Number/Type[a]	5H, 7V	2H	2H, 44V	2H, 53V	1H	2V+1V	71V	3V	42V
Maximum Capacity (m³/d)	322,000	87,000	132,000	128,500	76,000	5,300	40,000	6,000	150,000
Screen Zone Below Land Surface (m)	24 to 30	15 to 15.6	12 to 18	18 to 24	24 to 30	12 to 16.5	15 to 19	12 to 17	32 to 52
Screen Zone (m)	6	0.6	6	6	6	4.5	4	5	20
Distance to River (m)	0 to 75	3 to 18	<30 to >800	9 to 245	<30 to 84	46	80 to 180	100 to 150	300
Travel Time (days)	4.9	<1	<7 to >14	2 to 17	2 to 5	<55	25 to 50	50 to 100	80 to 300
River									
Discharge (m³/s)	<2.8 to >1,400	6,370 to 7,080	<50 to 3680	4 to 2,025	6,300 to >28,000	NA	120 to 2,000	120 to 2,000	120 to 2,000
Width/Depth[b] (m)	15 to 90/NA	4,000/3	300/1.5	225/2.5 to 3	600/10	12/NA	120/2	140/2	130/2
Bed Sediments	Sand	Sand/Silt	Sand	Sand	Sand	Gravel/Sand	Coarse Gravel	Gravel	Gravel
Aquifer									
Type	Unconfined	Unconfined, Some Leaky	Unconfined to Leaky	Unconfined to Confined	Unconfined	Leaky	Unconfined	Unconfined	Unconfined
K (m/s)	2.4×10^{-4} to 4.3×10^{-4}	3.7×10^{-3}	1.4×10^{-3}	1.5×10^{-4} to 1.1×10^{-3}	6×10^{-4}	4.3×10^{-4}	1×10^{-3} to 2×10^{-3}	1×10^{-3} to 2×10^{-3}	6×10^{-4} to 2×10^{-3}
Thickness (m)	8 to 26	15	21 to 26	15 to 20	0 to 40	15	10 to 13	15 to 20	40 to 55
Specific Yield (%)	NA	NA	15	10	NA	NA	20	20	20
Material	Sand and Gravel	Sand and Gravel	Sand and Gravel	Fine to Medium Sand on Coarse Gravel	Sand and Gravel with Silt and Clay	Sand with Some Gravel	Sand and Gravel	Medium and Coarse Sand	Medium and Coarse Sand
Heterogeneity	NA	Homogeneous	Few Clay Lenses	Silty Clay Lenses	Coarse Gravel and Pebbles	Fine Silt Lenses	Homogeneous	Few Fine Sand Lenses	Few Silt Lenses

a V = Vertical well. H = Horizontal collector well. b At mean river water level. NA = Not available. SCWA = Sonoma County Water District.

The production capacity of RBF systems that extract water beneath a riverbed is much higher than that of water extracted along a riverbank, if there is no or only low-clogging of the riverbed, but the advantages of RBF processes (e.g., mixing and equilibration) are widely missed in water extracted beneath the riverbed. Key aspects for planning water extraction beneath a riverbed include:

- Riverbed hydraulic conductivity.
- Riverbed hydraulics (erosion, deposition).
- River channel morphology.
- Flooding.
- Shipping and dredging uses.

Erosive conditions in the river are advantageous because they limit the formation of a clogging layer. Floods also remove the clogging layer, which might be enriched with heavy metals and adsorbed organic compounds; however, some floods may have a negative effect if the production wells or laterals of collector wells are located near the banks. Even if the wells are not flooded, the destruction of the clogging layer, changes in pore pressure, and higher flow velocities in the aquifer can lead to an increased transport of dissolved compounds, previously adsorbed particles, bacteria, and viruses, which could possibly result in breakthroughs. In some settings, moderate floods may be useful in removing the clogging layer.

For production wells and laterals of collector wells that are located at a distance from the riverbank, hydrogeological conditions and aquifer properties have a great impact on productivity and the quality of pumped water. For this type of RBF scheme, the following questions should be answered before planning:

- Which advantage of RBF is the most important?
- Which proportion of riverbank filtrate in pumped raw water should be achieved for water-quality purposes?
- Which length of the river or catchment area along the river can be used for water extraction?
- What amount of water has to be produced?
- Which drawdown is acceptable depending on aquifer thickness and land use?
- What amount of water can be continuously withdrawn from the river without ecological conflicts?

Table 14-3 gives examples for the siting and management of RBF schemes and their effects on the intended advantages of RBF.

In most cases, a long distance between the riverbank and production wells has a favorable effect on water quality. The common opinion is that the flow time of riverbank filtrate is the most important parameter for water quality; however, findings from a literature survey and field studies show that the surface area in contact with the infiltrated water is of higher importance (Nestler et al., 1998). Thus, the flow path length, together with the thickness of the aquifer and the infiltration area in the river, are the parameters to examine.

A long flow path further attenuates organic compounds in addition to eliminating easily biodegradable compounds, which already occurs in and near the riverbed. Bacteria removal is related to sorption and half-life. Commonly, a retention time of 50 days is considered adequate for this removal by most European utilities. That time, however, is not sufficient for fully removing viruses and pathogens. Furthermore, the processes to remove viruses and pathogens are not yet fully understood. For example, during floods and changes in flow velocity and pressure in the aquifer, a higher number of bacteria is observed at sampling points where surface water could not have been transported during a certain time of flooding (Schubert, 2000).

Table 14-3. Effects of Design Parameters on RBF Efficiency

Aim	Long Distance Between Bank and Wells	Low Pumpage along a Line Compared to High Pumpage at a Point	Siting of Wells in a Meander
Removal of Suspended Solids and Particles	0	0	0
Removal of Bacteria, Viruses	++	+	0
Elimination of Biodegradable Compounds	+	+	0
Equilibration of Temperature Changes	++	+	0
Equilibration of Changes in Concentrations of Compounds in Water	++	+	0
High Proportion of Riverbank Filtrate in Pumped Raw Water	- -	0	++

+ = Increase/improvement.   – = Decrease.   0 = No significant effect.

3. Design Options

Different scenarios for siting wells along a river are possible (see Table 14-3). The possible influence of well location on the proportion of riverbank filtrate that is extracted has been calculated by a simple, fictive groundwater flow model using MODFLOW (Harbaugh and McDonald, 1996). Table 14-4 gives a summary of the main model parameters. The size of the model domains was chosen equivalent to the necessary recharge area covering the landside groundwater flow to the wells.

Table 14-4. Parameters Used for MODFLOW Simulations

Parameter	Value	Parameter	Value
Aquifer Type	Unconfined	Aquifer Thickness	25 m
Hydraulic Conductivity	1×10^{-3} m/s	Number of Layers	1 Layer
Heterogeneity Issues	Homogeneous and Isotropic	Riverbed Conditions	No Clogging
Groundwater Flow	Perpendicular to the River	Well-Screen Placement	Fully Penetrating
Well Capacity	50 m³/h	Recharge	5 L/(s × km²)
Effective Porosity	0.25	Slope of River	0.03 Percent
Slope of Non-Pumping Groundwater	0.04 Percent	Distance Between Wells and Riverbank (L)	100 m
Boundary Type at Southern Edge	Fixed Head	Boundary Type for River	Fixed Head

When vertical wells are sited in a line parallel to the riverbank (Figure 14-1), the wells must be placed an appropriate distance apart to reduce interference with each other. This is often referred to as a well gallery. Another possibility includes placing wells in well groups (Figure 14-2). The calculated proportion of river water in pumped raw water for conditions as given in

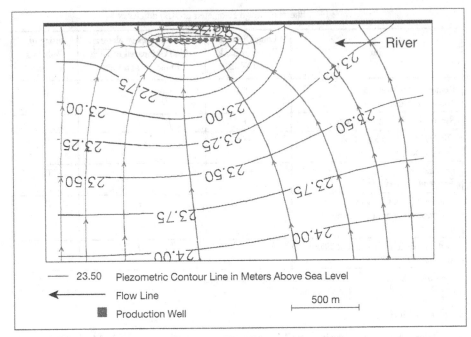

Figure 14-1. Piezometric contours and flow path lines around a well gallery parallel to a river.

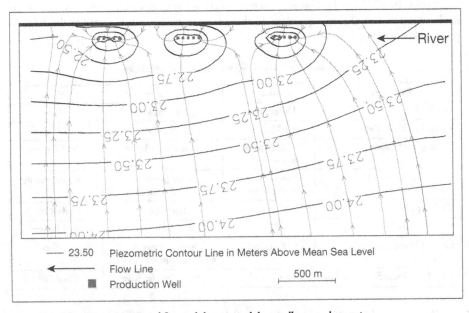

Figure 14-2. Piezometric contours and flow path lines around three well groups along a river.

Table 14-4 are 77 percent for a well gallery of 15 wells (see Figure 14-1) and 65 percent for three well groups of five wells each and a distance between the well groups of 400 to 500 m (see Figure 14-2). The higher the cone of depression, the higher the proportion of river water (if the infiltration is not limited by clogging). Calculations with different pumping rates of the five wells within a well group (but with the same total pumpage of 250 m³/h gave the same proportion of river water. More important is the distance between the well groups and the formation of a large cone of depression. If the aquifer is thin, well groups may have an advantage over a well gallery; if both settings (gallery versus groups) have the same number of wells, the well gallery would produce less water than the well groups because of less drawdown.

The higher the number of wells or the longer the distances between the wells, the higher the used aquifer volume and, therefore, the higher the reactive surface of the aquifer material in contact with the infiltrate. The distance between the river and the wells should be optimized based upon:

- Expected leakage rates of the river.
- Preferred flow path length.
- Preferred retention times.

If significant clogging in the riverbed is expected, then the wells must be placed closer to the bank to ensure that the planned proportion of riverbank filtrate is extracted.

The best place for pumping a large quantity of riverbank filtrate is on an island or within a meander, especially if the river has a steeper gradient than the groundwater in the connected aquifer and if the riverbed has high hydraulic conductivity. Another fictive groundwater flow model was used for evaluating four variants of siting wells in a meander (Figure 14-3 a-d). Five wells with pumping rates of 50 m³/h each and a distance of 40 m between the wells and 100 m from the riverbank (see Figure 14-3 a-c) are placed at different locations within a meander. The model parameters are similar to Table 14-4.

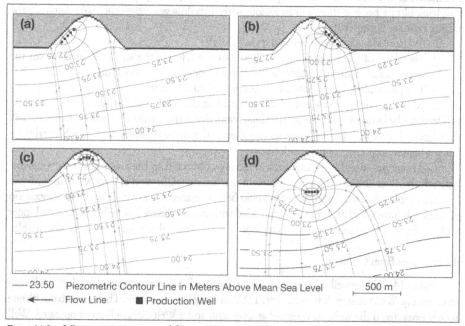

———— 23.50   Piezometric Contour Line in Meters Above Mean Sea Level   500 m
◄———— Flow Line   ■ Production Well

Figure 14-3 a-d. Piezometric contours and flow path lines around well groups at different locations in a meander.

Table 14-5 summarizes the calculated proportions of riverbank filtrate and groundwater in the raw water as well as the minimum retention times of riverbank filtrate and the lengths of infiltration areas in the river.

Table 14-5. Quantities of River Water, Flow Times, and Length of the Infiltration
for the Four Variants of RBF Well Configurations

Variant	River Water (%)	Landside Groundwater (%)	Minimum Flow Time (days)	Length of Infiltration Area in the River (m)
A	67	33	30	890
B	73	27	30	550
C	78	22	30	710
D	35	65	600	1,230

For the documented conditions, the highest proportion of riverbank filtrate is extracted in Variant C, where the well group is placed in the curve of the meander. The lowest proportion is obtained if the well group is outside the meander at a greater distance from the river (Variant D). The minimum flow time depends upon the distance between the wells and the riverbank under the documented conditions. Determining minimum flow times is helpful in designing monitoring programs and in estimating attenuation rates. The minimum travel time calculated here reflects the time between the riverbank and the well. If the river width is about 100 m or more, the average travel time would be much higher than the minimum, and the volume of water having that minimum retention time might be very low. Thus, the mean flow time of the pumped riverbank filtrate is more important even though it is not easy to calculate. The length of the infiltration area in the river is not helpful in characterizing the aquifer volume in contact with riverbank filtrate because there will be regions with high flow velocities that could dominate the quality changes during RBF.

Infiltration will occur naturally in the meander if:

- The slope of the hydraulic grade line from the river to the aquifer is higher than the groundwater gradient towards the river under non-pumping conditions.
- The piezometric head contours are not parallel to the river.

In addition, during floods, water will infiltrate at the bank with a higher water level and exfiltrate at the bank with a lower level. Along many rivers, the aquifer zone within meanders will show groundwater and surface water mixing due to the dynamic water level. Such conditions allow a higher proportion of riverbank filtrate.

Old branches, which are downstream and are connected to the river, can cause gradients between the river and the branch, resulting in the natural infiltration of river water towards the branch. Despite the fact that such zones would be advantageous for extracting a high proportion of riverbank filtrate, old branches are often filled with mud and have low bed conductivity. Under such conditions, the extraction of a high proportion of riverbank filtrate may not be sustainable.

4. Examples from Germany

Figure 14-4 shows well galleries sited along the Elbe River downstream of the City of Torgau in Germany. Well Gallery A is more than 1,000-m away from the river. There, riverbank filtrate is only extracted if the wells are operated at long-term and if Well Gallery B is not operated. Well Gallery B is placed very near the river. This results in low flow times and lower attenuation rates

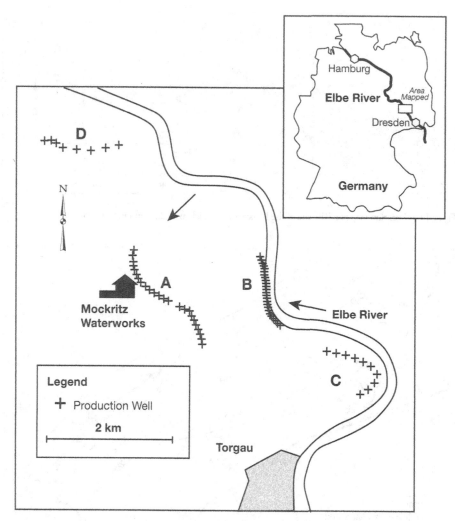

Figure 14-4. Location of well galleries along the Elbe River downstream of the City of Torgau in Germany.

of organic compounds compared to that of Well Gallery C. Well Gallery C was found to be the optimum for having a high proportion of riverbank filtrate (50 to 70 percent) and high attenuation rates for DOC.

Figure 14-5 shows single wells sited along the Elbe River upstream of the City of Meissen in Germany. Local geological conditions affect flow conditions. The river flows through a sharp valley of hardrocks, and alluvial deposits are only found along the river and up to 300 m south from the river. There is a very low proportion of landside groundwater; however, geological conditions allow groundwater flow beneath the river from the northeast to the wells in the south. Such undercurrent (the flow beneath a river) is also observed if the aquifer has high anisotropy or less permeable layers to promote such flow (Nestler et al., 1996).

An old, low-cost RBF scheme is shown in Figure 14-6. Three vacuum well galleries are each connected to a large well at the waterworks. No pumps are installed in the wells. In the 1980s,

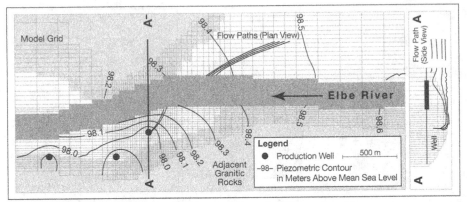

Figure 14-5. Location map of a RBF site of the Meissen-Siebeneichen Waterworks in Germany.

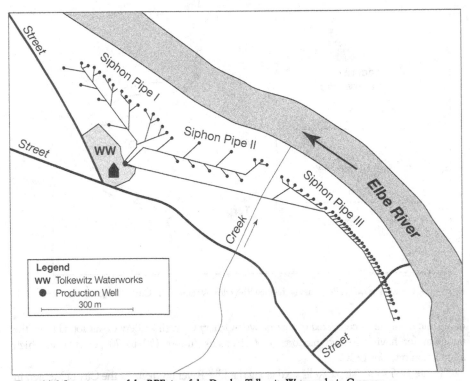

Figure 14-6. Location map of the RBF site of the Dresden-Tolkewitz Waterworks in Germany.

strong river-water pollution by organics (from pulp and paper mills) and high water extraction rate caused unsaturated conditions beneath the riverbed and problems with raw-water quality, especially bad taste and odor. After river-water quality was improved (1989 to 1993), the hydraulic conductivity of the riverbed increased. At present, the total pumpage is less than half of that from 1987, and selected wells are in operation to ensure longer flow times and better riverbank-filtrate quality.

The City of Düsseldorf in Germany maintains a series of wells that are located on the banks of the Rhine River. These are either vertical wells or horizontal filter wells. Vertical wells are connected to vacuum lines for pumpage. The horizontal filter wells are similar in design to collector wells; however, the laterals of these wells are away from the river and the laterals do not go beneath the riverbed. The river cross-section near Düsseldorf also shows a mixed set of conditions: part of the river sediment moves and part is clogged (Schubert, 2000).

5. Examples from Hungary

Some of the largest RBF systems have been built along riverbanks or on islands in rivers. Approximately 850 wells, both vertical and horizontal collector wells, are used by the Budapest Waterworks for water extraction along the Danube River in Hungary. Of these wells, approximately two-thirds are located on Szentendre Island, upstream of the City of Budapest. Water extracted from wells in this island does not undergo any further treatment with the exception of chlorination. Water is gravity-fed to the city distribution system. In addition, there are wells on Csepel Island, downstream of the City. Because the Danube River receives large amounts of pollutant load from Budapest and because the soil causes anaerobic conditions, water from the Csepel Island contains high concentrations of iron and manganese and must undergo further treatment. For many of the collector wells, the laterals go partly beneath the riverbed.

6. Example from the United States

Utilities within the United States mostly use collector wells that have laterals directly below the river. It is also possible to have landside laterals, and many of the old and newly installed wells have a small number of landward laterals. Ray (2001) conducted a simple modeling exercise for a collector well that had five laterals and was located on a highly pervious riverbank. The peak of a contaminant (ethylene dibromide) in the river was 4 µg/L. The concentration hydrograph is presented in Figure 10-19 of this book. The hydrograph was asymmetric, with an early peak and a tail end that lasted about a week. The duration of the contaminant pulse in the river was 9 days and, for 1 day, the concentration exceeded 3 µg/L. When all the laterals are directed towards the river (Figure 14-7 a), the peak concentration of ethylene dibromide at the caisson was estimated to reach 0.006 µg/L. If five other laterals (two parallel to the river and three on the landward side) were connected to the caisson at the same depth as the river laterals (Figure 14-7 b), the peak concentration at the caisson would be less than 0.001 µg/L. In the second case, more groundwater from the aquifer is pumped, thus diluting the impact of the concentration peak.

7. Conclusion

The siting of RBF schemes is an optimization task that needs a balance between the extracted volume of riverbank filtrate and the preservation of water quality due to attenuation and mixing processes during RBF. High extraction rates of riverbank filtrate and effective attenuation rates can be achieved if wells are placed on an island or in a meander. Furthermore, the siting and design of an RBF system is not only a function of hydrogeological factors, but also of technical, economical, regulatory, and land-use factors. Some utilities prefer to use horizontal collector wells because these wells require a small number of pumps and have low operation and maintenance costs compared to vertical wells; however, a complex cost analysis of vertical versus horizontal collector wells, including installation, operation, and maintenance under different hydrogeological conditions, is not available. Due to those site-specific conditions, no general

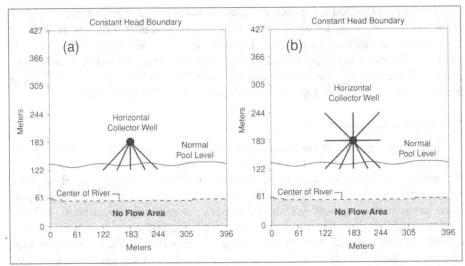

Figure 14-7 a-b. Layout of the horizontal collector wells with five laterals towards the river (a) and five additional laterals towards the land (b).

construction procedure can be defined. Investigations should focus on extended comparisons of existing RBF sites, as intended herein, and on classifying different features. More efforts should be geared toward understanding attenuation processes through site-specific investigations and on understanding how the design and siting of wells should address clogging and scouring issues.

References

Harbaugh, A.W., and M.G. McDonald (1996). "User's documentation for MODFLOW-96. An update to the U.S. Geological Survey Modular Finite-Difference Ground-Water Flow Model." *Open File Report 96-485*, U.S. Geological Survey, Reston, Virginia.

Nestler, W., J. Dehnert, P. Neitzel, and T. Grischek (1996). "Untersuchungen zur Unterströmung der Elbe" (Investigations into groundwater flow beneath the Elbe River). *Wasser und Boden*, 48: 53-58 (in German).

Nestler, W., W. Walther, F. Jacobs, R. Trettin, and K. Freyer (1998). "Water production in aquifers in the catchment areas of the River Elbe." *UFZ-Report 7*, 204 p. (in German).

Ray, C. (2001). "Modeling riverbank filtration systems to attenuate shock loads in rivers." *Proceedings, American Water Works Association Annual Meeting*, American Water Works Association, Denver, Colorado.

Schubert, J. (2000). "Entfernung von Schwebstoffen und Mikroorganismen sowie Verminderung der Mutagenität bei der Uferfiltration" (Removal of particles and microorganisms and decreasing mutagenicity during bank filtration). *Gas- und Wasserfach Wasser/Abwasser*, 141(4): 218-225 (in German).

Chapter 15. Natural Organic Matter Removal During Riverbank Filtration: Current Knowledge and Research Needs

Jörg E. Drewes, Ph.D.
Colorado School of Mines
Golden, Colorado, United States

R. Scott Summers, Ph.D.
University of Colorado
Boulder, Colorado, United States

1. Introduction

NOM in source waters can contain significant levels of DOC, which constitutes the reactive organic content and is responsible for the majority of reactions of interest in water-treatment processes (e.g., disinfectant demand, disinfection byproduct formation, biogrowth, coagulant demand). NOM control has become a focus of drinking-water treatment and is regulated for surface waters in the United States by the enhanced coagulation requirement of Stage 1 of the Disinfectant/Disinfection Byproduct Rule. Subsurface treatment has been shown to be an effective alternative to conventional and advanced drinking-water treatment. Two low-cost subsurface techniques are available:
- Surface spreading basins using treated or partially treated water.
- Infiltration systems with raw water from lakes, rivers, or reservoirs.

The latter approach, which is termed RBF, represents a low-cost element that can be integral to a multi-barrier concept of drinking-water treatment. Research studies conducted over the past 25 years in Europe (Sontheimer and Nissing, 1977; Sontheimer, 1980; Sontheimer, 1991; Kühn and Müller, 2000) and the past 5 years in the United States (Weiss et al., 2002; Wang, 2002) have demonstrated that RBF is an effective technology to remove and transform not only pathogens, but also organic matter, disinfection byproduct precursors, disinfection byproducts, and other trace organics.

2. NOM in Surface Water and Groundwater

NOM is commonly used to describe the complex matrix of organic material in both dissolved and particulate form that occurs ubiquitously in both surface water and groundwater. DOC is operationally defined by a single filtration step, most commonly through a 0.45-μm filter, whereas particulate organic matter represents colloidal and particulate matter. DOC consists of both humic (i.e., humic and fulvic acids) and non-humic components and is derived from allochthonous (external) as well as autochthonous (internal) sources.

According to MacCarthy (2001), "Humic substances comprise an extraordinary complex mixture of highly heterogeneous, chemically reactive yet refractory molecules, produced during early diagenesis in the decay of biomatter, and formed ubiquitously in the environment." Humic

303

C. Ray et al. (eds.), Riverbank Filtration, 303–309.
© 2002 *Kluwer Academic Publishers. Printed in the Netherlands.*

substances have been studied extensively because of their environmental significance. Due to the extreme complexity of humic substances, many research efforts have focused on:

- Molecular-size distribution.
- Functional group characterization.
- Metal complexation properties.
- Acidic properties.
- Reactivity with chlorine.

Fulvic acids represent the most water-soluble fraction of humic material and contribute 90 percent of the dissolved humic substances in most natural waters (Malcolm, 1985). Wilson (1959) reported that color in water is caused by fulvic acids. The molecular weight of fulvic acids typically ranges between 500 and 2,000 daltons (Thurman et al., 1982). Humic acids have molecular weights greater than 2,000 daltons. Humic molecules contain aromatic, carbonyl, carboxyl, methoxyl, and aliphatic units, with the phenolic and carboxylic functional groups providing most of the protonation and metal complexation sites.

The non-humic fraction of DOC includes hydrophilic acids, proteins, amino acids, amino sugars, and carbohydrates. All these groups of compounds are likely to be present in natural waters, although their absolute and relative concentrations are expected to vary from site to site. With respect to chemical properties and their implications for water treatment, the fraction of NOM designated as humic substances has, historically, been considered the most important (Thurman, 1985). Aquatic humic substances account for approximately 50 percent of the DOC present in most natural waters (Thurman, 1985; Owen et al., 1993). Humic substances impart a brown or yellow color to water, which is undesirable. The chlorination of humic substances has been shown to release humic sub-units through the formation of chlorinated and non-chlorinated products, some of which are known to be carcinogenic (Sonnenberg et al., 1989). Humic substances also compete with regulated target compounds for adsorption sites on activated carbon and can cause the fouling of high-pressure membranes.

Historically, the non-humic fraction was less understood as it was difficult to concentrate and to characterize; however, recent studies suggest that the reactivity of the non-humic fraction was shown to be similar to that of the humic fraction when disinfection byproduct formation was normalized on a DOC basis (Owen et al., 1993). In addition, the non-humic fraction may represent biodegradable organic carbon, which is important for the bacterial re-growth potential in distribution systems. NOM characterization has been a priority for the water-treatment industry, in part because such characterization holds the key to understanding, predicting, and perhaps controlling NOM reactivity under water-treatment conditions. Previous investigations of NOM from a wide variety of sources have led to some generalizations about NOM molecule characteristics in different environments (Owen et al., 1993).

NOM in lakes, reservoirs, and streams of moderate to high trophic status is often dominated by autochthonous material, whereas low-order rivers and streams usually carry more allochthonous NOM. Allochthonous NOM has large carbon-to-nitrogen ratios (near 100 to 1), is highly colored, and has significant aromatic carbon content, whereas autochthonous NOM has lower carbon-to-nitrogen ratios (near 10 to 1), is almost colorless, and has low aromatic carbon content.

3. Fate of NOM During Travel Through Subsurface/Porous Media

Studies conducted in the past on artificial groundwater recharge and RBF systems suggest that the removal and transformation processes of organic matter during travel through the subsurface are highly dynamic. Although findings generated at full-scale operations point to a site-specific removal behavior, some general statements with respect to removal of organic matter in RBF systems can be made.

Particulate Organic Matter

Particulate organic matter seems to be effectively removed during the initial phase of infiltration where river water penetrates into the subsurface. This removal is usually associated with coagulation and precipitation processes (Sontheimer and Nissing, 1977).

Dissolved Organic Carbon

Long-term measurements have indicated that RBF systems have a nearly constant performance in removing dissolved organic constituents in river water without significantly accumulating organic matter in the subsurface (Sontheimer, 1991). DOC removal during RBF at the Rhine River in Europe was constant and accounted for approximately 50-percent removal of organic matter in the river water (Sontheimer and Nissing, 1977; Kühn and Müller, 2000). The most recent studies conducted by Weiss et al. (2002) and Wang (2002) at RBF systems on the Ohio, Wabash, and Missouri Rivers in the United States confirmed previous findings that RBF can remove up to 50 to 60 percent of DOC.

The biological degradation of DOC under aerobic conditions is a major removal process of subsurface processes. Dissolved oxygen concentrations in the infiltration zone are maintained through infiltrated water and diffusion. Steady infiltration conditions in RBF systems generate a special bioactive filtration layer at the water/sediment interface (Matthess, 1990). Amy et al. (1993), Kivimaeki et al. (1998), and Drewes and Fox (1999) observed the highest removal of DOC and TOX within the first meter of infiltration. Wang (2002) also reported that the majority of NOM removal occurred within the first 15 m of infiltration. Size-exclusion chromatography with online DOC and ultraviolet-absorbance detection was used by Drewes et al. (in press) to evaluate changes in the biodegradable and refractory components of organic matter during the initial phase of soil-aquifer treatment. The size-exclusion chromatograms indicated a substantial removal of polysaccharides and protein-like non-humic organics during percolation through the upper vadose zone in soil-aquifer treatment systems using treated wastewater effluents. The preferred removal of non-humic biodegradable carbon, reflecting the re-growth potential for distribution systems, is consistent with observations made in RBF systems. Kühn and Müller (2000) reported that RBF could reduce the biological re-growth potential by more than 60 percent. The assimilable organic carbon level in riverbank filtrate determined by Wang (2002) was significantly lower than in river water, resulting in a significantly reduced re-growth potential.

In addition to findings pointing to a removal of non-humic material, Drewes et al. (in press) observed a significant removal of larger molecular weight fractions representing hydrophobic and transphilic acids and neutrals during soil-aquifer treatment. Based on advanced spectroscopic techniques using carbon-13 nuclear magnetic resonance spectroscopy (13C-NMR) and Fourier-transform infrared spectroscopy, carbohydrates of hydrophobic and transphilic acid isolates decreased during short-term soil-aquifer treatment of infiltrated wastewater while aliphatic carbon increased. The aromatic carbon of these isolates decreased significantly only during long travel times (2 to 8 years of transport) with relatively small changes in DOC concentration. Kivimaeki et al. (1998) and Gerlach and Gimbel (1999) also observed that the highest molecular weight fractions of organic matter were removed during the initial phase of infiltration because the collision efficiencies greatly increased with increasing molecular weight. These findings point to the long-term transformation of remaining organic matter during travel through the subsurface towards well-aged fulvic acids, while non-humic material is biodegraded and humic substances are adsorbed during the initial phase of infiltration. A fundamental understanding of the removal mechanisms for humic substances in RBF systems is still missing, although most studies suggest a combination of initial adsorption followed by biodegradation (Sontheimer, 1991; Gerlach and Gimbel, 1999). The accumulation of organic matter in RBF systems due to the continuous deposition of humic substances usually does not occur due to microbiologically induced degradation of sorbed humic substances (Gerlach and Gimbel, 1999). Gerlach (1998) reported that the oxygen consumption in RBF is two to three times greater than the stoichiometric oxygen requirements for DOC and ammonium ion (NH_4^+) mineralization. There is also evidence for a total increase of the molecular weight fraction of less than 1,000 daltons in RBF systems after a subsurface flow of nearly 100 m (Ludwig et al., 1997), which points to a partial degradation of humic substances.

Redox Conditions

The accumulation of organic compounds in sediments and the introduction of DOC into groundwater usually determine the predominant redox conditions in the subsurface (Matthess, 1990). Stuyfzand (1998) showed a strong relationship between the removal of organic constituents and the redox environment, which might represent conditions from suboxic (presence of dissolved oxygen and nitrate) to anoxic (absence of dissolved oxygen), toward deep anoxic (hydrogen sulfide [H_2S] in solution). Schwarzenbach et al. (1983) found strong evidence in their field-scale studies that certain organic micropollutants (e.g., 1,4-dichlorobenzene) were only biotransformed under aerobic conditions. The elimination of such compounds may, therefore, be hindered if anaerobic conditions prevail in the aquifer in the vicinity of the infiltration zone. In contrast to artificial groundwater recharge systems using surface spreading basins, redox conditions in RBF systems are not always constant. They heavily depend on:

- Flow regimes of the feeding river.
- Local geohydrological conditions.
- Amount of biodegradable carbon in river water.

At the Rhine River, riverbank filtrate in the early 1970s was characterized by anaerobic conditions, with dissolved oxygen concentrations less than 1 mg/L. Improved wastewater treatment and industrial pollution control programs in the watershed of the Rhine River resulted in decreased organic carbon loads in the river since the early 1980s (Kühn and Müller, 2000). This surface-water quality improvement resulted in aerobic redox conditions in the riverbank filtrate with dissolved oxygen concentrations reaching 3 mg/L.

Disinfection Byproduct Precursors

Weiss et al. (2002) conducted studies at water-treatment systems that employ RBF along major Midwestern rivers. The researchers observed a reduction of disinfection byproduct FP during RBF, which varied for THMs from 53 to 82 percent and for HAAs from 47 to 80 percent, respectively. The results also indicated a shift from chlorinated to more brominated disinfection byproduct species. This finding is important to public health since the brominated species are suspected to be more carcinogenic than the chlorinated species. In these studies and investigations by Wang (2002), the FP concentrations were reduced to a greater extent than DOC, suggesting a preferential removal of precursor material during RBF.

Micropollutants

Studies focusing on micropollutants in river water and riverbank filtrate indicate that RBF acts as a barrier for many substances (BMI-Fachausschuss, 1985; Mühlhausen et al., 1991; Sontheimer, 1991). Schwarzenbach et al. (1983) observed that the biological processes responsible for the removal of various micropollutants occurred predominantly within the first few meters of infiltration. The removal of organic compounds due to adsorption onto sediments seems to depend on the fraction of organic carbon (f_{oc}) of the subsurface and the hydrophobicity of individual compounds. Haberer et al. (1985) conducted adsorption experiments with sediments from RBF systems. Adsorption increased with the increasing organic carbon content of the sediments. During full-scale testing, no breakthrough of organic compounds occurred with the exception of 1,2-dichlorobenzene, 1,4-dichlorobenzene, and 1,2,4-trichlorobenzene. During field studies, Schwarzenbach et al. (1983) found that the retention of highly lipophilic compounds (such as hexachlorobenzene) was rather small in aquifer materials with low organic carbon content (less than 0.1 percent). Of all the compounds studied by Roberts et al. (1980), chlorobenzene was transported most rapidly in a groundwater recharge system. From the form of the concentration response at an observation well, chlorobenzene appeared to exhibit the properties of adsorption and dispersion without biodegradation. Dichlorobenzene isomers and 1,2,4-trichlorobenzene seemed to be more strongly adsorbed than chlorobenzene because their arrival at the observation wells was delayed longer.

Some micropollutants, however, show only partial or no significant removal during RBF, such as aromatic sulfonic acids, EDTA (a chelating agent), naphthalene-1,5-disulfonate (a concrete additive), clofibric acid (a lipid regulator drug), carbamazepine (an anti-epileptic drug), or X-ray contrast agents (Haberer and Ternes, 1996; Neitzel et al., 1998; Heberer et al., 1998; Kühn and Müller, 2000; Putschew et al., 2000). Among volatile organic compounds for which no evidence of biological transformation under any redox conditions was found were chloroform, 1,1,1-trichloroethane, trichloroethylene, and tetrachloroethylene (Schwarzenbach et al., 1983). The limited removal of these micropollutants requires additional post-treatment steps where riverbank filtrate is used for drinking-water supply.

4. Future Research Needs

Although there is evidence of a substantial removal of NOM, disinfection byproduct precursors, and organic micropollutants during percolation through subsurface systems, a lack of knowledge exists with respect to the relative changes of organic matter composition during RBF and factors responsible for those changes. Research is needed in the following areas:

- The most significant research need is a more complete description of removal processes for the non-humic and humic fractions of NOM.
- What is the role of dominating redox conditions?
- What is the role of travel distance/time?
- What is the fate of the biodegradable fraction of NOM? Why is there residual assimilable organic carbon or biodegradable organic carbon after passage?
- What is the impact on the speciation of disinfection byproducts that are formed in post-treatment disinfection?
- The potential role of organic nitrogen on nitrogen-containing disinfection byproducts should be investigated more fully.
- How important are redox conditions and NOM complexation for the removal of organic micropollutants?

References

Amy, G., G.L. Wilson, A. Conroy, J. Chahbandour, W. Zhai, and M. Siddiqui (1993). "Fate of chlorination byproducts and nitrogen species during effluent recharge and soil aquifer treatment (SAT)." *Water Environment Research*, 65(6): 726-734.

BMI-Fachausschuss (1985). *Kuenstliche Grundwasseranreicherung. Stand des Wissens und der Technik in der Bundesrepublik Deutschland. Bundesminister des Innern*. Erich Schmidt Verlag, Berlin, Germany.

Drewes, J.E., and P. Fox (1999). "Fate of natural organic matter (NOM) during groundwater recharge using reclaimed water." *Water Science and Technology*, 40(9): 241-248.

Drewes, J.E., D.M. Quanrud, G.L. Amy, and P.K. Westerhoff (2002). "Character of organic matter in soil-aquifer treatment systems." *Journal of Environmental Engineering*, in press.

Gerlach, M. (1998). "Zur Bedeutung von Huminstoffen bei der Trinkwassergewinnung aus Uferfiltrat." Ph.D. thesis, Institute of Chemical Engineering/Water Engineering, University of Duisburg, Duisburg, Germany.

Gerlach, M., and R. Gimbel (1999). "Influence of humic substance alteration during soil passage on their treatment behaviour." *Water Science and Technology*, 40(9): 231-239.

Haberer, K., M. Drews, H. Kussmaul, and D. Mühlhausen (1985). "Verhalten von organischen Schadstoffen bei der kuenstlichen Grundwasseranreicherung und Entwicklung von speziellen Methoden zu deren Ueberwachung." *Research Report 102 02 302/04 UBA-FB 83-053*, German Environmental Protection Agency, Berlin, Germany

Haberer, K., and T.A. Ternes (1996). "Bedeutung von wasserwerksgaengigen Metaboliten fuer die Trinkwassergewinnung." *Gass- und Wasserfach Wasser/Abwasser*, 137(10): 573-578.

Heberer, T., K. Schmidt-Bäumler, and H.J. Stan (1998). "Occurrence and distribution of organic contaminants in the aquatic system in Berlin. Part I: Drug residues and other polar contaminants in Berlin surface and groundwater." *Acta Hydrochima et Hydrobiolica*, 26: 272-278.

Kivimaeki, A.L., K. Lahti, T. Hatva, S.M. Tuominen, and I.T. Miettinen (1998). "Removal of organic matter during bank filtration." *Artificial Recharge of Groundwater*, J.H. Peters et al., eds., A.A. Balkema, Rotterdam, The Netherlands.

Kühn, W., and U. Müller (2000). "Riverbank filtration - An overview." *Journal of American Water Works Association*, 92(12): 60-69.

Ludwig, U., T. Grischek, W. Nestler, and V. Neumann (1997). "Behaviour of different molecular-weight fractions of DOC of Elbe River during riverbank filtration." *Acta Hydrochima et Hydrobiolica*, 25(3): 145-150.

MacCarthy, P. (2001). "The principles of humic substances: An introduction to the first principle." *Humic substances – Structures, models and functions*, E.A. Ghabbour and G. Davies, eds., Royal Society of Chemistry, Cambridge, England, 19-30.

Malcolm, R.L. (1985). "Geochemistry of stream fulvic and humic substances." *Humic substances in soil, sediment, and water: Geochemistry, isolation, and characterization*, G.R. Aiken, D.M. McKnight, R.L. Wershaw, and P. MacCarthy, eds., John Wiley and Sons, New York, New York.

Matthess, G. (1990). *Die Beschaffenheit des Grundwassers. Lehrbuch der Hydrogeologie Band 2*, Second Edition. Gebrueder Borntraeger, Berlin, Germany.

Mühlhausen, D., K. Zipfel, and U. Obst (1991). "Beurteilung der Langzeitdynamik in sandigen Grundwasserleitern bei Uferfiltration und kuenstlicher Grundwasseranreicherung." *Final Report 102 02 316 UBA-FB 91-056*, German Environmental Protection Agency, Berlin, Germany.

Neitzel, P.L., A. Abel, T. Grischek, W. Nestler, and W. Walther (1998). "Behaviour of aromatic sulfonic acids during bank infiltration and under laboratory conditions." *Vom Wasser*, 90: 245-271.

Owen, M.O., G. Amy, and Z.K. Chowdhury (1993). *Characterization of natural organic matter and its relationship to treatability*. American Water Works Association Research Foundation and American Water Works Association, Denver, Colorado.

Putschew, A., S. Wischnack, and M. Jekel (2000). "Occurrence of triiodinated X-ray contrast agents in the aquatic environment." *The Science of the Total Environment*, 255(1-3): 131-136.

Roberts, P.V., P.L. McCarty, M. Reinhard, and J. Schreiner (1980). "Organic contaminant behavior during groundwater recharge." *Journal of Water Pollution Control Federation*, 52(1): 161-172.

Schwarzenbach, R.P., W. Giger, E. Hoehm, and J.K. Schneider (1983). "Behavior of organic compounds during infiltration of river water to groundwater. Field studies." *Environmental Science and Technology*, 17(8): 472-479.

Sonnenberg, L.B., J.D. Johnson, and R.F. Christman (1989). "Chemical degradation of humic substances for structural characterization." *Aquatic humic substances: Influence on fate and treatment of pollutants*, I. H. Suffet and P. MacCarthy, eds., American Chemical Society, Washington, D.C.

Sontheimer, H., and W. Nissing (1977). "Aenderung der Wasserbeschaffenheit bei der Bodenpassage unter besonderer Beruecksichtigung der Uferfiltration am Niederrhein." *Gas- und Wasserfach Wasser/Abwasser*, 57(9): 639-645.

Sontheimer, H. (1980). "Experience with riverbank filtration along the Rhine River." *Journal American Water Works Association*, 72(7): 386.

Sontheimer, H. (1991). *Trinkwasser aus dem Rhein?* Academia Verlag, Sankt Augustin, Germany.

Stuyfzand, P.J. (1998). "Fate of pollutants during artificial recharge and bank filtration in the Netherlands." *Artificial Recharge of Groundwater*, J.H. Peters et al., eds., A.A. Balkema, Rotterdam, The Netherlands.

Thurman, E.M., R.L. Wershaw, R.L. Malcolm, and D.J. Pinckney (1982). "Molecular size of aquatic humic substances." *Organic Geochemistry*, 4: 27-35.

Thurman, E. (1985). *Organic geochemistry of natural waters*. Nijhoff/Junk Publishers, Durdrecht, The Netherlands.

Wang, J. (2002). "Riverbank filtration case study at Louisville, Kentucky." *Riverbank filtration: Improving source-water quality*, C. Ray, G. Melin, and R.B. Linsky, eds., Kluwer Academic Publishers, Dordrecht, The Netherlands.

Weiss, W.J., E.J. Bouwer, W.P. Ball, C.R. O'Melia, H. Arora, and T.F. Speth (2002). "Reduction in disinfection byproduct precursors and pathogens during riverbank filtration at three Midwestern United States drinking-water utilities." *Riverbank filtration: Improving source-water quality*, C. Ray, ed., Kluwer Academic Publishers, Dordrecht, The Netherlands.

Wilson, A.L. (1959). "Determination of fulvic acids in water." *Journal of Applied Chemistry*, 9(10): 501-510.

Chapter 16. Research Needs to Improve the Understanding of Riverbank Filtration for Pathogenic Microorganism Removal

Philippe Baveye, Ph.D.
Laboratory of Environmental Geophysics
Cornell University
Ithaca, New York, United States

Philip Berger, Ph.D.
Ijamsville, Maryland, United States

Jack Schijven, Ph.D.
National Institute of Public Health and the Environment
Microbial Laboratory for Health Protection
Bilthoven, The Netherlands

Thomas Grischek, Ph.D.
Institute for Water Chemistry
Dresden University of Technology
Dresden, Germany

1. Introduction

After more than a century of RBF operation in Europe and over a decade of detailed scientific research, a significant body of empirical observations has been accumulated, as evinced by previous chapters in this book. Most available evidence pertains to characteristics of the input (river) and output (well) waters only and relates primarily to the removal of chemicals and some particles (e.g., turbidity), treating the riverbank itself as a black box. In spite of these limitations, the results obtained to date have appeared encouraging enough to some water utilities to further investigate RBF as a means to remove pathogenic microorganisms. In some cases, high pathogen removals are observed (e.g., aerobic spores of *Bacillus subtilus*), thereby leading to claims of high removal rates. At this stage, further research appears necessary to determine if these high removal claims are warranted.

This chapter will outline several potentially fruitful avenues for research on microorganism removal by RBF, including:

- Improved assay methods for pathogen enumeration.
- The use of surrogate organisms as indicators of pathogen transport.
- Direct observation of transport pathways in riverbanks.
- The effects of riverbank heterogeneity (particularly due to clogging and microbial activity near the river/aquifer interface) on pathogen transport and removal.

311

C. Ray et al. (eds.), Riverbank Filtration, 311–319.
© 2002 *Kluwer Academic Publishers. Printed in the Netherlands.*

2. Improved Assay Methods for Pathogen Enumeration – Research Needs

Average pathogen removal at a particular site is usually difficult to evaluate in practice because, typically, there are insufficient numbers of pathogens in both raw-river water and riverbank-filtered water to measure removal with sufficient accuracy, given the assay methods now available. This situation is particularly critical for *Cryptosporidium* oocysts and *Giardia* cysts.

Typically (Zanelli et al., 2000), the enumeration of *Cryptosporidium* oocysts in water samples relies on:

- A concentration step leading to a reduction of the volume from 10 to 1,000 L to 1 to 40 mL.
- A purification step that separates the oocysts from any interfering particles.
- A detection step, which is routinely performed by epifluorescent microscopy after the water concentrate is stained with anti-*Cryptosporidium* fluorescein-isothiocyanate (FITC)-labeled monoclonal antibodies (Smith et al., 1995).

This last part of the technique requires the observation of all the microscopic fields of a 25-mm diameter membrane. Often, oocyst detection is further hampered by the presence of contaminating debris (Smith et al., 1995) and autofluorescing particles, like algae; therefore, oocyst enumeration via this traditional approach is tedious, time-consuming, and often inaccurate.

Fortunately, a number of new techniques have recently been developed, which, in the relatively near future, promise to greatly facilitate the enumeration of oocysts in water samples. Zanelli et al. (2000), for example, describe the use of a technique that involves obtaining water concentrates by cartridge filtration or flocculation, followed by analysis either without purification or after immunomagnetic separation or flotation on percoll-sucrose gradients. Oocyst enumeration is subsequently performed using ChemScan® RDI, a solid-phase cytometry equipment that enables a rapid, automated analysis of an entire 25-mm diameter membrane within 3 minutes. After laser scanning, the results are displayed on a scan map, which identifies the position of the presumed oocysts on the membrane surface. An epifluorescent microscope can then be automatically positioned on the "events" detected by the apparatus to confirm the identification of *Cryptosporidium*. Zanelli et al. (2000) report recoveries yielding close to 100 percent in most cases (average 125 percent, ranging from 86 to 467 percent) for all the concentration/purification techniques tested. Compared with direct microscopic determinations, counting times via solid-phase cytometry are four to six-fold shorter.

Another approach (Esch et al., 2001a) to accurately and rapidly detect viable *Cryptosporidium parvum* oocysts in environmental water involves a microfluidic chip that detects RNA amplified by nucleic acid sequence based amplification (NASBA). The mRNA serving as the template for NASBA is produced by viable *Cryptosporidium parvum* as a response to heat shock. The chip uses sandwich hybridization by hybridizing the NASBA-generated amplicon between capture probes and reporter probes in a microfluidic channel. The reporter probes are tagged with carboxy-fluorescein-filled liposomes. These liposomes, which generate fluorescence intensities not obtainable from single fluorophores, allow the detection of very low concentrations of targets. A variant (Esch et al., 2001b) of this technique, using a single-use visual-strip assay, involves extracting *Cryptosporidium parvum*'s mRNA coding for heat-shock protein hsp70, followed by amplification using NASBA.

These and other emerging approaches to the enumeration of *Cryptosporidium* oocysts in water samples should be adopted in the near future in research efforts dealing with pathogen removal efficiency through riverbanks. With these new technologies, removal estimates will most likely become significantly more reliable than at present. Also, the increased ability that these technologies will afford to rapidly monitor *Cryptosporidium* oocysts in river and well waters will facilitate the selection of effective surrogate or indicator organisms, which are easier and faster to enumerate than oocysts.

3. Use of Surrogate Organisms — Research Needs

As a general rule, in ideal sand filters, particle mobility is greatest for particles of about 1 μm in diameter; for smaller particles and particles several microns larger in diameter, mobility is reduced (e.g., Mackie and Bai, 1993). Some pathogenic microsporidia are approximately 1 μm in diameter, so there is a need to consider the mobility of these sized particles. In general, particles of other sizes have differing mobility or differing physical removal processes:

- For particles less than 1 μm, diffusion enhances the transport to a sorption site at a collector grain, reducing overall mobility.
- For particles greater than 1 μm, sedimentation enhances transport to a collector grain.
- For particles of oocyst size, the relative importance of sedimentation versus straining is unclear.

Although bulk density measurements are available for both oocysts and some indicators, such as anaerobic spores, the theoretical results are sensitive to the density of the particle, so more density measurements are necessary. Most recently, Metge and Harvey (2001) have identified populations of dwarf oocysts (2 to 3 μm in diameter) that are more likely to be found in the eluant after passage through a porous media column. Being smaller, these oocysts may undergo insignificant straining and other removal processes and, thus, may be more mobile in porous media. Similar intra-population variability has been found to be significant for some bacterial transport studies in porous media (Bolster et al., 2000). Because surface charge and the hydrophobicity of oocysts depends on age and the method of purification from feces (Brush et al., 1998), it is recommended that the characteristics of oocysts released from manure, and their subsequent removal when passing through soil, be studied in more detail. More research is needed on the size range of oocysts and other factors that govern oocyst transport in porous media.

Indicator or surrogate organisms are very useful because, typically, they occur in higher concentration in both raw and riverbank-filtered water and, typically, have standardized assay methods with lower detection limits. One research need is to determine the most suitable indicators for each of the pathogens in groundwater flowing through porous media. It is likely that the most appropriate indicators used in studies of surface water or high velocity (rapid sand) filters in surface-water treatment plants may not be the most appropriate indicators for use in porous-media flow involving low water velocity. One criterion for suitability might be whether that indicator has hydrodynamic properties similar to the pathogen of concern. For example, the nearly spheroid form of *Cryptosporidium* oocysts suggests that a suitable indicator should be similar in shape (as well as in size and density). Although some rapid sand filtration experiments have been conducted with aerobic spores, each of these studies was conducted using coagulants together with rapid sand granular filtration or was conducted with dual media, such as sand and anthracite. As a result, the performance of the system was not applicable to RBF.

A study of suitable pathogen indicators and their transport in granular experimental systems should be undertaken and the results should be compared with the transport of pathogens in those same systems. Additional research is needed to compare the relative mobility of aerobic versus anaerobic spores and the relative mobility of spores versus oocysts. Both aerobic and anaerobic spores have been suggested as oocyst indicators, yet little is known about their relative mobility vis-à-vis oocysts or each other. It is difficult to find published data on the size of *Clostridium perfringens* spores, but it is likely that they are about the same size as *Clostridium bifermentans* spores and slightly larger than aerobic spores of *Bacillus subtilus*.

It has been observed in various field studies that high initial virus removal often takes place due to the presence of more favorable attachment sites within the first few meters of soil passage (Schijven, 2001). It is recommended that the quantitative relationship between virus attachment, pH, and the fraction of favorable attachment sites on the surface of sand grains be further evaluated. Future experiments will be designed to identify and quantify the removal mechanisms (inactivation, adsorption, and physical straining) operating on a range of differing-sized organisms. Adsorption and straining rates may be different due to variations in the sizes of microorganisms. Also, more work is needed on straining, ripening, and biological activity in natural systems.

Viruses, due to their size and survival times in the environment, are generally regarded as the most critical microorganisms where the effectiveness of soil passage for removing pathogens from source water is concerned; however, fecal indicator bacteria have been observed, under some environmental conditions, to penetrate into an aquifer as far as viruses and may, therefore, be useful indicators of fecal contamination in those conditions. In many situations, it is not feasible to predict if a groundwater well is adequately protected or if soil-aquifer treatment is sufficient due to the lack of knowledge (uncertainty) on the relevant characteristics of a given situation. It would, therefore, be very useful to have a sensitive, inexpensive, and rapid method for detecting a model or indicator organism (e.g., bacteriophage) that identifies groundwater at risk of viral and fecal contamination. In that respect, MS-2 bacteriophage and naturally occurring F-specific RNA-bacteriophages have already proven to be useful.

MS-2 bacteriophage is an icosahedral phage with a diameter of 27 nm and a low isoelectric point of 3.5. MS-2 may be considered as a relatively conservative tracer for virus transport in saturated sandy soils with a low organic carbon content in the pH-range of 6 to 8. Under those conditions, where both soil grains and MS-2 have a net negative surface charge, MS-2 has shown little or no adsorption. In most soils, MS-2 attachment is less than or equal to that of most other viruses. Possibly due to the presence of multivalent cations, MS-2 may attach more than the less negatively charged ϕX-174 bacteriophage. MS-2 is less stable than several pathogenic viruses and is inactivated faster at temperatures of 10 to 25°C. But, at temperatures lower than 7°C, its inactivation rate is very low. Under unsaturated conditions, MS-2 is not a good choice as a relatively conservative virus tracer because of its strong sensitivity to air/water interfaces and, consequently, strongly enhanced inactivation.

F-specific RNA bacteriophages have similar physical properties to enteroviruses, especially with respect to size. MS-2 belongs to Group I of F-specific RNA bacteriophages. As naturally present viruses, F-specific RNA bacteriophages are good candidates to represent enteroviruses in various treatment processes of surface water, including soil passage. Before entering a treatment like soil passage, enteroviruses and F-specific RNA bacteriophages have largely followed the same path (i.e., both have passed the sewerage system, followed by sewage treatment, discharge into surface water, and some kind of pretreatment). It may be reasoned that along this path, from the sewerage system to the point of recharge into an aquifer, viruses that are less stable, or that adsorb readily to solid surfaces, have disappeared already. This suggests that a selection of very stable and poorly adsorbing viruses (i.e., worst-case viruses) has taken place. This selection has been the same for F-specific RNA bacteriophages and enteroviruses. In surface water, F-specific RNA bacteriophages occur in numbers of 100 to 10,000 times greater than enteroviruses; therefore, it has been possible to show 4- to 6-log removal of F-specific RNA bacteriophages by RBF. Removal of F-specific RNA bacteriophages and MS-2 has been shown to be similarly low at field-scale.

To predict virus removal in a particular field situation, detailed knowledge of the soil properties, virus characteristics, and environmental and hydraulic conditions is required. At this time, there is insufficient knowledge of the quantitative relationships among these factors to enable independent *a priori* predictions of virus behavior at field sites. A quantitative relationship between microorganism removal and physico-chemical properties of the aquifer (grain sizes, porosity, pH, redox parameters) is needed to predict removal at field-scale.

The physical processes that govern the probability of indicator attachment to porous media grains have received some research attention in recent years. More research is needed for many of the pathogens as compared with their indicators. In particular, research is needed on the reversibility of attachment and whether the attachment process promotes or inhibits pathogen (or indicator) die-off. Some longer-lived pathogens may remain infectious despite temporary attachment. The contribution of irreversible adsorption is important because this is an actual removal mechanism. Reversible adsorption is not a removal mechanism, but causes retardation and, therefore, allows more time for inactivation. Although the hazard is primarily defined by those pathogens that take the fastest path and arrive first at the wellhead, there may be a hazard associated with those pathogens that arrive later, if they are long-lived.

A significant research effort is underway at the Bolton Well Field in Cincinnati, Ohio, to evaluate flow and transport during induced infiltration. Uniquely, the U.S. Geological Survey has installed a slant well that bottoms just under the riverbed. The slant well is used to measure changes in water-quality parameters after passage through the riverbed, but without disturbing the bed. To date, the slant well has been used primarily for studies on temperature and specific conductance (Sheets et al., 2001). Funding limitations have prevented studies on the removal of particulates, especially biological particulates, such as pathogens and their indicator organisms. More studies at this site are needed to take maximum advantage of the slant well as a research tool. Other types of sub-riverbed monitoring were installed in the Elbe River (Macheleidt et al., 2000) and Enns River (Ingerle et al., 1999) in Europe. Studies are needed to determine the best monitoring tools for representative sampling of infiltrate without perturbing the system such that the colloids desorb during pumping.

4. Direct Observations of Transport Pathways in Riverbanks – Research Needs

Recent studies (Brown et al., 2002) of bacterial transport in laboratory columns show that bacterial filtration is dominated by small grains and, especially, by the small dimension of the oblong grains. These data suggest that transport may be especially sensitive to the distribution and percentage of the fine material, although that feature may have less significance on water passage through the medium. More research is needed on the effects of heterogeneity on flow and transport. The methods available to measure streambed properties include both direct measurement methods and secondary methods based on textural analysis, models, aquifer pump tests, and water-balance models. An up-to-date compilation of streambed permeability determination methods and results was published by Calver (2001). Recently, Landon et al. (2001) compared several differing instream measurement methods at various locations within the Platte River watershed in Nebraska. More studies like this are needed elsewhere to evaluate other streambed matrices and measurement techniques. The freeze-core method (e.g., Palcsak, 1995) should be applied at RBF sites to measure particle gain sizes in the streambed.

Traditionally, a major impediment to such site characterization is that conventional sampling or borehole techniques for measuring subsurface hydraulic parameters are costly, time-consuming, and invasive. Fortunately, geophysical observations can complement direct characterization data by providing multi-dimensional and high resolution subsurface measurements in a minimally invasive manner. Several techniques have been developed in the preceding decade to use joint geophysical-hydrological data to characterize the hydraulic properties of the subsurface and their spatial correlation structures. One of these approaches (Hubbard and Rubin, 2000) consists of using high-resolution tomographic data together with limited borehole data to infer the spatial correlation structure of log permeability, which can be used within stochastic simulation techniques to generate parameter estimates at unsampled locations.

Using these techniques, Hubbard et al. (2001) have evaluated the importance of heterogeneities in controlling the field-scale transport of bacteria injected into the ground for remediation purposes. Geophysical data, collected across a range of spatial scales, include "surface" ground-penetrating radar, radar cross-hole tomography, seismic cross-hole tomography, cone penetro-meter, and borehole electromagnetic flowmeter measurements. These data were used to:

- Interpret the subregional and local stratigraphy.
- Provide high-resolution hydraulic conductivity estimates.
- Provide information about the log conductivity spatial-correlation function.

The information from geophysical data was used to guide and assist field operations and to constrain the numerical bacterial transport model. Although more field work of this nature is necessary to validate the usefulness and cost-effectiveness of including geophysical data in the characterization effort, qualitative and quantitative comparisons between tomographically obtained flow and transport parameter estimates with well bore and bromide breakthrough measurements suggest that geophysical data can provide valuable, high-resolution information. This information, traditionally only partially obtainable by performing extensive and intrusive well bore sampling, may help to reduce the ambiguity associated with hydrogeological heterogeneity that is often encountered when interpreting field-scale transport data.

Research similar to that described above, but located at RBF sites, might improve the under-standing of the effects of lithological and other types of heterogeneities on the subsurface transport of water and the removal of pathogens.

5. Effect of Riverbank Heterogeneity on Pathogen Transport and Removal – Research Needs

Clogging can significantly affect RBF well yields. For example, Heeger (1987) and Wang et al. (2001) document mechanical clogging that decreased well yield. In contrast, at sites along the Danube River in Slovakia and Hungary, iron and manganese precipitation has contributed to clogging. At some sites along the Rhine and Elbe Rivers in Germany, clogging was increased by poorly degradable organic compounds.

But there remains insufficient information for identifying the importance of the organic load in river water (TOC/DOC) for RBF applicability. Investigations into the effects of DOC, its biodegradability, and redox conditions on hydraulic conductivity of the riverbed will start in Germany in 2002. At present, there is no agreed upon conceptual model to predict clogging effects at future RBF sites.

Some work has been carried out to demonstrate the importance of geological heterogeneity on the efficiency of RBF (e.g., Maxwell and Welty, 2001). Beyer and Banscher (1976) and Grischek (2002) determined that the most significant clogging in the Elbe riverbed occurs in the upper 3 to 5 cm. Schubert (2001) has investigated the effect of physical and chemical clogging at the river/aquifer interface on the removal of xenobiotics and particulates from percolating water. Laboratory investigations of bioclogging of sand and soil columns (see Baveye et al. [1998] for a comprehensive review) also suggest that the river/aquifer interface should be a biologically active zone and could, therefore, have markedly different hydraulic and transport characteristics than the bulk of the riverbanks. This has been verified in the field by Battin and Sengschmitt (1999).

The formation of gas bubbles, microbial growth, and exopolymer production are among the mechanisms that can change the hydraulic conductivity of the active layer (Heeger, 1987). High pathogen removals might be associated with low groundwater velocity. Differentiating the exact mechanism is important because lower velocity zones at the river/aquifer interface may be quickly reestablished after flood scour whereas biological activity may take longer to reestablished. In other situations where the biological active layer forms quickly (in only about 2 to 3 days after scouring [Macheleidt et al., 2000; Grischek et al., 1994]), the opposite may occur. Depending on the kinetics of these different processes, there may be a more or less protracted time period with less than optimal pathogen removal by the riverbanks and, therefore, increased risk of well-water contamination.

Clogging at the river/aquifer interface may also induce water transport instabilities and lead to preferential flow pathways, similar to those found in unsaturated soils and responsible for the uncharacteristically fast transport of colloids and pathogens like *Cryptosporidium* oocysts (Darnault et al., 2001). In the absence of preferential transport, the thin, biologically active zone at the river/aquifer interface may be responsible for much of the filtration/attenuation of biocolloids, xenobiotics (like endocrine disruptors), or heavy metals present in river water. To date, none of these issues have been investigated. More research is needed to find out if there might be simple and cost-effective ways to engineer the river/aquifer interface in the area of the riverbank that feeds the collector wells to minimize scouring or the occurrence of preferential transport and to maximize the filtration/attenuation efficiency of the sediments. To reach that objective, it is necessary to better understand the dynamics, filtration/attenuation capacity, and spatial heterogeneity of the biologically active zone at the river/aquifer interface.

A variety of groundwater collection devices are used to provide drinking water in riverbank-filtered systems. The choice of a vertical or horizontal collector well maybe dictated by factors other than pathogen removal efficiency. Because some wells may be designed to provide large water quantities, there remains a question as to whether such wells also provide suitable pathogen removal efficiencies. For example, a horizontal collector well may be designed to collect water primarily from higher hydraulic conductivity zones in the center of the alluvial channel, but may also receive a higher proportion of direct surface-water recharge because the fine-grain bed material is thinner or absent in the high velocity zone at the center of the stream channel. Such wells may be more at risk for pathogens because of the greater component of recent recharge.

The simulation of horizontal collector wells is difficult. Analytical element models are suited for the task, but more research is needed to apply the method. Heroic measures have been used to apply three-dimensional finite element models (Ray, 2001; Eckert, 2000; Schafer, 2000), but it is not clear whether the simulation adequately addresses the unique character of horizontal collector wells. An adequate prediction of pathogen transport in porous media aquifers yielding water to horizontal collector wells relies upon an adequate simulation of the groundwater flow field. More research is needed on simulating horizontal collector wells in shallow alluvial aquifers adjacent to surface water so that predictive pathogen transport models can be used to simulate these complex, three-dimensional groundwater flow fields.

6. Conclusions

In summary, predicting pathogenic microorganism removal by RBF requires a sophisticated understanding of the flow and transport of biological particles within a porous media ecotope. The ecotope is biologically complex and difficult to study in the field. Similarly, the much more accessible schmutzdecke layer overlying a slow sand filter is poorly understood. Despite these problems, there have been significant recent improvements in understanding due to work in Germany, The Netherlands, and the United States. Nevertheless, more work remains necessary, in particular along the direction outlined in this chapter.

References

Battin, T.J., and D. Sengschmitt (1999). "Linking sediment biofilms, hydrodynamics, and river bed clogging: Evidence from a large river." *Microbial Ecology*, 37: 185-196.

Baveye, P., P. Vandevivere, B.L. Hoyle, P.C. DeLeo, and D. Sanchez de Lozada (1998). "Environmental impact and mechanisms of the biological clogging of saturated soils and aquifer materials." *Critical Reviews in Environmental Science and Technology*, 28: 123-191.

Beyer, W., and E. Banscher (1976). "Zur Erkundungsmethodik der Uferfiltratgewinnung" (Exploring techniques for riverbank filtration sites). *Zeitschrift Angewandte Geologie*, 22: 149-154 (in German).

Bolster, C.H., A.L. Mills, G. Hornberger, and J. Herman (2000). "Effect of intra-population variability on the long-distance transport of bacteria." *Ground Water*, 38: 370-375.

Brown, D.G., J.R. Stencel, and P.R. Jaffe (2002). "Effects of porous media preparation on bacteria transport through laboratory columns." *Water Research*, 36: 105-114.

Bush, C.F., M.F. Walter, J.L. Anguish, and W.C. Ghiorse (1998). "Influence of pretreatment and experimental conditions on electrophretic mobility and hydrophobicity of *Cryptosporidium parvum* oocysts." *Applied and Environmental Microbiology*, 64 (11): 4439-4445.

Calver, A. (2001). "Riverbed permeabilities: Information from pooled data." *Ground Water*, 39: 546-553.

Darnault, C., P. Garnier, K.L. Oveson, T.S. Steenhuis, J.Y. Parlange, P. Baveye, M. Jenkins, and W.C. Ghiorse (2001). "Effect of preferential flow on the fate and transport of *Cryptosporidium parvum* through the vadose zone." *Journal of Contaminant Hydrology*, submitted.

Eckert, P., C. Blomer, J. Gotthardt, S. Kamphausen, D. Liebich, and J. Schubert (2001). "Correlation between the well field catchment area and transient flow conditions." *Proceedings, International Riverbank Filtration Conference*, W. Julich and J. Schubert, eds., Internationale Arbeitsgemeinschaft der Wasserwerke im Rheineinzugsgebiet (IAWR), Amsterdam, The Netherlands, 103-113.

Esch, M.B., L.E. Locascio, M.J.Tarlov, and R.A.Durst (2001a). "*Cryptosporidium parvum* using DNA-modified liposomes in a microfluidic chip." *Analytical Chemistry*, 73: 2952-2958.

Esch, M.B., A.J. Baeumner, and R.A. Durst (2001b). "Detection of *Cryptosporidium parvum* using oligonucleotide-tagged liposomes in a competitive assay format." *Analytical Chemistry*, 73: 3162-3167.

Grischek, T. (2002). "Zur Bewirtschaftung von Uferfiltratfassungen an der Elbe" (Management of river bank filtration along the Elbe River). Unpublished Ph.D. dissertation, Dresden University of Technology, Dresden, Germany.

Grischek, T., W. Nestler, J. Dehnert, and P. Neitzel (1994). "Groundwater/river interaction the Elbe River basin in Saxony." *Ground water ecology*, J.A. Stanford and H.M. Vallett, eds., American Water Resources Association, Herndon, Virginia, 309-318.

Heeger, D. (1987). "Untersuchungen zur Kolmationsentwicklung in Fliessgewaessern. En Beitrag zur Methodik der Hydrogeologischen Erkundung" (Investigations into clogging of river beds). Unpublished Ph.D. dissertation, Bergakademie Freiberg, Freiberg, Germany.

Hubbard, S.S., and Y. Rubin (2000). "Hydrogeological parameter estimation using geophysical data: A review of selected techniques." *Journal of Contaminant Hydrology*, 4: 3-34.

Hubbard, S.S., J.S. Chen, J. Peterson, E.L. Majer, K.H. Williams, D.J. Swift, B. Mailloux, and Y. Rubin (2001). "Hydrogeological characterization of the South Oyster bacterial transport site using geophysical data." *Water Resources Research*, 37: 2431-2456.

Ingerle, K., B. Wett, and H. Jarosch (1999). "Geohydraulische Verhaeltnisse" (Geohydraulic Conditions). Forschungsprojekt Uferfiltrat. Biogeochemie und mikrobielle Oekologie eines Oberflaechenwasser-Grundwasser-Oekotons in einem Stauraum der Enns (Research project bank filtration. Biogeochemical and microbial ecology of as surface water – Ground water ecotone in a reservoir of the River Enns), C. Hasenleithner, ed., *Schriftenreihe Forschung im Verbund*, 60: 43-77 (in German).

Landon, M.K., D.L. Rus, and F.E. Harvey (2001) "Comparison of instream methods for measuring hydraulic conductivity in sandy streambeds." *Ground Water*, 39: 870-885.

Macheleidt, W., T. Grischek, and W. Nestler (2000). "The crucial role of sub-river bed monitoring in water-quality assurance." *Proceedings, International Riverbank Filtration Conference*, W. Julich and J. Schubert, eds., Internationale Arbeitsgemeinschaft der Wasserwerke im Rheineinzugsgebiet, (IAWR), Amsterdam, The Netherlands, 293-295.

Mackie, R.I., and R. Bai (1993). "The role of particle size distribution in the performance and modeling of filtration." *Water Science and Technology*, 27:19-34.

Maxwell, R.M., and C. Welty (2001). "Simulation of the impact of geologic heterogeneity on colloid transport in riverbank filtration." *Proceedings, International Riverbank Filtration Conference*, W. Julich and J. Schubert, eds., Internationale Arbeitsgemeinschaft der Wasserwerke im Rheineinzugsgebiet (IAWR), Amsterdam, The Netherlands, 241-250.

Metge, D.W., and R.W. Harvey (2001). "Potential for *Cryptosporidium* oocyst migration within riverbank filtration systems." *American Water Works Association Water Quality Technology Conference*, American Water Works Association, Denver, Colorado (unpublished abstract).

Palcsak, B.B. "Using the freeze-core method to collect streambed samples for determination of particle-size distribution." *U.S. Geological Survey Open-File Report 95-466*, U.S. Geological Survey, Denver, Colorado, 14 p.

Ray, C. (2001). "Modeling riverbank filtration systems to attenuate shock loads in rivers." *American Water Works Association Annual Meeting*, American Water Works Association, Denver, Colorado.

Schafer, D.C. (2001). "Groundwater modeling in support of riverbank filtration for Louisville Water Company." *Proceedings, International Riverbank Filtration Conference*, W. Julich and J. Schubert, eds., Internationale Arbeitsgemeinschaft der Wasserwerke im Rheineinzugsgebiet (IAWR), Amsterdam, The Netherlands, 241-510.

Schijven, J.F. (2001). "Virus removal from groundwater by soil passage: Modeling, field and laboratory experiments." Ph.D. thesis, Delft University of Technology, Delft, The Netherlands.

Schubert, J. (2001). "How does it work? Field studies on riverbank filtration." *Proceedings, International Riverbank Filtration Conference*, W. Julich and J. Schubert, eds., Internationale Arbeitsgemeinschaft der Wasserwerke im Rheineinzugsgebiet (IWR), Amsterdam, The Netherlands, 41-55.

Sheets, R. (2001). "Results of continuous monitoring at a riverbank filtration site." *American Water Works Association Water Quality Technology Conference*, American Water Works Association, Denver, Colorado (unpublished abstract).

Smith, H.V., L.J. Robertson, and J.E. Ongerth (1995). "*Cryptosporidiosis* and *Giardiasis*: The impact of waterborne transmission." *Journal of Water SRT-Aqua*, 44: 258-274.

Wang, J.Z., S.A. Hubbs, and M. Unthank (2001). "Factors impacting the yield of riverbank filtration systems." *American Water Works Association Water Quality Technology Conference*, American Water Works Association, Denver, Colorado (unpublished abstract).

Zanelli, F., B. Compagnon, J.C. Joret, and M.R. de Roubin (2000). "Enumeration of *Cryptosporidium* oocysts from surface water concentrates by laser-scanning cytometry." *Water Science and Technology*, 41: 197-202.

Chapter 17. Organic Chemical Removal Issues

Ingrid M. Verstraeten, Ph.D.
United States Geological Survey
Baltimore, Maryland, United States

Thomas Heberer, Ph.D.
Institute of Food Chemistry
Technical University of Berlin
Berlin, Germany

1. Current State of Knowledge

To illustrate the importance of riverbank-filtered water in Europe, it is estimated that about 90 percent of the volume of the drinking-water supply in Hungary is groundwater, about 50 percent of which originated from river water. The City of Berlin, Germany, has a total population of about 4-million people and uses 100-percent groundwater in drinking-water production (Worch, 1997); however, about 75 percent of the drinking water originates from surface water mainly produced by bank filtration (from lakes) and, to a lesser extent, by artificial groundwater recharge (SENSUT, 1997).

In the United States, more than 50 percent of the total population uses groundwater (Hallberg et al., 1987). In the Midwest and Great Plains regions, more than 50 percent of the population relies upon groundwater (Hallberg et al., 1987). The total volume of riverbank-filtered water used for drinking water, however, has never been quantified in the United States. Thus, the effects of the presence of micropollutants, pesticides and their degradates, and endocrine disrupters in drinking water in the United States remains unknown. In the future, contaminated surface water may become a more commonly used source of drinking water because of the growing world population and the limitations of existing drinking-water reservoirs.

Organic Contaminants of Interest

Poorly degradable organic compounds are relevant in evaluating RBF because of their potential to reach the wells used for drinking-water production. As shown in Chapter 9, groundwater and drinking-water studies indicate that organic compounds, such as pesticides, pharmaceuticals, complexing agents, amines, sulfonamides, and aromatic sulfonates, can be present in drinking water and, thus, are relevant to RBF and drinking-water supply. Some organic compounds, such as aromatic sulfonates (which are typical industrial chemicals), are widespread in the aquatic environment, and several sulfonates are very persistent (Knepper et al., 1999). They can pass through the treatment processes in sewage-treatment plants and then can be transported from rivers into groundwater. As reported by Knepper et al. (1999) and Verstraeten et al. (1999), some compounds are not completely removed in drinking-water treatment plants. Metabolism is often accompanied by the introduction of polar moieties, such as hydroxyl or carboxyl groups, into the parent molecules, making the parent molecules much more soluble and, thus, more easily transportable.

C. Ray et al. (eds.), Riverbank Filtration, 321–330.
© 2002 *Kluwer Academic Publishers. Printed in the Netherlands.*

The presence of numerous organic compounds and their removal rates during natural bank filtration, chemical treatment, or physical treatment have not been studied sufficiently and, in many cases, remain totally unknown. Often, their biodegradability, toxicological risk to humans, or ecotoxicological risk are not understood completely. Standardized protocols for the testing of many compounds do not exist. Of those protocols that do exist, the analytical methods sometimes are not sensitive enough to confirm the presence of compounds in untreated water or drinking water.

a. Endocrine Disrupting Chemicals

The potential presence of endocrine disrupting chemicals in river water is a topic of increasing interest (Barber et al., 2000), and the potential transport of endocrine disrupting chemicals into riverbank-filtered water presents a major concern. Endocrine disrupting chemicals are present in industrial and domestic wastewater, both of which are significant sources of river water. In the United States, the mean per capita wastewater flow rate is 200 to 500 L per person per day (Barber et al., 1996). Most municipal wastewater is treated by activated sludge or filtering methods that rely on sorption and biodegradation to remove compounds. Strongly sorbed compounds can be removed by 25 percent (after primary treatment) to more than 98 percent (after tertiary treatment) (Barber et al., 1996). Some rivers are wastewater dominated (Barber et al., 2000). Concentrations of wastewater-derived chemicals in river water are influenced by the discharge of the river (dilution) and the flow volume of the wastewater-treatment plant. The presence of surfactants in rivers is thought to be a good indicator of the presence of domestic waste in rivers (Barber et al., 1996). But several other compounds, such as some pharmaceutical residues or synthetic musk compounds, have been reported as excellent indicators of the presence and percentages of municipal sewage effluents in surface water (Heberer et al., 1998, 1999). Several polar drug residues have proven to be much more reliable as municipal sewage indicators than some of the longer known indicator compounds, such as caffeine or coprostanole (Heberer, in press). Because of their leaching behavior, some of these compounds, or compounds such as EDTA, also may be used to identify sewage influences on groundwater (Brauch et al., 2000; Heberer, in press).

Endocrine disrupting chemicals (such as natural and synthetic hormones), degradation products of nonionic surfactants, and plasticizers can have an adverse impact on aquatic organisms (Colborn et al., 1993; Colborn et al., 1996; Beer, 1997; U.S. Environmental Protection Agency, 1997; Purdom et al., 1994) as well as on people (Timm and Maciorowski, 2000). Until now, no known research has been conducted in the United States on the transport of these chemicals from a river into a drinking-water supply using riverbank-filtered water. Surface water used for drinking water has been shown to be contaminated (Kolpin et al., in press); therefore, the U.S. Geological Survey has proposed performing research at a RBF site to evaluate the potential impact of river-water endocrine disrupting chemicals on drinking water. Endocrine disrupting chemicals of main interest include not only pesticides, pharmaceuticals, and degradates or byproducts of treatment, but also alkylphenolpolyethoxylate-derived compounds, 17-β-estradiol, bisphenol A, nonylphenol, and octylphenol. EDTA and nonylphenol carboxylates were found to be the most abundant in wastewater (<1 to 439 μg/L), and they persisted for considerable distance downstream in river water (Barber et al., 2000).

A study of the occurrence of endocrine disrupting chemicals along the Meuse and Rhine Rivers in Europe (Ghijsen and Hoogenboezem, 2000) led the authors to suggest that the presence of endocrine disrupting chemicals in drinking water is minor and that the removal of these contaminants is generally substantial during drinking-water treatment processes, unless high concentrations of these contaminants are released in surface water and RBF is used as the main drinking-water treatment process.

b. Pharmaceuticals

It is important to investigate the occurrence and to understand the fate of pharmaceutical compounds in surface water and their removal rates during natural treatment or other chemical or physical treatment. The direct effects on aquatic organisms and the possible indirect effects on human health are, generally, unknown. Thus, the use of antibiotics may have resulted in the emergence of multi-resistant bacteria or the resistance of known pathogenic bacteria transmitted in the aquatic environment at unknown concentrations (Mons et al., 2000).

c. Synthetic Compounds

Synthetic compounds other than pharmaceuticals have been found in rivers, including polychlorinated biphenyls and hexachlorobenzene, generally associated with suspended sediments in streams. Polycyclic aromatic hydrocarbons as well as volatile and semi-volatile organic compounds also have been found in streams. Many of these contaminants are biodegraded in the rivers, with a half-life of less than 1 day; others resist biodegradation and have half-lives on the order of years. For example, DDT and its degradates frequently are found in the aquatic environment and, as shown in Chapter 9, the polar DDT metabolite, DDA, also may be present in riverbank-filtered water.

Analytical Techniques

According to Notenboom et al. (1999), it is not feasible to compare the detections of pesticides or other organic compounds between nations because of inconsistencies in:
- Data sets (spatial and temporal variabilities).
- Sampling design (e.g., targeted sampling) (Kuhlmann and Zullei-Seibert, 1998).
- Collection methods.
- Analytical techniques (analytical detection limits, variations in analytical methods, and use of stricter quality-assurance quality-control methods).
- Environmental settings (Kolpin et al., 1994, 1995, 2000; Barbash and Resek, 1996).

Kolpin et al. (1995) noted a large increase in pesticide detections when using a reporting level as much as 20 times lower than a previous reporting level and when the number of degradates analyzed were increased threefold. Atrazine detections more than doubled when the reporting limit was modified from 0.10 to 0.01 µg/L.

Moreover, the fact that atrazine and its degradates are more commonly detected than other herbicides, and that herbicides are more frequently detected than other pesticides, may reflect a bias in the emphasis of monitoring programs for priority substances. The list of priority substances is reevaluated regularly by some nations (e.g., The Netherlands) (Van Genderen et al., 2000). Other compounds may be more commonly present in the environment or may be present in even larger concentrations, but these compounds may be less emphasized in the monitoring programs or not analyzed at all. In the United States, degradates of pesticides generally remain unregulated, which decreases the monitoring efforts searching for the presence of these compounds, because available funding is limited for non-priority pollutants. In Europe, pesticide degradates are included in the drinking-water guidelines. The European regulations for pesticides and several other parameters are not based on toxicological aspects, but on a "precautionary principle." Thus, the maximum tolerance levels for pesticide residues in drinking water has been set to 0.10 µg/L for concentrations of the individual parent compounds and their degradates after drinking-water treatment, and to 0.50 µg/L for the sum of all pesticides and their degradates present in drinking water in the European Union (European Economic Commission, 1980; European Communities, 1998). In the new European regulations and, consequently, in Germany's new drinking-water regulations (which will be in effect beginning January 1, 2003), there will no longer be a limitation on the maximum tolerance levels for toxicologically relevant degradates of pesticides (European Union, 1998).

Even though several analytical techniques exist to detect pharmaceutically active compounds, personal care products, and endocrine disrupting chemicals, these techniques need to be adapted to analyze samples in complex matrices with very low detection levels. Recently, new techniques have been developed to determine endocrine disrupting chemicals, such as chromatographic techniques combined with sensitive immunochemical methods (Snyder et al., 2000; Van Emon et al., 2000); however, the endocrine disrupting chemicals have no common structural characteristics, thereby requiring a variety of analytical procedures at a high cost, which makes analyses expensive and labor intensive.

Drinking-Water Treatment Techniques

As early as 1965, Robeck et al. (1965) were concerned about the presence of pesticides in drinking water because of the contamination of surface water with organochlorine and other compounds. Although RBF has the ability to reduce the amounts of parent pesticides in water, the process may not be sufficient to produce water of potable quality when river water is heavily polluted by organic contaminants; therefore, RBF in combination with other treatment methods, such as the use of biological treatment in addition to activated organic carbon treatment (Robeck et al.,1965; Miltner et al., 1989; Verstraeten et al., in press) or the use of membrane filtration techniques (Najm and Trussell, 1999) should be considered not only as a preventive tool by wastewater utilities, but also as part of the treatment process by drinking-water utilities when contamination by persistent organic contaminants can be expected.

A study done on glyphosate (Speth, 1993) indicates that powdered activated carbon, ultrafiltration, and 0.45-µm filters were ineffective, and suggests the use of ozonation and chlorination as treatment methods. Other suitable methods for the removal of organic compounds, including those with small molecular sizes, may be highly sophisticated membrane-filtration techniques (e.g., nanofiltration or reverse osmosis) (Najm and Trussell, 1999). Such techniques may, however, be energy and cost intensive. The large-scale application of reverse osmosis is limited because of great operational pressures increasing the cost of energy. Nevertheless, membrane-filtration techniques and, especially, nanofiltration are becoming increasingly popular because of both the decreasing costs of membrane devices and operational improvements (e.g., cross-flow techniques). In some cases, even reverse osmosis provides some striking advantages compared to conventional purification methods, such as active charcoal filtration. For example, reverse osmosis is used now in mobile drinking-water units for military and civilian catastrophe protection purposes in Germany. It substitutes for conventional purification units that use large quantities of chemicals and charcoal filtration. Membrane filtration has been proven an effective, reliable, and relatively inexpensive alternative to these drinking-water units (Heberer et al., 2001). Nanofiltration remains superior to reverse osmosis for use with large drinking-water purification facilities because of its lower operational costs.

2. Implications

Data Limitations

The results of studies completed during the last 20 to 30 years illustrate that RBF may not be an adequate process for the total removal of organic compounds and micropollutants present in river water (Kühn and Müller, 2000). This is true not only when organic compounds have a low molecular weight and are hydrophilic (Barbash and Resek, 1996), but also for the removal of other compounds because of variable factors, such as when:

- Preferential flowpaths exist.
- Travel time is short.
- Aquifer media are coarse.
- Amount of organic matter is low.
- Well construction is unfavorable (well type, laterals extending under the riverbed).
- Pumping rate is such that the percentage of surface water induced in the wells is great.

Analyzing only for parent compounds precludes a full understanding of the fate and transport of herbicides and their impact on the ecosystem and human health. Although significant amounts of parent compounds can be removed by RBF and ozonation, the toxicity of the remaining byproducts is of concern unless activated carbon is used in conjunction with other treatment processes to remove the byproducts as well as parent compounds (Verstraeten et al., in press).

Special attention also should be given to recently discovered contaminants (e.g., the residues of pharmaceuticals and personal care products) for which little or no data are available. For a long time, these residues were not suspected to occur in the aquatic environment or, if present, occurred only at negligible concentrations; however, the annual amounts of pharmaceutical chemicals that are prescribed in Germany are on the same order of magnitude as the amounts of pesticides applied in agriculture (Heberer and Stan, 1998). Pesticides are usually sprayed onto the ground during one or two seasons of the year and are more widely dispersed in the environment than pharmaceuticals. Pharmaceutical residues, on the other hand, are continuously discharged from sewage-treatment plants, which act as point sources; therefore, it can be expected that:

- The concentrations of these residues in surface water and riverbank-filtered water may, under unfavorable conditions, exceed the residue levels of pesticides.
- The presence of these chemicals is maintained throughout the year.

For example, concentrations of pharmaceuticals could exceed concentrations of herbicides when surface water has large percentages of municipal sewage in densely populated municipal areas (see Chapter 9).

Analytical and Cost Limitations

The combination of immunoassay techniques with more HPLC-MS/MS and GC/MS or GC/MS-MS techniques, also called "effect-related analysis," potentially can aid in cutting analytical costs; however, immunoassay analyses have limited the selectivity of antibodies. To address this issue, research is being conducted to improve existing immunoassay techniques for a wider spectrum of compounds. Nevertheless, at this time, immunoassay techniques cannot replace the more sensitive (parts per trillion level) and more expensive alternative methods of analyses.

It is necessary to:
- Evaluate the presence of organic compounds and their degradates in drinking water.
- Develop additional analytical methods.
- Develop standardized methods.
- Determine the toxicological character of the most common organic compounds in drinking water.

It is also necessary to use efficient, adapted, and selective repeatable pre-concentration methods together with rapid, reliable, sensitive, and wide-spectrum analytical techniques in the laboratory, combined with screening techniques, such as immunoassay, in the field (Jeannot and Sauvard, 2000). Today, the presence of degradation products in treated and untreated water is better known than before, but:

- Analytical techniques commonly do not exist.
- Pre-concentration methods are being improved.
- Screening methods, such as immunoassays, are being developed.
- Analytical methods are being improved by using coupled analyses (e.g., GC with MS and HPLC with MS-MS).

Ultimately, however, the contamination problems need to be addressed at the source rather than at the drinking-water intake.

The City of Lyon in France has established a rigorous monitoring program to assess the presence and concentrations of herbicides in riverbank-filtered water and in the City's drinking water. The program will help define the technical solutions needed to meet European drinking-water criteria for pesticides (Mazounie et al., 2000). The program emphasizes that the main compounds detected in drinking water in 1997 (e.g., atrazine, deethylatrazine, simazine, terbutylazine, and other triazines, and their temporal variations) be monitored. The analytical cost of pesticides and other organic analyses has increased the cost of water to citizens. Preferred remedies other than costly water treatment might include wellhead protection, an educational campaign to reduce pesticide loading onto fields, or changes in the type of pesticides applied.

Drinking-Water Treatment Limitations

Differences in the physical, chemical, and biological parameters of water make it difficult to predict the removal or attenuation of any organic compound during RBF. Varying conditions during chemical and physical treatment procedures in the treatment plant, including flow rate, filter rate, and chemical dose, also affect the presence of organics in drinking water.

Currently, the best-known methods for the natural or active treatment of organics are:
- Increasing RBF travel times (e.g., distance from the river, depth of well completion, and removal and filtration capacities of the sediments), selecting RBF locations with high organic matter content and an active microbial population at the river/aquifer interface, and avoiding RBF locations with sediments containing mostly large particle sizes.
- Encouraging a high pH in combination with oxidation processes or aerobic conditions.
- Using granular activated carbon or powdered activated carbon to adsorb the contaminants.
- Promoting microbial degradation either during RBF or active treatment.
- Remediating at the source, such as best management practices by individuals, industry, and the wastewater community.

In addition, although compounds occur in lower concentrations in riverbank-filtered water, it is possible that these lower concentrations do not indicate true removal, but a result of the mixing of surface water and groundwater and of dispersion during transport, ultimately resulting in longer exposure times at lower levels (Kühn and Müller, 2000). Because riverbank-filtered water tends to have a lower DOC content than river water, the life of activated-carbon filters as a means of reducing the pollutants in drinking water is extended.

Existing and Future Drinking-Water Regulations on Pharmaceutical Residues

Unlike prescribed tolerances for pesticide residues in drinking water set by the European Union (European Economic Commission, 1980; European Union, 1998), no such values exist for pharmaceutical residues yet. New regulations concerning the assessment of potential risks posed by pharmaceutical products for human use in the environment were suggested in the European Union draft guideline III/5504/94 (European Union, 1994), but these regulations have not been promulgated by the European Union at this time. Such regulations already exist for veterinary pharmaceutical products (European Union, 1996). New European Union guidelines refer to the registration of new pharmaceutical chemicals only. Guidelines for the environmental assessment of already registered pharmaceuticals have not been planned (Stan and Heberer, 1997; Heberer and Stan, 1998).

3. Remedies

Improving Data Availability

Existing data on the occurrence of organic chemicals in the environment are diverse and poorly distributed over the globe. The information on RBF is sporadic. Sometimes, water-supply companies even keep occurrence data confidential, fearing any consequences that publication may produce. In addition, some governments do not acknowledge the interaction between surface water and groundwater. For example, in the State of Nebraska in the United States, this interaction was accepted only during the last decade and mainly focused on water-quantity and not on water-quality issues. Compounding the problem, groundwater generally has not been defined as originating from nearby streams and has not been identified as riverbank-filtered water. Furthermore, water from wells receiving induced river water has not always been recognized as being influenced by surface water. These misconceptions need to be resolved before the effect of RBF on the presence of organics in water through induced infiltration can be assessed.

The occurrence of pharmaceutically active compounds, personal care products, and endocrine disrupting chemicals in the environment needs to be evaluated. These chemicals are natural or synthetic hormones or other compounds causing androgenic, anti-androgenic, estrogenic or anti-estrogenic effects, and include several DDT derivatives, nonylphenols, bisphenol A, and tributyltin (Beer, 1997). Although some effects of endocrine disrupting chemicals on invertebrates and on several vertebrates have been demonstrated, the effects on humans are less clear. In addition, evidence exists of the temporal fluctuations of concentrations in rivers, but information has not been developed on the removal of these contaminants during drinking-water treatment. In the future, wastewater-treatment plants and drinking-water treatment plants may be required to treat their water to a much higher standard. Also, the pharmaceutical industry may formulate organic chemicals for a wide variety of purposes that are less persistent in the environment.

Not only should analytical efforts emphasize the analysis of water samples for the presence of organic compounds, but should also include an analysis of sewage sludges, sediments, suspended solids, soil, and other potential sources. Improved information is needed on the occurrence of organic compounds in the environment, their sources, and the variables and processes that are important in removing contaminants from water during RBF and other treatment steps.

Additional information should be gathered on the removal of emerging contaminants at the river/aquifer interface to broaden our understanding of removal processes. More information should be collected about the toxicology of compounds, not only focusing on parent compounds, but also on degradates.

Developing New Analytical Methods

More analytical methods need to be developed and improved. For example, the lack of monitoring for degradates of pesticides may understate the effect of organics present in the aquatic environment on human health, especially in drinking water derived from rivers that, at times, contain large amounts of pesticides. As shown in Chapter 9, by the example of the DDT metabolite DDA, degradates of organic compounds may be much more relevant than the parent compounds in RBF processes. Sometimes, these degradates may originate from industrial chemicals not expected in the environment, and samples may not be analyzed for them. These compounds often are only found by chance when analyzing for other target compounds. This phenomenon also has been demonstrated in Chapter 9 (see the discussion on N-[phenylsulfonyl]-sarcosine).

Future of RBF as a Drinking-Water Treatment Technique

The objective of an optimized RBF process is to provide a drinking-water supply based on a natural treatment process. To achieve this objective, the use or release of persistent organic substances into the environment of interest to drinking-water production should be limited. Additional purification techniques may be necessary to remove recalcitrant organic compounds if concentrations exceed the existing or potential future maximum tolerance levels. In general, regardless of which organic compound is of concern, the use of multi-barrier drinking-water treatment processes, which could include RBF and ozonation with an adsorption step using frequently reactivated or exchanged activated carbon, is probably the best approach to remove these chemicals (Lange et al., 2000). RBF also may be used in combination with membrane-filtration techniques, such as nanofiltration, which has the ability to reduce or, at times, even remove residual contaminants not totally removed by RBF. Because most drinking-water treatment processes are not optimized for organic compound removal, there is a need to focus on reducing or, if possible, eliminating the release of organic compounds into surface water. Nevertheless, although RBF may not remove all organics or reduce organics to acceptable levels in drinking water, the use of RBF as a natural treatment step now can significantly reduce the cost of surface-water treatment to a drinking-water utility and, ultimately, reduce the cost of water to the consumer.

References

Barbash, J.E., and E.A. Resek (1996). "Pesticides in ground water: Distribution, trends, and governing factors." *Pesticides in the hydrologic system*, Volume 2, R.J. Gilliom, ed., U.S. Geological Survey Water Quality Assessment, Ann Arbor Press, Inc., Ann Arbor, Michigan, 588 p.

Barber, L.B., G.K. Brown, and S.D. Zaugg (2000). "Potential endocrine disrupting organic chemicals in treated municipal wastewater and river water." *Analysis of Environmental Endocrine Disruptors*, L.H. Keith, T.L. Jones-Lepp, and L.L. Needham, eds., American Chemical Society, Washington, D.C., 97-123.

Barber, L.B., J.A. Leenheer, W.E. Pereira, T.I. Noyes, G.K. Brown, C.F. Tabor, and J.H. Writer (1996). "Organic contamination of the Mississippi River from municipal and industrial wastewater." *Contaminants in the Mississippi River, 1987-1992*, R.H. Meade, ed., U.S. Geological Survey Circular, Reston, Virginia, 115-127.

Beer, A.J. (1997). "Something in the water." *Biologist*, 44(2): 296.

Brauch, H.J., F. Sacher, E. Denecke, and T. Tacke (2000). "Wirksamkeit der Uferfiltration fuer die Entfernung von polaren organischen Spurenstoffen" (Efficiency of bank filtration for the removal of polar organic tracer compounds). *Gas- und Wasserfach Wasser/Abwasser*, 14(1/4): 226-234.

Colborn, T., D. Dumanoski, and J.P. Myers (1996). *Our stolen future*. Plume/Penguin, New York, New York, 316 p.

Colborn, T., F.S. Vom Saal, and A.M. Soto (1993). "Developmental effects of endocrine-disrupting chemicals in wildlife and humans." *Environmental Health Perspectives*, 101: 378-384.

European Economic Commission (1980). "Drinking Water Guideline." 80/779/EEC, EEC No. L229/11-29, European Economic Commission, Brussels, Belgium.

European Union (1998). Council Directive 98/83/EG on the quality of water for human consumption. *Official Journal of the European Communities L330/32*, 5 December 1998.

European Union (1994). *European Union draft guideline III/5504/94 (Draft 4): Assessment of potential risks posed by medicinal products for human use (excluding products containing genetically modified organisms)*. European Commission, Directoral-General Industry, Pharmaceuticals Service III/E/3. The European Agency for the Evaluation of Medical Products (EMEA), London, England.

European Union (1996). *Note for guidance: Environmental risk assessment for veterinary medicinal products other than GMO-containing and immunological products*. European Agency for the Evaluation of Medicinal Products (EMEA)/ Committee for Veterinary Medicinal Products (CVMP)/055/96, London, England.

Ghijsen, R.T., and W. Hoogenboezem (2000). *Endocrine disrupting compounds in the Rhine and Meuse Basin – Occurrence in surface, process, and drinking water*. Sub-project of the National Research Project on the occurrence of endocrine disrupting compounds, Association of River Waterworks (RIWA), Amsterdam, The Netherlands, 96 p.

Hallberg, G.R., R.D. Libra, K.R. Long, and R.C. Splinter (1987). "Pesticides, groundwater, and rural drinking water quality in Iowa." *Proceedings, Conference on pesticides and groundwater: A health concern for the Midwest*, Freshwater Foundation, Navarre, Minnesota.

Heberer, T. (2001). "From municipal sewage to drinking water: Occurrence and fate of pharmaceutical residues in the aquatic system of Berlin." Article in "Attenuation of groundwater pollution by bank filtration," T. Grischek and K. Hiscock, eds., *Journal of Hydrology*, in press.

Heberer, T., and H.J. Stan (1998). "Arzneimittelrückstände im aquatischen system" (Drug residues in the aquatic system). *Wasser & Boden*, 50(4): 20-25.

Heberer, T., S. Gramer, and H.J. Stan (1999). "Occurrence and distribution of organic contaminants in the aquatic system in Berlin. Part III: Determination of synthetic musks in Berlin surface water applying solid-phase microextraction (SPME)." *Acta Hydrochimica et Hydrobiolica*, 27: 150-156.

Heberer, T., K. Schmidt-Bäumler, and H.J. Stan (1998). "Occurrence and distribution of organic contaminants in the aquatic system in Berlin. Part I: Drug residues and other polar contaminants in Berlin surface and groundwater." *Acta Hydrochimica et Hydrobiolica*, 26: 272-278.

Heberer T., D. Feldmann, K. Reddersen, H. Altmann, and T. Zimmermann (2001). "Removal of pharmaceutical residues and other persistent organics from municipal sewage and surface waters by applying membrane filtration." Sponsored by the Universities Council on Water Resources, Carbondale, Illinois, Water Resources Update, Issue 120: 18-29.

Jeannot, R., and E. Sauvard (2000). "Les dosages des substances phytosanitaires et de leurs produits de dégradation dans les eaux: Situation, tendances, et perspectives" (Analysis of pesticides and their degradation products in water: Status, trends, and prospects). *Hydrogéologie*, 1: 13-29.

Knepper, T.P., F. Sacher, F.T. Lange, H.J. Brauch, F. Karrenbrock, O. Roerden, and K. Lindner (1999). "Detection of polar organic substances relevant for drinking water." *Waste Management*, 19: 77-99.

Kolpin, D.W., M.R. Burkart, and E.M. Thurman (1994). "Herbicides and nitrate in near-surface aquifers in the midcontinental United States, 1991." *U.S. Geological Survey Water Supply Paper 2413*, U.S. Government Printing Office, Washington D.C.

Kolpin, D.W., D.A. Goolsby, and E.M. Thurman (1995). "Pesticides in near-surface aquifers: An assessment using highly sensitive analytical methods and tritium." *Journal of Environmental Quality*, 24: 1125-1132.

Kolpin, D.W., E.M. Thurman, and S.M. Linhart (2000). "Finding minimal herbicide concentrations in ground water? Try looking for their degradates." *The Science of the Total Environment*, 248: 115-122.

Kolpin, D.W., E. Furlong, M.T. Meyer, E.M. Thurman, S. Zaugg, L. Barber, H. Buxton, and M. Lindsey (2002). "Pharmaceuticals, hormones, and other emerging contaminants in U.S. streams, 1999-2000: A national reconnaissance." *Environmental Science and Technology*, in press.

Kühn, W., and U. Müller (2000). "Riverbank filtration – An overview." *Journal American Water Works Association*, December 2000: 60-69.

Kuhlmann, B., and N. Zullei-Seibert (1998). "Pesticide monitoring: Measuring the right substances at the right place and time." *Artificial recharge of groundwater*, J.H. Peters et al., eds., A.A. Balkema, Rotterdam, The Netherlands.

Lange, F.T., R. Furrer, and H.J. Brauch (2000). *Polar aromatic sulfonates and their relevance to Waterworks*. Association for River Waterworks (RIWA), Amsterdam, The Netherlands, 70 p.

Mazounie, P., D. D'Arras, and E. Brodard (2000). "Surveillance et contrôle des pesticides: Le point de vue d'un exploitant, Lyonnais des Eaux" (Monitoring and control of pesticides: The operation approach of Lyonnaise des Eaux). *Hydrogéologie*, 1: 7-11.

Miltner, R.J., D.B. Baker, and T.F. Speth (1989). "Treatment of seasonal pesticides in surface water." *Journal American Water Works Association*, 81(1): 43-52.

Mons, M.N., J. Van Genderen, and A.M. Van Dijk-Looijaard (2000). *Inventory on the presence of pharmaceuticals in Dutch water*. Association for River Waterworks (RIWA), Vereniging van Exploitanten van Waterleidingberdrijven in Nederland, Kiwa, Nieuwegein, The Netherlands, January 2000.

Najm, I., and R.R. Trussell (1999). "New and emerging drinking water treatment technologies." *Identifying future drinking water contaminants*, National Academy Press, Washington D.C., 220-243.

Notenboom, J., A. Verschoor, A. Van der Linden, E. Van de Plassche, and C. Reuther (1999). "Pesticides in groundwater: Occurrence and ecological impacts." *RIVM Report 601506002*, National Institute of Public Health and the Environment, Bilthoven, The Netherlands.

Purdom, C.E., P.A. Hardiman, V.J. Bye, N.C. Eno, C.R. Tyler, and J.P. Sumpter (1994). "Estrogenic effects of effluents from sewage treatment works." *Chemistry and Ecology*, 8: 275-285.

Robeck, G.G., K.E. Dostal, J.M. Cohen, and J.F. Kreissl (1965). "Effectiveness of water treatment processes in pesticide removal." *Journal American Water Works Association*, 57(2): 181-199.

Senatsverwaltung für Stadtentwicklung, Umweltschutz und Technologie (SENSUT) (1997). *Berlin digital atlas*, Second Edition, Kulturbuchverlag, Berlin, www.sensut.berlin.de/sensut/umwelt/uisonline/dua96/html/edua_index.shtml.

Snyder, S.A., E. Snyder, D. Villeneuve, K. Kurunthachalam, A. Villalobos, A. Blankenship, and J. Giesy (2000). "Instrumental and bioanalytical measures of endocrine disruptors in water." *Analysis of environmental endocrine disruptors*, L.H. Keith, T.L. Jones-Lepp, and L.L. Needham, eds., American Chemical Society, Washington, D.C.

Speth, T.F. (1993). "Glyphosate removal from drinking water." *Journal of Environmental Engineering*, 119(6): 1139-1157.

Stan, H.J., and T. Heberer (1997). "Pharmaceuticals in the aquatic environment." Article within a special issue on "Dossier water analysis," M.J.F. Suter, ed., *Analusis*, 25: M20-23.

Timm, G.E., and A.F. Maciorowski (2000). "Endocrine disruptor screening and testing: A consensus strategy." *Analysis of environmental endocrine disruptors*, L.H. Keith, T.L. Jones-Lepp, and L.L. Needham, eds., American Chemical Society, Washington D.C.

U.S. Environmental Protection Agency (1997). "Special report on environmental endocrine disruption: An effects assessment and analysis." *EPA/630/R-96/012*, U.S. Environmental Protection Agency, Washington, D.C.

Van Emon, J.M., C.L. Gerlach, K.L. Bowman (2000). "Analytical challenges of environmental endocrine disruptor monitoring." *Analysis of environmental endocrine disruptors*, L.H. Keith, T.L. Jones-Lepp, and L.L. Needham, eds., American Chemical Society, Washington D.C.

Van Genderen, J., M.N. Mons, and J.A. Van Leerdam (2000)." *Inventory of toxicological evaluation of organic micropollutants*. Association of River Waterworks (RIWA), Amsterdam, The Netherlands.

Verstraeten, I.M., J.D. Carr, G.V. Steele, E.M. Thurman, and D.F. Dormedy (1999). "Surface-water/ground-water interaction: Herbicide transport into municipal collector wells." *Journal of Environmental Quality*, 28(5): 1396-1405.

Verstraeten, I.M., E.M. Thurman, E.C. Lee, and R.D. Smith (2002). "Degradation of triazines and acetamide herbicides by bank filtration, ozonation, and chlorination in a public water supply." Article in "Attenuation of groundwater pollution by bank filtration," T. Grischek and K. Hiscock, eds., *Journal of Hydrology*, in press.

Worch, E. (1997). *Wasser und Wasserinhaltsstoffe. Eine Einführung in die Hydrochemie*. B.G. Teubner Verlagsgesellschaft, Leipzig, Germany.

Conclusion

Chittaranjan Ray, Ph.D., P.E.
University of Hawaii at Mānoa
Honolulu, Hawaii, United States

Gina Melin
National Water Research Institute
Fountain Valley, California, United States

Ronald B. Linsky
National Water Research Institute
Fountain Valley, California, United States

The current and state-of-the-art applications of RBF technology, as it is presently practiced in both the United States and Europe, are described within this book. The three main topics of interest include: the mechanisms behind RBF, its ability to remove contaminants from surface water, and critical research needs.

1. Major Findings

- While RBF technology is popular in Europe, it is beginning to gain interest in the United States due to pending federal regulations that will impact the use of surface water as drinking water.
- Many utilities in the United States are unaware that their wells may be classified as RBF wells (wells that may not be under the direct influence of surface water).
- The mixing of riverbank-filtered water with native groundwater in an aquifer can increase the quantity of supply as well as dilute contaminants.
- High extraction rates of riverbank filtrate are best achieved when wells are placed on an island or meander.
- Under optimal conditions, RBF can remove up to 8 logs of virus over a distance of 30 m in about 25 days. Greater removal efficiencies may be expected for bacteria, protozoa, and algae under the same conditions.
- RBF can reduce biological regrowth potential by more than 60 percent.
- Biodegradation and the physical removal of particulate matter at the river/aquifer interface are the primary mechanisms for removing NOM and other contaminants. For instance, NOM removal occurs within the first 15 m of infiltration.
- RBF can remove more than 50 percent of NOM and disinfection byproduct precursors from surface water.
- RBF can remove between 35 and 67 percent of TOC and DOC from surface water, which also significantly reduces disinfection byproduct formation potentials by 53 to 82 percent (THM) and 47 to 80 percent (HAA).
- RBF has greater reductions in theoretical cancer risk (28 to 45 percent reduction) due to removing THM than conventionally treated water (11 to 47 percent).

331

C. Ray et al. (eds.), Riverbank Filtration, 331–333.
© 2002 *Kluwer Academic Publishers. Printed in the Netherlands.*

- Organic matter and redox conditions affect the sorption of bacteria and viruses in the subsurface.
- Some micropollutants showed only partial or no significant removal during RBF, including aromatic sulfonic acids, EDTA, naphthalene-1,5-disulfonate, clofibric acid, carbamazepine, or X-ray contrast agents. Additional post-treatment steps are needed to remove these micropollutants (such as activated carbon adsorption and ozonation) when riverbank filtrate is used for drinking-water supply.
- Pollutants infiltrating at different places in the river will take different pathways through the aquifer and will finally converge at the RBF well. As a result, the shock-load concentration is effectively attenuated. Moving the well farther from the river could help further attenuate river-water contaminants.
- The contaminant travel time in an RBF system is significantly greater than in a direct intake system because of phenomena such as sorption and degradation. In addition, there is a lag time between the peaks observed between pumped water and surface water. This lag time in RBF allows water utilities to respond to an emergency faster within an RBF system than within a direct intake system.
- The installation of an early warning system and the presence of back-up wells can help alleviate the problems associated with shock loads during flood conditions.
- The cleaning frequencies of nanofiltration membranes are significantly lower with RBF-pretreated water than with conventionally pretreated water, meaning that conventionally pretreated water has higher fouling rates.
- Flux loss during the operation of nanofiltration membranes is lower with RBF-treated water (12 to 24 percent) than with conventionally treated water (36 to 50 percent).
- RBF significantly reduces mutagenic activity. For instance, the number of induced revertants in well water (20/L) was much lower than in river water (about 250/L).

2. Research Needs

- Detailed life-cycle cost comparisons between horizontal collector wells and vertical wells for a given production capacity at selected utilities.
- Estimates of surface-water contributions to wells for utilities located on riverbanks.
- A means to classify RBF wells as wells under the direct influence of surface water.
- Reliable data to determine whether the rehabilitation frequency for horizontal filter wells is lower than for vertical wells.
- Data on capital investment, operation, and maintenance costs of wells and well fields to help determine the unit cost of finished water of different well types.
- Process-level studies to understand the mechanisms that contribute to pathogen retention in soils and aquifer sediments.
- The effects of the presence of low (microgram to nanogram per liter) levels of pesticides, personal care products, and pharmaceuticals found in RBF filtrate.
- A greater understanding of the infiltration behavior of the streambed under low-flow and flood conditions.
- The dynamics of the formation and loss of sediment (or clogging) layer at the river/aquifer interface and its impact on infiltration rates.
- The impact of the loss of the clogging layer on pathogen transport to the aquifer.
- The impact of the clogging layer on the redox zone.
- The removal processes of non-humic and humic fractions of NOM during RBF.

- The fate of the biodegradable fraction of NOM during RBF.
- The impact of site hydrogeology and the associated redox conditions on NOM removal.
- A means to identify and characterize the microbial community at the river/aquifer interface and their role in contaminant degradation.
- The flow and transport of biological particles in porous media.
- A means to improve methodologies for determining pathogen concentrations, especially for protozoa
- The use of surrogates to study the transport behavior of target pathogens (e.g., phages for viruses).
- Methodology to directly measure or observe microbial pathogens and their pathways in the riverbank.
- The effect of riverbank material heterogeneity on pathogen removal.
- Better methodologies to analyze and remove organic chemicals, such as insecticides, personal care products, and pharmaceuticals, from water.
- Laboratory studies on the fate and transport of pharmaceuticals to evaluate their persistence and movement in the subsurface and to incorporate into transient simulation models.
- An evaluation of the occurrence of pharmaceutically active compounds, personal care products, and endocrine disrupting chemicals in the environment.
- Validated three-dimensional models for RBF sites to help utility managers plan for emergencies and other unforeseen events.
- Modeling efforts to estimate the enhancement of interface hydraulic conductivity during flooding and its impact on contaminant transport.
- Optimization and cost studies of RBF with other pretreatment methods to show that RBF is a reliable and cost-effective treatment mechanism.
- Data on river stage at the RBF site during flood events.

3. Summary

RBF is an applicable strategy for treating drinking-water supplies. There is sufficient experience in both Europe and the United States that validates RBF in an array of environmentally different settings and under conditions that have rigorously tested the technology. Those who are contemplating RBF should move forward and utilize its principles because RBF will not only add value to the water supplies, but will also enhance the sustainability of water supplies for future generations.

Glossary

Absorption: The transfer process of a gas, liquid, or dissolved substance to the surface of a solid, where it is bound.

Abstraction: See *Extraction*.

Adsorption: The attraction and adhesion of molecules of a gas, vapor, or dissolved substance to a surface. Adsorption is generally a passive and reversible process. Granular or powdered activated carbon is often used as an adsorption medium.

Age Stratification: Technique used to differentiate the age of geologic materials or the age of water present in aquifers.

Alluvial Aquifer: A water-bearing geologic unit composed of sedimentary material deposited by flowing rivers, streams, or melted snow. The size distribution of the material is controlled to some extent by gravity and fluid properties. An alluvial aquifer is usually a good source of easily exploited groundwater when the aquifer is adjacent to a flowing stream.

Ames Test: A test of mutagenic activity using a series of genetically engineered strains of *Salmonella*. Also called the "*Salmonella typhimurian* Microsomal Mutagenicy Assay."

Aqueous Phase: A contaminant can remain in air or water based upon its Henry's law constant. In water, it is often referred to as the "aqueous phase," and air or other media (such as oil or solvent) are referred to as "non-aqueous phases." Volatile contaminants partition between these phases based upon their Henry's law constants.

Aquifer: Layers of sand and gravel that contain (store) water underground and are sufficiently permeable to transmit water to wells and springs.

Aquifer Grain: Sand grains of aquifer material.

Aquitard: A low-permeability geologic unit that can store groundwater. It transmits groundwater slowly from one aquifer to another.

Artificial Groundwater Infiltration: The process of intentionally adding water to an aquifer by injection or infiltration. Dug basins, injection wells, or the simple spreading of water across the land surface are all means of artificial groundwater infiltration. Also referred to as "artificial recharge" and "artificial groundwater recharge."

Attachment: Process in which microbes stick to fine sediment, grams, or particles.

Attenuation: The process by which a compound is reduced in concentration over time through absorption, adsorption, degradation, dilution, and/or transformation, or is "killed," in the case of biological organisms.

Atrazine: An herbicide and plant-growth stimulator used primarily on corn and soybeans. It is slightly soluble in water and is regulated by the U.S. Environmental Protection Agency. Also known as "2-chloro-4-ethylamino-6-isopropyl amino-1,3,5-triazine."

Bacteriophage: A group of viruses that infect and grow in bacteria, such as coliphages that grow in *E. coli*. Following replication of the host bacterium, new bacteriophages are released by lysis of the host cell. Bacteriophages can be used as surrogates or models in place of human enteric viruses during water treatment testing and can be potential indicators of pathogens. Examples of common bacteriophages are MS-2 coliphages and F-specific phages. A bacteriophage is also called a "phage" or "coliphage."

Bank Filtration: A generic term that refers to water derived or drawn through the banks of lakes and other surface-water bodies (such as reservoirs or artificial recharge into spreading basins). See *Riverbank Filtration*.

Bank Storage: A natural process in which groundwater is temporarily stored in sediments adjacent to a stream channel as a result of a rise in stream elevation during flooding.

Bed Load: Sediments such as soil, rocks, particles, or other debris that rest on or near the riverbed. These sediments may be moved along the riverbed by flowing water.

Bed Load Transport: The movement (rolling, skipping, or sliding) of sediment, such as soil, rocks, particles, or other debris, along or very near the riverbed by flowing water.

Bend: A curve in a river. Bends generally grow into windings or turns, which are known as "meanders."

Biodegradation: The breakdown or decomposition of organic matter by microorganisms.

Biofouling: The presence or growth of organic matter in a water-treatment system. This phenomenon can compromise the performance of a unit process. For example, biofouling can occur in a membrane process when microorganisms attach and grow on the membrane surface, resulting in a premature and excessive loss of flow rate, or "flux," through the membrane.

Biological Filtration: The process of passing water through a filter medium that has been allowed to develop a microbial biofilm ("mat") that assists in the removal of fine particulate and dissolved organic materials.

Biomass: (1) The total weight of biological matter, including any attached extracellular polymeric materials. (2) A material that is or was a living organism or was excreted from a microorganism.

Biomass Distribution: Refers to the presence of biomass in control filter columns. The biomass amount could vary initially within the column and, later, may reach a steady-state condition. Filter biomass helps degrade organic contaminants present in water.

Borehole: A hole drilled into the earth, often to a great depth, as a prospective well or for exploratory purposes.

Brackish Water: Water that has a salinity lower than that of sea water, but higher than that of fresh water. Brackish water generally contains 1,000 to 10,000 milligrams per liter of total dissolved solids. Waters in estuaries are often brackish.

Breakthrough: The time it takes for a chemical to move through a measured column of simulated soil material.

Caisson: A vertical, large-diameter concrete pit where pumps are located and where water from the laterals of collector wells enter. Also known as a "concrete wet well caisson."

Cake Filtration: Filtration classification for filters where solids are removed on the entering face of the granular media.

Capture Time: The time needed for a water particle to travel from surface water to the laterals of horizontal wells. It also refers to the travel time of a water particle from the river/water interface to the screen zones of a vertical well.

Capture Zone: The up-gradient and down-gradient areas of an aquifer that drain into a particular well. The delineation of capture zones is used extensively in wellhead protection planning and in contaminant recovery.

Catchment Area: The contributing area for a pumping well in an aquifer. It may also refer to all areas that contribute to flow in a river at a given location. Also called a "catchment basin" or "zone of contribution."

Charcoal Filtration: The process of sorption using activated carbon.

Chemical Clogging: An impediment of flow to a well or porous media due to chemical precipitates, such as calcium carbonate.

Chlorination: The addition of chlorine to water or wastewater, usually for the purpose of disinfection. In chlorination, chlorine oxidizes microbiological material, organic compounds, and inorganic compounds. It is the principal form of disinfection for water supplies in the United States.

Clay-Bearing Formations: Geologic formations in an aquifer that contain clay or fine-grained material.

Clogging: An impediment of flow, typically as a result of particle blockage (small particles enter the pore space of a coarse-grained material and block the movement of fluid through the large pores of the material). In rapid sand filtration (typically employed in water-treatment plants), clogging can occur as a result of excessive particle removal prior to backwash or mudball formation. Also referred to as "plugging."

Coagulation: The process of destabilizing charges on particles in water by adding chemicals (coagulants). Natural particles in water have negative charges that repel other material and, thereby, remain in suspension. In coagulation, positively charged chemicals are added to neutralize or destabilize these charges and allow the particles to clump together and be removed by physical processes, such as sedimentation or filtration. Commonly used coagulants include aluminum and iron salts and cationic polymers.

Coliform Bacteria: Bacteria commonly found in the intestinal tracts of humans and other warm-blooded animals and that is shed in fecal material. The presence of coliform bacteria in water indicates that the water has received contamination of an intestinal origin. In sanitary bacteriology, these organisms are defined as aerobic and facultative anaerobic, gram-negative, non-spore-forming, rod-shaped bacteria that ferment lactose with gas and acid formation within 48 hours at 95°Fahrenheit (35°Celsius).

Coliform Count: The number of coliform bacteria present in a given amount of water.

Collector Well: See *Horizontal Collector Well*.

Collision Efficiency Factor: A measure of the efficiency of the collision of destabilized particles in forming larger particles in the flocculation process. After particles become destabilized in the coagulation process, they can be brought together to agglomerate into larger particles via flocculation.

Colloid: A small, discrete solid particle in water that is suspended (not dissolved) and will not settle by gravity because of molecular bombardment.

Colloidal Filtration: A filtration process that remove colloids, which are suspended solids with a diameter <1 micron that cannot be removed by sedimentation alone. Colloidal filtration allows the removal of these particles during collision with a collector (e.g., sand grain) and subsequent attachment.

Column Study: An experiment in which undisturbed or packed sand/soil is placed in columns for evaluating the one-dimensional transport of chemicals and microbes under various experimental conditions.

Concrete Wet Well Caisson: See *Caisson*.

Cone of Influence: The depression, roughly conical in shape, produced in a water table or other piezometric surface by the extraction of water from a well at a given rate. The volume of the cone will vary with the rate and duration of withdrawal of water. Also called a "cone of depression."

Control Filter: An experimental setup in which a series of packed columns are used to examine the degradation of chemicals or attenuation of microorganisms under a set of controlled experimental conditions

Convection: The process of mass movement within a fluid medium, like water, which results in the transport and mixing of properties (such as heat, chemicals, or particles) of that medium. In physics, it refers to the transfer of heat by fluid motion caused by the force of gravity and by differences in density resulting from non-uniform temperature; thus, the process moves both the fluid and the heat, and the term "convection" is used to signify either or both. The terms "convection" and "advection" are often used synonymously.

Conventional Treatment Plant: A system that employs coagulation, flocculation, sedimentation, filtration, and disinfection as sequential unit processes for water treatment and/or drinking-water production.

Cryptosporidium: A protozoan about 3.5 micrometers in diameter that can survive in water and, when ingested can cause illness.

Deep Well Injection: The disposal of raw or treated waste (such as sewage or brine) by pumping into a deep well discharging to an aquifer that is not used for water supply.

Degradate: The degradation product or metabolite of a chemical. For organic chemicals such as pesticides, microorganisms breakdown the parent compound to secondary compounds through biochemical mechanisms. Some time after pesticide application, a portion of the original compound may be present as a degradation product.

Deposition: (1) The laying down of potential rock-forming material, such as layers of sediment. (2) The material that collects on the inside surface of a distribution system pipe as a result of suspended material in water.

Desiccation: A process used to remove virtually all moisture.

Detachment: Opposite of attachment. This is a process in which attached microorganisms can be removed from solid surfaces.

Dewatering: The process of partially removing water. It may refer to the draining or removal of water from an enclosure, such as a basin, tank, reservoir, or other storage unit, or to the extraction/separation of water from aquifer material, sludge, or a slurry.

Diatom: A microscopic, single-celled algae that is commonly found in freshwater and marine environments. Thousands of species, which are characterized by a cell wall composed of polymerized silica (exoskeleton), have been identified. Diatoms secrete siliceous frustules in a great variety of forms that may accumulate in sediments in enormous numbers. Large deposits of diatoms are mined as diatomaceous earth, which is used in specific situations as filter media in water treatment. In addition, certain types of diatoms can contribute to taste-and-odor problems in drinking-water supplies.

Dichlorodiphenyl Trichloroethane: A chlorinated hydrocarbon insecticide banned in many countries, including the United States, because of its persistence in the environment and accumulation in the food chain. Also known as "DDT."

Diffusion: The movement of suspended or dissolved particles from a more concentrated to a less concentrated area. Molecular diffusion is quantified by Fick's law.

Dilution: The lowering of a chemical concentration by adding or mixing with it water that contains no chemicals or the same chemical at very low concentrations.

Direct Intake: Water that is pumped from rivers or lakes.

Discharge: A generic term used to represent the flow rate (volume per unit time) of water coming from a pipe, moving in rivers/streams, or moving through a section of an aquifer. It is the volume of water flowing past a specific point in a water system in a given period of time. Typical units are cubic meters per second, gallons per minute, million gallons per day, and cubic feet per second. It can also refer to the release of any pollutant, by any means, into the environment.

Discharge Hydrograph: A graphic representation of the relationship of the flow, stage, or velocity of a stream or conduit at a given point as a function of time.

Disinfection: The selective destruction and/or inactivation of disease-causing (pathogenic) organisms, such as bacteria, viruses, fungi, and protozoa, by either chemical or physical means. In water treatment, water is exposed to chemical disinfectants — chlorine, chloramines, chlorine dioxide, iodine, or ozone — for a specified time period to eliminate these organisms.

Disinfection Byproduct: Compounds formed as a result of a series of complex reactions between commonly used water treatment disinfectants (including chlorine, chloramines, chlorine dioxide, and ozone) and organic compounds during the disinfection of water.

Disinfection Byproduct Formation Potential: The amount of disinfection byproducts formed during a test in which a source or treated water is dosed with a relatively high amount of disinfectant (normally chlorine — $HOCl$, OCl^-) and is incubated under conditions that maximize disinfection byproduct production (e.g., natural to alkaline pH, warm water temperature, contact time of 4 to 7 days). This value is not a measure of the amount of disinfection byproducts that would form under normal drinking-water treatment conditions, but rather an indirect measure of the amount of disinfection byproduct precursors in a sample. If water has a measurable level of disinfection byproducts prior to the formation potential test (e.g., in a prechlorinated sample), then the formation potential equals the terminal value measured at the end of the test minus the initial value.

Disinfection Byproduct Precursor: A substance that can be converted into a disinfection byproduct during disinfection. Typically, most of these precursors are constituents of natural organic matter. The bromide ion (Br^-) is a precursor material.

Dispersion: The phenomenon in which a solute flowing in groundwater is mixed with uncontaminated water and becomes reduced in concentration. Dispersion is caused by differences in pore velocity and differences in flow paths at a small scale in the aquifer. A similar phenomenon is caused by turbulence in surface-water systems.

Diving Bell: A large, open-bottomed vessel for underwater work that is supplied with air under pressure. It is often used in rivers to make visual observations and for collecting samples. Also known as a "diving cabin."

Drawdown: (1) The drop in the water table when water is pumped by a well. (2) The drop in the water level of a tank or reservoir. (3) The amount that the water level in a well will drop once pumping begins. In this case, drawdown equals the static water level minus the pumping water level.

Dredging: To remove sediments or sludge from a river or lake bottom using a dragline or similar mechanical equipment.

Drinking Water: Water that is safe for human consumption or that may be used in the preparation/cleaning of food or beverages when it meets or exceeds all applicable federal, state, and local requirements concerning safety. Also called "potable water."

Dug Well: See *Pit Well.*

Dune Recharge: The process of replenishing groundwater near coastal sand dunes in The Netherlands with surface water. Part of the recharged water can be pumped back for drinking purposes.

Early Warning Water Quality Monitoring: Monitoring program meant to detect incidents of environmental damage and pollutant releases.

Enhanced Coagulation: A modified coagulation process relying on the addition of excess coagulants to achieve increased removals of natural organic matter. It can also be used to remove arsenic during the coagulation process.

Enteric Bacteria: Gram-negative, non-spore forming, facultative anaerobic, rod-shaped bacteria (bacilli) of the family *Enterobacteriaceae* found in the intestinal tracts of animals. This family of bacteria is broadly divided into three groups based on lactose utilization: the lactose fermenters, the coliforms (the genera *Escherichia*, *Enterobacter*, and *Klebsiella*); the lactose nonfermenters (the genera *Salmonella*, *Shigella*, and *Proteus*); and the slow lactose fermenters, the paracolon bacteria (organisms of the Bethesda-Ballerup and Arizona groups [genus *Citrobacter*], the Hafnia, and the Providencia).

Enteric Pathogens: Pathogens that inhabit the gastrointestinal tract of animals.

Enteric Virus: Any virus that inhabits the alimentary and gastrointestinal tracts of animals. Many such viruses are stable in feces and wastewater and can be transmitted through contaminated water supplies.

Erosion: A natural process by which rock material is loosened/dissolved and removed from the earth's surface. It includes the processes of weathering, solution, corrosion, and transportation.

Eutrophication: Nutrient enrichment of water, causing excessive growth of aquatic plants and the eventual deoxygenation of water.

Excystation: The biological process in which sporozoites or trophozoites emerge from a protective shell (a *Cryptosporidium* oocyst or *Giardia* cyst).

Exfiltration: The removal of water from a given location. In the context of RBF, when the river level drops, the water stored in bank areas will exfiltrate back to the river.

Extraction: The process of drawing forth or obtaining a substance by chemical or mechanical action, as by pressure or distillation. It is a separation technique used to increase the concentration of a solute, remove interferences from a matrix, or both. Also referred to as "abstraction."

Filter Bed: (1) A tank for water filtration that has a false filter bottom covered with granular media. (2) A pond with sand bedding, as in a sand filter or slow sand filter. (3) A type of bank revetment consisting of layers of filtering medium such that the particles gradually increase in size from the bottom upward. Such a filter allows the groundwater to flow freely, but it prevents even the smallest soil particles from being washed out.

Filter Well: Well in which laterals with sand filters are used to bring water to a central caisson. These laterals typically are placed away from the river (not directly under the river).

Filtrate: The liquid remaining after it has passed through a filter. Pumped water from RBF wells is often referred to as the "filtrate."

Filtration Improvement Period: See *Ripening Period*.

Fine-Grain Formation: A geologic formation with fine-grained (0.3 to 0.6 mm in diameter) material soil consisting mostly of clay and silt.

Flocculation: A water-treatment process following coagulation that uses gentle stirring to bring suspended particles together so they will form larger, more settleable clumps called "floc."

Flood Scour: Loss of riverbed or riverbank material due to the shearing action of water during flood periods.

Flood Wave: A rise in stream flow to a crest in response to runoff generated by precipitation, as well as the subsequent recession of the stream's flow after precipitation ends.

Flow: (1) The movement of a stream of water or other fluid from place to place; movement of silt, water, sand, or other material. (2) A fluid that is in motion. (3) The quantity or rate of movement of a fluid; the discharge; the total quantity carried by a stream. (4) To issue forth a discharge. (5) The liquid or amount of liquid per unit time passing a given point.

Flow Rate: A measure of the volume (quantity) of water moving past a given point in a given period of time. Also known as "flux."

Formation Samples: Geologic samples collected from an aquifer during drilling operations.

Gas Chromatography: An analytical technique used to determine the molecular composition and concentrations of various chemicals in water and soil samples. In most cases, an extract of a water sample is injected into a gas chromatograph (an instrument used to separate organic compounds at trace concentrations). Analytes are volatilized in an injector port and migrate as a gas through a chromatography column. The speed with which the analytes migrate depends on the relative affinity of the analyte for the stationary phase in the chromatography column. Compounds are identified based on their retention time in the column.

Giardia: The generic name for a group of single-celled, flagellated, pathogenic protozoans found in a variety of vertebrates, including mammals, birds, and reptiles.

Glacial Aquifer: Aquifers created during the recession of the glaciers and the movement of melt waters. Many of the buried-valley aquifers in northern United States were created during the retreat of the last continental glacier.

Grain-Size Distribution: A presentation of the size of distribution of sedimentary materials as a function of their diameter. The x-axis is generally the diameter of the grain and the y-axis represents the fraction of particles that are finer than a given diameter.

Granular Activated Carbon: A form of particulate carbon manufactured with increased surface area per unit mass to enhance the adsorption of soluble contaminants. Granular activated carbon is used in fixed-bed contactors in water treatment and is removed and regenerated (reactivated) when the adsorption capacity is exhausted. In some applications, granular activated carbon can be used to support a biological population for stabilizing biodegradable organic material. The other type of carbon most frequently used in water treatment is powdered activated carbon.

Gravel Pack: Gravel surrounding a well intake screen, artificially placed ("packed") to aid the screen in filtering sand out of the aquifer. Gravel packs are usually needed in aquifers containing large proportions of fine-grained material.

Groundwater: Subsurface water contained in porous rock strata and/or soil.

Groundwater Collector Well: Horizontal collector wells constructed solely within aquifers (away from rivers) to enhance well yield. When groundwater collector wells are close to rivers, they act as RBF wells.

Groundwater Disinfection Rule: A regulation of the U.S. Environmental Protection Agency to establish disinfection requirements for public-water systems using groundwater.

Groundwater Protection Zone: An area where activities are controlled to reduce the contamination of groundwater. This area normally covers the contributing zone of a pumping well.

Groundwater Recharge: The process of adding water to the aquifer (often by percolation, the injection of tertiary treated wastewater, or other means) to replenish groundwater that has been pumped from the aquifer. Also known as "groundwater replenishment."

Groundwater Under the Direct Influence of Surface Water: Water defined by the U.S. Environmental Protection Agency in the Surface Water Treatment Rule as any water beneath the surface of the ground that has a significant occurrence of insects or other microorganisms, algae, organic debris, or large-diameter pathogens like *Giardia lamblia*, or significant and relatively rapid shifts in water characteristics — such as turbidity, temperature, conductivity, or pH — that closely correlate with climatological or surface-water conditions. It is a legal definition that implies that groundwater pumped from a well has been affected by recently infiltrated surface water.

Hard-Rock Tunnel: A tunnel constructed in rock that requires drilling and blasting for its economical removal.

Head: The energy per unit weight of a liquid. In practical terms, head is the pressure at any given moment in a water system, and is calculated as the pressure exerted by a hypothetical column of water standing at the height to which the free surface of water would rise above any point in a hydraulic system. Head is often measured in pounds per square inch or kilopascals, and is sometimes expressed as the height of a column of water in feet or meters that would produce the corresponding pressure; this measurement may be called hydrostatic head. Head is also called "pressure head" or "velocity head."

Head Gradient: The gradient of hydraulic pressure (head). It is calculated by taking the difference in heads over a given distance and dividing this value with the distance. It is a dimensionless term.

Head Loss: A reduction of water pressure (head) in a hydraulic or plumbing system. Head loss is a measure of (1) the resistance of a medium bed (or other water-treatment system), a plumbing system, or both to the flow of the water through it, or (2) the amount of energy used by water in moving from one location to another. In water-treatment technology, head loss is basically the same as "pressure drop."

High Performance Liquid Chromatography: A technique that is able to separate compounds in the liquid state as they migrate through a chromatographic column. Analytes migrate through the column at a rate based on their relative affinity for the stationary phase versus the mobile phase. The technique has many applications in the analysis of organic compounds that may not be amenable to separation by gas or ion chromatography.

Horizontal Collector Well: A circular central collection caisson sunk into the ground with horizontal lateral well screens pushed out into unconsolidated aquifer deposits, in many cases into alluvial deposits beneath a river or lake. It is typically used by United States water utilities to produce drinking-water supplies from groundwater sources or from riverbanks through filtration. Also referred to as "radial collector well" or "collector well."

Horizontal Directional Drilling: Wells drilled horizontally or in an angular way rather than standard vertical drilling. This technique has recently been used for removing contaminated water and introducing nutrients to contaminated zones where such actions using vertical wells are not feasible. For thin aquifers or aquifers with low hydraulic conductivity, this technique is generally used to install horizontal wells.

Horizontal Setback Distance: The proximity of a vertical well to the riverbank. This distance is used by regulatory agencies (i.e, U.S. Environmental Protection Agency) to calculate filtration credits for RBF.

Horizontal Well: Wells that are often used for environmental applications where vertical wells are not suitable, such as for removing contaminants under buildings or developed areas (parking lots and roads). These wells are ideal in low permeable, shallow formations. Horizontal wells are typically single perforated pipes that are placed in desired locations using horizontal directional drilling technology. The screen lengths of horizontal wells are much larger than vertical wells. Also referred to as "directionally-drilled horizontal wells."

Hydraulic Conductivity (K): A coefficient describing how fast or slow water can move through a permeable medium. It is the constant of proportionality, K, in Darcy's law of linear seepage:

$$Q = -KA\ grad(h)$$

Where Q is the total discharge; A is the cross-sectional area of flow; -grad(h) is the hydraulic gradient, dimensionless; and K is the hydraulic conductivity. Typical units of hydraulic conductivity are feet per day, gallons per day per square foot, or meters per day (depending on the units chosen for the total discharge and the cross-sectional area).

Hydraulic Gradient: The rate of change of pressure (head) per unit of distance flow at a given point and in a given direction. For a stream, it is the slope of the energy grade line, or slope of line representing the sum of kinetic and potential energy along the channel length. It is equal to the slope of the water surface in steady, uniform flow. For a pipe, it is the change in static head (pressure) per unit of distance in a pipeline in which water flows under pressure. The slope of the hydraulic grade line indicates the change in pressure per head unit of distance. The gradient of hydraulic head determines the direction of flow. Water moves in the direction of higher head to lower head. It is used synonymously with "Head Gradient."

Hydraulic Monitoring: The process of checking or tracking hydraulic parameters in the aquifer and in the river stage at an RBF site.

Hydraulics: The branch of science or engineering that deals with water or another fluid in motion.

Hydrodynamic Dispersion: This is the spreading process of a contaminant in a porous media (e.g., aquifer) due the heterogeneities of media and is a combination of two terms: mechanical mixing due to heterogeneities and Fickian diffusion due to the concentration gradient within the aquifer.

Hydrodynamics: The study of the motion of, and the forces acting on, fluids.

Hydro-Mechanical Dispersion: This is the mechanical mixing component of hydrodynamic dispersion. It is dependent upon the dispersivity of the porous media (which is dependent upon scale and direction) and the velocity of water flowing in the pore spaces.

Hydrophilic: Having a tendency or affinity to bind with water molecules or absorb water.

Hydrophobic: Having a tendency to repel water molecules.

Immunoassay: The identification of a substance based on its capacity to act as an antigen (a substance capable of stimulating an immune response).

Inactivation: The process that renders viruses unable to reproduce or infect. Usually, a specific percentage of a population is affected over time. This effect can be accomplished by a variety of methods, such as heat, chemicals, or ultraviolet light. In contrast, the term "kill" is applied to other forms of life, such as bacteria, cysts, or algae.

Induced Infiltration: The flow or movement of water from an adjacent water source, such as a stream or lake, into pumping wells or through the soil surface and into the aquifer under pumping stresses. Also referred to as "induced recharge."

Induced Riverbank Filtration: See *Riverbank Filtration.*

Infiltrated Water Supplies: Water supplies that rely upon induced infiltration to aquifers, especially at RBF settings.

Infiltration: The flow or movement of water downward through soil.

Infiltration Gallery: A horizontal underground conduit of screens, perforated pipes, or porous material that collect percolating water, often under a riverbed.

Inflow: Surface and subsurface water or stormwater that moves towards a collection center, such as a well. For a storm sewer, the term "inflow" refers to groundwater entering the sewers through cracks and joints.

Inflow Area: The area that contributes to flow, similar to a catchment area.

Information Collection Rule: A U.S. Environmental Protection Agency rule requiring water utilities serving more than 10,000 customers to conduct monitoring that will aid in the gathering of data for use in the Stage 2 Disinfectants-Disinfection/Disinfection Byproduct Rule (a proposed rule to limit the maximum contaminant level of trihalomethanes) and the Enhanced Surface Water Treatment Rule (a rule under development that will set maximum contaminant level goals for *Cryptosporidium* in public-water systems using surface-water sources or groundwater under the direct influence of surface water). The Information Collection Rule was promulgated on May 14, 1996.

Injection Well: A hole drilled below the ground surface into which wastewater or treated effluent is discharged.

Intake: (1) The works or structures at the head of the conduit into which water is diverted. (2) The process or operation by which water is absorbed into the ground and added to the aquifer or groundwater basin (recharge). (3) The flow of rate into a canal, conduit, pump, stack, tank, or treatment process before treatment.

Ionic Strength: A measure of solution strength based on both the concentrations and valences of ions present.

Isoelectric Point: A pH at which the net charge on a compound is neutral. In electrophoretic separation, amphoteric compounds have equal positive and negative charges at the isoelectric point and fail to migrate. This phenomenon is also used in the separation of proteins by precipitation because the solubility of proteins is lowest at the isoelectric point. Good coagulation also occurs at the isoelectric point of the coagulant.

Large Diameter Bucket Auger Borings: Wells drilled by augers (tools used to bore holes) that are several feet (often more than 1 meter) in diameter; the auger is often called a "bucket auger." The inside of the auger is smooth and the cutting edges are located at the bottom of the auger, pointing outward.

Laterals: Screened pipes that extend horizontally out of a vertical well caisson.

Leaching: (1) A process by which soluble materials are washed out of soil, ore, or buried waste, and into a water source. (2) To wash or drain by percolation. (3) The dissolution of solids and chemicals into water flowing through a porous sample.

Leakage: The uncontrolled loss of water from one aquifer to another. The leakage may be natural, as through a semi-impervious confining layer, or artificial, as through an uncased well.

Leakance: A measure of the ease of flow between the river/aquifer interface.

Life-Cycle Cost: A method of expressing cost in which both capital costs and operation and maintenance costs (of equipment, etc.) are considered for comparing different alternatives. This includes the salvage value. Typically, the amortized annual cost of the capital investment, based on a fixed interest rate and design period, is added to annual operations and maintenance cost to arrive at a total annual cost.

Load: Almost any quantity of water carried by a conduit. For chemical contamination, it is referred to as the amount or concentration of the chemical in the river or aquifer.

Log: The exponent in the representation of a number as a power of 10. For example, the common logarithm of 100 is 2, because 10 raised to the power of 2 equals 100. Also referred to as "common logarithm."

Log Removal: A shorthand term for $\log_{10}$ removal, used in reference to the Surface Water Treatment Rule and the physical-chemical treatment of water to remove, kill, or inactivate pathogenic organisms such as *Giardia lamblia* and viruses. A 1-log removal equals a 90-percent reduction in density of the target organism; a 2-log removal equals a 99-percent reduction; a 3-log removal equals a 99.9 percent reduction; and so on.

Log Removal Credit: A regulatory term used in the United States that expresses the amount of pathogens that a water utility has removed from water using technologies like slow sand filtration and RBF. For example, some water utilities that employ RBF may receive 1-log removal credit. That means the RBF process has removed 90 percent of initial concentration of pathogens; however, if the target removal is 99.9 percent (3 logs), the utility must remove an additional 2 logs using conventional filtration or other alternative techniques.

Low-Flow Period: Periods of the year when the river flow is close to its minimum. This low-flow period is sometimes referred to as "base flow."

Mass Spectrometry: A method of chemical analysis in which compounds emerging from a gas chromatograph are fragmented and ionized by bombardment with a beam of electrons. An electromagnetic field separates the ions according to their individual mass-to-charge ratios into a characteristic mass spectrum for each molecule. An analog computer analyzes the spectra and makes it possible to identify the molecules even in cases of poor separation on the chromatography column, hence the advantage of mass spectrometry compared to selective chromatograph detectors.

Maximum Contaminant Level: The maximum permissible level (concentration) of a contaminant in water delivered to the end user of a public-water system. Maximum contaminant levels are the legally enforced standards in the United States.

Mean Flow: The arithmetic average of the discharge at a given point or station on the line of flow for a specified period of time.

Mean Sea Level: The average height of the sea for all stages of the tide.

Mean Velocity: The average velocity of a stream flowing in a channel or conduit at a given cross-section or in a given reach. It is equal to the discharge divided by the cross-sectional area of the section or by the average cross-sectional area of the reach. Also known as "average velocity" or "mean flow velocity."

Mechanical Clogging: Clogging that is induced due to the physical movement of particles to pore spaces by forces like gravity and convection.

Mesophilic Bacteria: Bacteria that grow best at temperatures between 25 and 40°Celsius. Also known as "mesophiles."

Micropollutants: Pollutants such as pesticides, polychlorinated biphenyls, volatile organics, and pharmaceutically active substances that are detected at the microgram-per-liter to nanogram-per-liter ranges.

Microscopic Particulate Analysis: The use of any one of several methods outlined by the U.S. Environmental Protection Agency to identify and size particles in water. Several versions of analytical methods are available that can identify *Giardia*, *Cryptosporidium*, algae, nematodes, and other microorganisms. Particles from large volumes of water are isolated onto a cartridge filter with a typical pore size of 1 micrometer. Microscopic particulate analyses are used to assess the performance of water filtration plants, as well as to help identify groundwater that may be under the influence of surface water.

MODFLOW: A modular three-dimensional flow model used by the U.S. Geological Survey that uses a finite difference approach to solve the groundwater flow equation. It calculates hydraulic heads in a flow domain.

MODPATH: A groundwater flow and particle tracking program used by the U.S. Geological Survey that tracks the flow path of a water or non-reactive contaminant particle solely due to velocity gradients. The effects of dispersion and retardation are not accounted for in this process. MODPATH uses hydraulic heads computed in MODFLOW to calculate travel path and times.

Molecular Diffusion: A process whereby mobile compounds (dissolved or suspended in another compound) move from areas of high concentration to areas of low concentration. Molecular diffusion is described mathematically by Fick's first law of diffusion.

Monitoring Well: A well (normally 2 to 6 inches in diameter) drilled primarily for monitoring water levels or water quality in an aquifer on a temporal basis. Such wells are distinct and different from production wells that are drilled to produce quantities of water for distribution to water users.

Most Probable Number: Statistical analysis technique based on the number of positive and negative results when testing multiple portions of equal volume. The most probable number is commonly used to estimate the density of coliform bacteria in a water sample.

MT3D: A commonly used solute transport model that uses the MODFLOW grid to solve the advection-dispersion-reaction equation. Also called "Mass Transport in Three-Dimension."

Mutagenesis: The process of producing a change in the base sequence of deoxyribonucleic acid (DNA) in a cell that is capable of reproducing in the organism. A variety of tests are available to determine whether a chemical or one of its metabolites is capable of inducing a DNA change that can be passed on to the progeny of the affected cell. The most commonly applied assays are designed to detect point mutations (e.g., the Ames Test).

Mutagenic: A chemical or agent with properties that cause mutation.

Mutation: An abrupt change in the genotype of an organism, not resulting from recombination. In a mutation, the base sequence of a nucleic acid molecule may undergo qualitative and quantitative alteration or rearrangement.

Nanofiltration: A pressure-driven membrane separation process that generally solutes larger than approximately 1 nanometer (10 angstroms) in size. Its separation capability is controlled by the diffusion rate of solutes through a membrane barrier and by sieving and is dependent on the membrane type. In drinking-water treatment, nanofiltration is typically used to remove nonvolatile organics larger than the 200- to 500-dalton molecular weight cutoff (e.g., natural and synthetic organics, color, disinfection byproduct precursors) and multivalent inorganics (for softening).

Natural Organic Matter: A term used to describe the organic matter (substances containing carbon compounds, usually of animal or vegetable origin) present in natural waters. Natural organic matter contributes to the color of water. Humic substances (e.g., fulvic acid) represent a significant fraction of natural organic matter in surface-water sources. Natural organic matter is derived from sources such as farmlands and forests.

Nonpoint Source: Sources of pollution that cannot be accurately identified (e.g., farmlands are nonpoint sources of agricultural chemicals).

Non-Purgeable Organic Carbon: The fraction of total organic carbon removed after purifying a sample with an inert gas.

Observation Well: A well placed near a production well to monitor changes in the hydraulic head of the aquifer.

Oocyst: An outer shell that protects an organism from the environment.

Open-End Caisson Method: A technique in which sections of the caisson are formed and poured at ground surface and are sunk into place by excavating soils from within the caisson. The sections (also called "lifts") are tied together by reinforcing steel.

Organic Carbon: Carbon derived from living organisms.

Organic Compound: A carbon-containing compound — such as hydrocarbon, alcohol, ether, aldehyde, ketone, carboxylic acid, or carbohydrate — that is derived from living organisms.

Organic Mat: Biofilm that is present at the river/aquifer interface.

Oscillation: A periodic movement back and forth, or up and down.

Oxic Zone: An area in a reservoir, lake, or treatment process in which a dissolved oxygen concentration can be measured.

Ozonation: The process of applying ozone that has disinfection properties similar to chlorine to water or wastewater for oxidation, disinfection, or odor control.

Parent Compounds: This term refers to the original chemical before it is broken down into secondary products that are known as "metabolites."

Pathogen: An infectious organism that can cause disease in a host.

Peak Capacity: In the context of RBF, it is the maximum (highest) capacity of the pumping equipment.

Peak Concentration: Maximum (highest) concentration

Peak Demand: It is usually the maximum (highest) water demand at a given time in a day. The peak demand can be broken down on the basis of time such as peak 1-hour demand, peak 24-hour demand, and so on, and is expressed as water needs per person per day during that period of time.

Peak Flow: (1) A flow rate of a given magnitude that is sustained for a specified period of time. Because it is difficult to compare numerical peak flow rate values from different treatment plants, peak flow rate values are normalized by dividing the long-term average flow rate. The resulting ratio is known as a "peaking factor." (2) Excessive flows experienced during hours of high demand, usually determined to be the highest 2-hour flow expected to be encountered under any operational conditions. Also known as "Maximum Flow."

Perforated Pipe Well Screen: A screen that permits water to enter a well through holes that have been punched or cut into the screen. These screens prevent sand from entering the well.

Permeability: A measure of the relative ease with which water flows through a porous material. A sponge is very permeable; concrete is much less permeable. Permeability is sometimes called "perviousness."

Piezometer: An instrument for measuring pressure (head) in a conduit, tank, or soil by determining the location of the free water surface.

Piezometric Measurement: Measurement of hydraulic pressure (head) using piezometers

Piezometric Surface: A surface that represents the position to which water would rise in wells anywhere in the aquifer. A groundwater-level contour map is a method of depicting the piezometric surface of an aquifer.

Piston Pump: A reciprocating pump in which the cylinder is tightly fitted with a reciprocating piston.

Pit Wells: Shallow, large-diameter wells that, in most instances, are manually dug into the ground (pit wells are either constructed by excavating with power machinery or by hand tools rather than drilling or driving). Typically, a pit well is constructed for an individual residential water supply. Pit wells are also known as a "dug wells" in the United States.

Point Source: The origin or source of a contaminant that can be identified accurately. Discharge from sewage-treatment plants is known as a "point source."

Pore Settling: Settlement of a suspension within the pore space. For example, bacteria can settle in the pore spaces of aquifers since it has a density that is slightly higher than water.

Pore-Size Distribution: Pore spaces divided into various size groups (similar to *Grain-Size Distribution*).

Pore Throat: This the opening area between adjoining sand or silt in the aquifer. The minimum (open) cross-sectional area available for water flow or chemical/microorganism transport is known as a "pore throat."

Porosity: The ratio of the aggregate volume of interstices (pores) in a rock or soil to its total volume. It is usually stated as a percentage.

Potable Water: See *Drinking Water*.

Powdered Activated Carbon: A highly adsorbent form of carbon that is composed of fine particles and provides a large surface area for adsorption. Powdered activated carbon is typically added as a slurry on an intermittent or continuous basis to remove taste-and-odor causing compounds or trace organic contaminants and is not reused.

Pressure Transducer: A device (or sensor) used to measure the hydraulic head or water pressure at a given location in the subsurface. It is frequently used to measure hydraulic heads in observation wells during groundwater investigations. Pressure transducers can be connected to data loggers for automated collection of hydraulic head or pressure data.

Primacy Agency: The organization that is responsible for administering and enforcing regulations.

Production Rate: The amount of water that is pumped from a well over a given period of time. It is expressed in cubic meters per second. Also known as "pumping rate."

Production Well: A well from which water is pumped for use. Also known as "pumping well."

Protozoa: Small, one-celled microorganisms, including amoebae, ciliates, and flagellates.

Protozoa Breakthrough: If protozoa (such as *Cryptosporidium* and *Giardia*) are present in filtered water of RBF, it is said that there is a "breakthrough" of protozoa through the porous media.

Psychrophilic Bacteria: Bacteria that grow best in temperatures between 54 to 64°Fahrenheit (12 to 18°Celsius). Also known as "psychrophiles."

Pump House: (1) A structure that houses the pump(s). (2) A shelter for housing pumping equipment at the top of a well.

Pump Station: A chamber that contains pumps, valves, and electrical equipment necessary to pump water or wastewater.

Pumping Unit: Individual pumps.

Pumping Well: See *Production Well*.

Purgeable Organic Carbon: The fraction of the total organic carbon removed from an aqueous solution as by gas stripping under specified conditions. Purgeable organic carbon is also referred to as "volatile organic carbon," which includes the carbon content of chemicals that are volatile under the specified stripping conditions of the test.

Ranney Collector Well: See *Horizontal Collector Well*.

Raw Water: Untreated surface or groundwater.

Recirculation Filter: A filter or a sand column in which the effluent is pumped back to the filter or column in a continuous manner. The filters or the columns operate under upflow conditions where the feed solution is pumped to the bottom of the column. The recirculation process continues until the desired removal of the contaminant is achieved.

Redox: The study of reactions in which electron transfers occur. Also known as "oxidation-reduction (redox) chemistry."

Redox Zones: Areas where the oxygen concentration is very low and reduced species of chemicals are present.

Retardation Factor: An index that measures the relative mobility of a contaminant particle with respect to water. If the retardation factor of a chemical is 1, it travels at the same velocity as water. If the retardation factor is 10, the chemical moves at one-tenth of the velocity of water.

Retention Tanks: (1) A basin that stores a chemical at a chemical plant in the event of a spill. (2) A basin that is used for temporary storage of water-treatment plant residuals.

Retention Time: (1) The average length of time a drop of water or chemicals remains in porous media while moving from the river. (2) The time spent by a drop of water or suspended particle in a tank or chamber. Mathematically, it is the volume of water in the tank divided by the flow rate through the tank. Also known as "detention time" or "residence time."

Reverse Osmosis: A pressure-driven membrane separation process that removes ions, salts, and other dissolved solids and nonvolatile organics from a liquid. The separation capability of the process is controlled by the diffusion rate of solutes through a membrane barrier and by sieving; it is dependent on the membrane type. In drinking-water treatment, reverse osmosis is typically used for desalting, specific ion removal, and natural and synthetic organic removal.

Revertants: Modified strains of the bacteria *Salmonella typhimurium* that revert to their natural state and form colonies in the presence of mutagenic-active compounds. The bacteria are modified so that they cannot produce Histidin, which is an amino acid necessary for growth. Mutagenic active compounds, however, "switch on" the ability of bacteria to produce Histidin. The Ames Test method uses revertants to determine the presence of mutagenic compounds in water samples.

Right of Way: The land dedicated to use by transportation agencies and utilities for the operation and maintenance of roads and power lines.

Ripening Period: The time taken for a filter medium to achieve normal operational efficiency. In early times, particulates and other materials passed through filter beds.

River/Aquifer Interface: The boundary between river water and the underlying riverbed sediments.

River Basin: The land area drained by a river and its tributaries.

River System: The principal stream and all its tributaries.

Riverbank Filtration: The process of collecting water in an infiltration gallery located near the bank of a river to allow river water to pass through the soil in the riverbank. Riverbank filtration provides particle removal, as well as partial or nearly complete removal of organic compounds and pathogenic organisms.

Sand Filtration: The oldest and most basic filtration process, which generally uses two grades of sand (coarse and fine) for turbidity and particle removal. A sand filter can serve as a first-stage roughening filter or prefilter in more complex processing systems. Sand filtration is typically used in conventional water-treatment plants.

Schmutzdecke: The thin organic mat (or layer of solids and biological growth) that forms on top of a slow sand filter, allowing the filter to remove turbidity effectively without chemical coagulation.

Scouring: Erosion and the subsequent wash-off of the material on the sides or bottom of a waterway, conduit, or pipeline by the movement of water. See *Flood Scour.*

Scouring Velocity: The minimum velocity (speed) required to carry away material accumulations in a waterway, conduit, or pipeline by a fluid in motion.

Screen Zone: The screened area of a well. This is the zone where water enters a production well or a monitoring well.

Seawater (Beach) Collector Wells: Horizontal collector wells that are installed in beach areas to collect filtered seawater.

Sedimentation: The removal of settleable suspended solids from water or wastewater by gravity in a quiescent basin or clarifier.

Shock Load: River water with a temporary and unusual amount of pollutants (such as algae, colloidal matter, color, suspended solids, turbidity, etc.).

Siphon: A hydraulic process in a closed system in which low enough pressure created at the siphon discharge permits a fluid to flow upward and be transferred across a higher elevation than the hydraulic grade of the system at the siphon starting point. For example, a siphon will transport water up and over the edge of a container as long as the discharge point is below the water surface in the container and the system is closed (i.e., no air gaps occur in the siphon).

Siphon Tubes: Pipes that connect one well from another and act as a siphon. Thus, pumping one well induces water from other wells to move to the pumped well as long as the suction is below 1 atmospheric pressure.

Slow Sand Filtration: A purification process in which raw water is passed downward through a filtering medium that consists of a layer of sand 0.6- to 1-meter thick. The filtrate is removed by an underdrainage system, and the filter is cleaned by scraping off the clogged sand and eventually replacing the sand. A slow sand filter is characterized by a slow rate of filtration (commonly 0.1- to 0.2-meters per hour). Its effectiveness depends on the biological mat, or schmutzdecke, that forms on the top few millimeters.

Slug: A temporary, abnormally high concentration of an undesirable substance that appears in the product or distributed water. For example, a slug of iron rust might appear because of the shearing action of a high-demand flow that loosens a previously deposited iron precipitate.

Slug Test: An aquifer test, conducted in wells, in which a known volume of tracer (the "slug") is added to groundwater to determine hydraulic properties of the aquifer or overlying soils.

Soft-Soil Tunnel: A tunnel excavated in soft (unconsolidated) material.

Soil Passage: The movement of water at a RBF site in the subsurface media, such as in soil or sand and gravel. This is sometimes synonymously used as "subsoil passage."

Sorption: The process of taking up and holding onto a soil, as by absorption or adsorption, or a combination of the two.

Spores: A propagative unit that is typically unicellular, is often uninucleate, and may be formed with (sexually) or without (asexually) a change in ploidy. Most types of nonmotile spores are dormant and are more resistant to environmental change than are vegetative cells. The principal groups of spore-forming bacteria are those in the genera *Bacillus* and *Clostridium*.

Stage: The elevation of a water surface above its minimum or above or below an established low-water plane or datum of reference. The height of water above a given datum in a river is known as "river stage."

Sterilized Filter Media: Filter media that has been sterilized by chemical or heat.

Stormwater: Water that is collected as runoff from a rainfall event.

Straining: A filtration process in which particles are separated from a fluid stream by sieving. The removal rate is dependent upon the size and shape of the strainer openings and the physical characteristics of the particles.

Surface Water: Water from sources open to the atmosphere, such as lakes, reservoirs, rivers, and streams.

Surfactant: A substance – such as a detergent, wetting agent, or emulsifier – that, when added to water, lowers surface tension and increases the "wetting capabilities" of water. Reduced surface tension allows water to spread and penetrate fabrics or other substances, enabling them to be washed or cleaned. Also called "surface-active agents" or "wetting agents."

Surrogate Compound: A substance that is not usually found in the environment but is chemically similar to an environmental contaminant. Such similarities allow the use of surrogate compounds as internal standards in organic analytical procedures. Surrogate compounds are used, for example, to determine recoveries and adjust the retention times of gas chromatographic methods.

Suspended Solids: Solid organic and inorganic particles that are held in suspension in water and are not dissolved. Suspended solids are retained on a standard glass fiber filter or a 0.45-micrometer pore diameter membrane filter after filtration of a well-mixed sample. The residue is dried at 21.7 to 221°Fahrenheit (103 to 105°Celsius). Also known as "suspended matter" or "suspended sediment."

Synthetic Organic Chemicals: Manmade organic chemicals, some of which are volatile while others tend to stay dissolved in water instead of evaporating. Some synthetic organic chemicals are contaminants in drinking water and are regulated by the U.S. Environmental Protection Agency. The regulated synthetic organic contaminants include volatile organic chemicals, pesticides, herbicides, polychlorinated biphenyls, selected treatment chemicals (e.g., acrylamide), and polynuclear aromatic hydrocarbons. Synthetic organic chemicals are also referred to as "synthetic organic compounds," "synthetic organic contaminants," or "synthetic organic matter."

Thermophilic Bacteria: Bacteria that grow best at temperatures between 45 and 60°Celsius. Also known as "thermophiles."

Tile Drains: Subsurface (not open) drains that are used in clay-rich farmlands to lower the water table by draining water to a nearby stream or open ditch. This draining process allows for root development and plant growth in such soils.

Tracer: (1) A foreign substance that is mixed with or attached to a given substance to determine the location or distribution of the foreign substance. (2) An element or compound that has been made radioactive so that it can easily be followed (traced) in biological and industrial processes. Radiation emitted by the radioisotope pinpoints its location.

Trapping: A process used to prevent a material flowing or carried through a conduit from reversing its direction of flow or movement or from passing a given point.

Travel Time: The time for a water particle to travel from one location to another along a flow line.

Trend Monitoring: Long-term measurement programs at an RBF site that tracks the water level and water quality of the river and aquifer.

Turbidity: A measure of the muddiness or cloudiness (caused by the presence of suspended matter) of water. It is an analytical quality, usually reported in nephelometric turbidity units, determined by measurements of light scattering. The turbidity of finished water in the United States is regulated by the U.S. Environmental Protection Agency.

Ultrafiltration: A pressure-driven membrane process that separates submicron particles (down to 0.01-micrometer size or less) and dissolved solutes (down to molecular weight cut-off of approximately 1,000 daltons) from a feed stream by using a sieving mechanism that is dependent on the pore-size rating of the membrane.

Ultraviolet Absorbance at 254 Nanometers: The absorbance of electromagnetic radiation at wavelength 254 nanometers by a liquid through a 1-centimeter path. The value is an indirect measure of compounds containing double bonds (including, but not limited to, aromatic compounds); therefore, this measurement has been considered representative of the humic content of natural organic matter, as well as acting as a surrogate for disinfection byproduct precursors.

Ultraviolet Light: Light rays beyond the violet region in the visible spectrum, invisible to the human eye.

Ultraviolet Light Disinfection: A process for inactivating microorganisms by irradiating them with ultraviolet light. The ultraviolet waves disrupt reproductive capability, rendering them inactive and incapable of reproduction. The ultraviolet light does not leave a disinfectant residual, however, so a form of chlorine disinfection must be applied if a residual is desired. To allow the irradiation to reach the organisms effectively, the water to be disinfected must be relatively free of particles, as in a filtered water.

Uniformity Coefficient: Method of characterizing filter sand where the uniformity coefficient is equal to the sieve size in millimeters that will pass 60 percent of the sand divided by that sieve size passing 10 percent. It is a measure of size distribution of sediments. A small value of the uniformity coefficient means all materials are of same size.

Vertical Setback Distance: The distance from the bottom of the streambed to the lateral of a collector well under the streambed.

Vertical Wells: A tubular well that is drilled vertically downward into a water-bearing stratum or under the bed of a lake or stream.

Viability: (1) An indication of whether a water system has the technical, financial, and managerial capabilities to provide drinking water that complies with federal and state drinking-water regulations. (2) The ability to live, grow, and develop.

Viscosity: The degree to which a fluid resists flow under an applied force. Viscosity is the ratio of the shear stress to the rate of shear strain (equation included). Water at 20°Celsius has a viscosity of 0.0208-pounds force seconds per foot squared (1.002×10^{-3} pascal-seconds, or 1.002 centipoise).

Volatile Organic Compound: These are trace chemicals, such as solvents (pesticides), that can easily be transferred from the aqueous phase to the air phase (accordingly, they have large Henry's constants). Also referred to as "purgeable organic carbon"

Waterworks: The collective set of features of a system that provides drinking water to the public, including the source-water facilities, treatment plant, and water distribution network. This term is typically used to designate all aspects of a given system. It is also referred to as a "water utility."

Well Casing: The non-perforated riser pipe that connects a well to the surface.

Well Development: The process of cleaning the fines and drilling residue from a new well to improve its yield and quality for subsequent use in water production and water-level monitoring.

Well Field: A group of wells treated as a single entity for administrative, production, and treatment purposes. Hydraulic models frequently treat several wells in the same geographic area as a single well with a discharge equivalent to that of all the wells in a well field.

Well Gallery: A number of wells connected together by an underground conduit, such as siphon tubes or pressurized conduits (pipes).

Wellhead: The surface appurtenance (top) of a particular well. The wellhead is the location of the pump motor (unless the motor and pump are both submersible), the concrete slab that surrounds the well casing, and any plumbing dedicated to that particular well. A group of wells that are geographically close may share a storage tank, but each has its own wellhead. A group of wells is often referred to as a "well field."

Wellhead Protection: Control of activities around the areas that contributes to flow of water to the wells. The aim is to guard against potential groundwater contamination from chemical and microbial sources.

Well Screen: A sleeve with slots, holes, gauze, or wire wrap placed at the end of a well casing to allow water to enter the well. The screen prevents sand from entering the water supply.

Well Water: Water from a hole bored, drilled, or otherwise constructed in the ground to tap an aquifer.

Wet Weather Flow: The flow in a combined sewer (which carries sanitary sewage and stormwater) during snowmelt and rain events.

Wetted Area: The length of wetted contact area between a stream of flowing water and the channel that contains it.

Wetted Perimeter: The length of wetted contact between a stream of flowing water and its containing conduit or channel, measured in a plane at right angels to the direction of flow.

Wire-Wound Well Screens: Screens that are made of vertical bars with wires wound around these vertical bars.

XAD: A resin of polystyrene or polyacrylic ester, which is used as adsorbent for the analysis of non- or weakly polar organic substances.

Yield: The quantity of water – expressed as a rate of flow – that can be collected for a given use or uses from surface water or groundwater sources on a watershed. The yield may vary based on the use proposed, the plan of development, and economic considerations.

Zone of Contribution: The volume of the aquifer that contributes water to the well.

Index

Water Science and Technology Library

17. V.P. Singh: *Dam Breach Modeling Technology.* 1996   ISBN 0-7923-3925-8

18. Z. Kaczmarek, K.M. Strzepek, L. Somlyódy and V. Priazhinskaya (eds.): *Water Resources Management in the Face of Climatic/Hydrologic Uncertainties.* 1996   ISBN 0-7923-3927-4

19. V.P. Singh and W.H. Hager (eds.): Environmental Hydraulics. 1996   ISBN 0-7923-3983-5

20. G.B. Engelen and F.H. Kloosterman:
    Hydrological Systems Analysis. Methods and Applications. 1996   ISBN 0-7923-3986-X

21. A.S. Issar and S.D. Resnick (eds.): *Runoff, Infiltration and Subsurface Flow of Water in Arid and Semi-Arid Regions.* 1996   ISBN 0-7923-4034-5

22. M.B. Abbott and J.C. Refsgaard (eds.): *Distributed Hydrological Modelling.* 1996
                                                                          ISBN 0-7923-4042-6

23. J. Gottlieb and P. DuChateau (eds.): *Parameter Identification and Inverse Problems in Hydrology, Geology and Ecology.* 1996   ISBN 0-7923-4089-2

24. V.P. Singh (ed.): *Hydrology of Disasters.* 1996   ISBN 0-7923-4092-2

25. A. Gianguzza, E. Pelizzetti and S. Sammartano (eds.): *Marine Chemistry.* An Environmental Analytical Chemistry Approach. 1997   ISBN 0-7923-4622-X

26. V.P. Singh and M. Fiorentino (eds.):
    Geographical Information Systems in Hydrology. 1996   ISBN 0-7923-4226-7

27. N.B. Harmancioglu, V.P. Singh and M.N. Alpaslan (eds.):
    Environmental Data Management. 1998   ISBN 0-7923-4857-5

28. G. Gambolati (ed.): CENAS. *Coastline Evolution of the Upper Adriatic Sea Due to Sea Level Rise and Natural and Anthropogenic Land Subsidence.* 1998 ISBN 0-7923-5119-3

29. D. Stephenson: *Water Supply Management.* 1998   ISBN 0-7923-5136-3

30. V.P. Singh: *Entropy-Based Parameter Estimation in Hydrology.* 1998   ISBN 0-7923-5224-6

31. A.S. Issar and N. Brown (eds.):
    Water, Environment and Society in Times of Climatic Change. 1998   ISBN 0-7923-5282-3

32. E. Cabrera and J. García-Serra (eds.):
    Drought Management Planning in Water Supply Systems. 1999   ISBN 0-7923-5294-7

33. N.B. Harmancioglu, O. Fistikoglu, S.D. Ozkul, V.P. Singh and M.N. Alpaslan:
    Water Quality Monitoring Network Design. 1999   ISBN 0-7923-5506-7

34. I. Stober and K. Bucher (eds): *Hydrogeology of Crystalline Rocks.* 2000 ISBN 0-7923-6082-6

35. J.S. Whitmore: *Drought Management on Farmland.* 2000   ISBN 0-7923-5998-4

36. R.S. Govindaraju and A. Ramachandra Rao (eds.):
    Artificial Neural Networks in Hydrology. 2000   ISBN 0-7923-6226-8

37. P. Singh, V.P. Singh: *Snow and Glacier Hydrology.* 2001   ISBN 0-7923-6767-7

38. B.E. Vieux: *Distributed Hydrologic Modeling Using GIS.* 2001   ISBN 0-7923-7002-3

39. I.V. Nagy, K. Asante-Duah, I. Zsuffa:
    Hydrological Dimensioning and Operation of Reservoirs. 2002   ISBN 1-4020-0438-9

40. I. Stober, K. Bucher: *Water-Rock Interaction.* 2002   ISBN 1-4020-0497-4

Kluwer Academic Publishers ~ Dordrecht / Boston / London